On the Boundary of Two Worlds: Identity, Freedom, and Moral Imagination in the Baltics
35

Baltic Eugenics

Bio-Politics, Race and Nation in Interwar Estonia, Latvia and Lithuania 1918-1940

Edited by
Björn M. Felder & Paul J. Weindling

Rodopi

Amsterdam - New York, NY 2013

Cover photo : Jēkabs Prīmanis measuring skulls (from Jēkabs Prīmanis (1937), Ievads antropoloģijas metodikā. Riga: Valters un Rapa.)

Cover background map: racial map from the article "How the European peoples are formed " (Kā cēlušās Eiropas Tautas), in the illustrated Latvian journal "Atputa" (issue no.759, 19 May 1939)

The paper on which this book is printed meets the requirements of "ISO 9706:1994, Information and documentation - Paper for documents - Requirements for permanence".

ISBN: 978-90-420-3722-9
E-Book ISBN: 978-94-012-0976-2

Printed in the Netherlands

JWH

Table of Contents

Acknowledgments 3

Introduction: Eugenics, Sterilisation and the Racial State:
The Baltic States and Russia and the Global Eugenics Movement
Björn M. Felder 5

Part 1: Eugenics in the Baltics

Race, Eugenics and National Identity in the Eastern Baltic:
From Racial Surveys to Racial States
Paul J. Weindling 33

The Application of Eugenics in Estonia 1918-1940
Ken Kalling 49

Racial Identity and Physical Anthropology in Estonia 1800-1945
Ken Kalling and Leiu Heapost 83

"God forgives - but Nature never will" – Racial Identity,
Racial Anthropology, and Eugenics in Latvia 1918-1940
Björn M. Felder 115

Latvian Psychiatry and Medical Legislation of the 1930s
and the German Sterilisation Law
Vladimirs Kuznecovs 147

"Over-Latvianization in Heaven" - Attitude towards
Contraception and Abortion in Latvia 1918-1940
Ineta Lipša 169

Eugenics against State and Church:
Juozas Blažys (1890-1939), Eugenics, Abortion
and Psychiatry in Interwar Lithuania 1918-1940
Björn M. Felder and Arūnas Germanivičius 203

Part 2: Eugenics in the Baltic Sea Region

World War One and National Characterology
in East-Central Europe
Maciej Górny 235

Soviet Eugenics for National Minorities: Eradication
of Syphilis in Buriat-Mongolia as an Element of Social
Modernisation of a Frontier Region 1923-1928
Vsevolod Bashkuev 261

Sterilisations in the Swedish Welfare State: A Gender Issue?
Maija Runcis 287

Eugenic Concerns, Scientific Practices: International
Relations and National Adaptations in the Establishment
of Psychiatric Genetics in Germany, Britain, the USA,
and Scandinavia 1910-1960
Volker Roelcke 301

Contributors 335

Acknowledgments

The majority of the papers included in this volume were presented at the workshop *Eugenics, Race and Psychiatry in the Baltic States: a Trans-National Perspective*, held at the Goethe Institute Riga between 7 and 8 May 2009. The workshop was organized by the Working Group in the History of Race and Eugenics in Europe, the Department of History at Oxford Brookes University, the Nordost-Institute in Lüneburg and the Estonian University of Life Science in Tartu and generously sponsored by the Gerda-Henkel-Foundation and the Fritz Thyssen foundation. First I should express my gratitude to Andreas Lawaty, who as the director of the Nordost-Institute encouraged me in organizing the workshop and to prepare this conference volume. Without him and his generous support the workshop and this volume would not have been possible. I also have to thank to Ulrich Everding, director of the Goethe Institute in Riga for his hospitality and the financial and logistical support. Thanks are also due to Erki Tammiksaar, director of the Karl Ernst von Baer House at the Estonian University of Life Science in Tartu.

As the publication of this volume took some time I first have to thank the contributors for their patience. Firstly, I owe gratitude to the Gerda-Henkel-Foundation, which supported this volume with a publication grant. Manfred Hildermeier welcomed me at his chair for Eastern European History at the Seminar for Medieval and Modern History at the Georg-August University of Göttingen and supported my research. Thanks to his open-mindedness and his support, this volume could be completed. There were several persons who helped this volume coming into existence in some way. Ken Kalling was never tired to answer any of my questions. Konrad Mayer and Katrin Steffen were reviewing chapters and gave important advices. Marius Turda also encouraged me to follow the trace of Baltic eugenics and to work on this volume. I have especially wish to thank Eva-Lotta Kalz, Katharina Sewening and Julian Nieding, who worked on the text and did a great job with the layout and the notes. Finally I am very thankful to Leonidas Donskis and Eric van Broekhuizen who accepted this volume to the Rodopi series "On the Boundary of Two Worlds" and kindly facilitated the process of publication.

Finally, I have to thank the Deutsche Forschungsgemeinschaft for generously supporting me to conduct research at the Georg-August University in Göttingen as well as enabling me to finish this volume.

Björn Felder Göttingen, Spring 2013

I wish to add my grateful thanks to the above mentioned persons, to Margit Berner and Maria Teschler-Nicola, in her capacity as Director of the Department of Anthropology at the Natural History Museum Vienna for crucial access to library resources, and to acknowledge the Wellcome Trust for Grant Number 082808.

Paul Weindling Oxford, Spring 2013

Introduction:
Eugenics, Sterilisation and the Racial State: The Baltic States, Russia, and the Global Eugenics Movement

Björn M. Felder

In 1938, the annual conference of the International Federation of Eugenics Organisations was scheduled to be held in the Estonian town of Pernu on the Baltic Sea. The host was the Estonian Association for Eugenics and Genealogy (*Eesti Eugeenika ja Genealoogia Selts*), which had been a member of the federation since 1928, and which was headed by Hans Madissoon. Estonia would be in the spotlight of eugenics researchers from throughout the world. However, the conference was postponed because of the impending war, and it would never be rescheduled for Estonia.[1] This episode shows that while the Baltic states were on the periphery of Europe geographically, they were scientifically directly in the centre of the eugenics debate and applied eugenics. International cooperation was not limited to conferences and communications. Since the beginning of the 20th century, scholars from the Baltics travelled to the West, in particular to Germany, to study bio-medical science, including eugenics. During the 1930s, students and scholars went abroad to visit well-known researchers, especially in Germany. Madissoon, for example, visited Fritz Lenz at the Kaiser Wilhelm Institute for Anthropology, Human Heredity and Eugenics in Berlin, and Otmar von Verschuer at the University of Frankfurt/Main in 1936. In addition to the Kaiser Wilhelm Institute in Berlin, one of the Kaiser Wilhelm Institutes in Munich, the *Deutsche Forschungsanstalt für Psychiatrie* headed by Ernst Rüdin was also frequently visited by eugenics pilgrims from the Baltic states.[2] This volume also considers crucial aspects of transnational science, the transfer of scientific knowledge, and scientific interactions in the field of eugenics in Europe. Thus the volume includes not only contributions on eugenics in the Baltics but also in its second part chapters on eugenics and race in the influential neighbouring countries, as Germany, Sweden, Soviet Union and Poland. The rationale for this is that Baltic eugenics has to be investigated in the European and international context.

Estonia probably had the strongest eugenics movement in the Baltics, even before World War One. A eugenics society was established in the 1920s, and the first law for applied eugenics was enacted in 1936, including mandatory sterilisation and abortion. Though Estonia may have been the country with the broadest social support for eugenics, neighbouring

Latvia and Lithuania also had active eugenics debates and movements that led to the enacting of a eugenics law, including non-compulsory sterilisation, in Latvia in 1937. In Catholic Lithuania, a eugenics counselling bureau was opened for the public in 1934. Further, the application of eugenics in the Baltic states was far more than a simple expression of the eugenics *Zeitgeist*. Applied eugenics in the Baltics was part of national eugenics projects planned and carried out by Baltic governments. The goals of these projects went far beyond "improving" the "nation's vitality"; they were clear forms of bio-politics, as described by Michel Foucault: a state power that seeks to control and form the individual body, specially reproduction.

Considering the strong ethno-nationalism that defined the nation as a biological group, which was fostered by authoritarian regimes established in Lithuania in 1926 and in Estonia and Latvia in 1934, the eugenics projects in general were designed to establish a nation in biological terms: not only to homogenise the nation ethnically but also genetically and racially to form a racial state. This can be defined as: A non-democratic state that defined its population in biological terms and followed eugenics as a main agenda. Achieving the biological rebirth of the nation and to construct it as a "pure" and "original" race was seen as a fulfilment of the national task, the national regeneration that carried strong elements of secular religion. This agenda was also a common phenomenon among bio-medical elites and politicians in Central Europe in the 1930s and 1940s.[3] Even if national regeneration was still a utopian dream of eugenicists in the Baltics in the 1930s, and the means of eugenics were only one step in that direction, the debate on race and eugenics was clearly directed toward forming a racial state: "Who owns the soil should also sow" was a phrase of Latvian dictator Kārlis Ulmanis that was used by eugenicist Jēkabs Prīmanis to describe the main agenda of national eugenics.[4] But unlike many Central European states as well as the Swedish example, ethnic minorities were not directly affected by eugenic policies in the Baltics. The Baltic states remained small but relatively peaceful harbours for Jewish refugees until the Nazi invasion of 1940.[5]

While eugenics and bio-politics in Scandinavia and Central Europe have been studied in depth,[6] it is quite astonishing that the Baltics' bio-politics and eugenics agendas remain outside the broader focus of historians. Such studies would contribute fundamentally to a different evaluation of the authoritarian regimes, the social debates, and the social and political implications of eugenics projects during the 1930s. Research on Estonia has been conducted by Ken Kalling in recent years, Björn Felder has written about the Latvian national eugenics project, and Arūnas Germanivičius has recently dealt with Lithuanian psychiatry, focusing on the connections to eugenics.[7] Because the Baltic states are situated at the European crossroads of eugenics-related ideas, and are exposed to many influences, the study of Baltic eugenics is of particular interest in regard to the specialised history of eugenics and bio-politics. Due to the close proximity to Germany, there were close connections with the Baltic-German middle class and scientists, as well as strong contacts with Scandinavia. In the case of Lithuania, eugenics was influenced by Catholicism. The Russian debate over Darwin and "degeneration" was also received in the Baltic region, as it was a part of Tsarist Russia until 1918 as I will show later.

Reasons for the late start of researching Baltic eugenics is clearly due to the years of Soviet occupation. Studies on the "nationalist and fascist regimes" of the 1920s and 1930s were not tolerated by the ruling communist parties in general. After 1991, Baltic historical science had to struggle with transforming history from being a political tool of the Communist Party back to an academic science. Beyond the typical problems of post-Soviet science, Russian and Soviet eugenics and genetics since the 1960s have been described by Russian scholars as victims of Stalinist terror and "Lysenkoism", in contrast to the perception of eugenics in Western Europe during this period.[8] Another important factor is the fact that racial (physical) anthropology was a normal academic discipline at Baltic and Russian universities referring to the racial framework until the early 1990ies.

The contemporary history of medicine in the Baltics as a discipline has failed to develop fundamental methodologies for analysing and criticising its own discipline and its "national heroes" who were invented in the last decades. Further, historical science in the Baltics in general is trapped in what could be called the "national historical discourse", where history is used to create national identity instead of producing scientific knowledge. One example could be the 2003 conference in Riga where Gustav Backman, the first anatomy professor in Latvia, was remembered and praised by a gathering of the Pauls Stradiņš Museum of the History of Medicine. Backman was the founding father of Latvian racial anthropology and an ardent eugenicist.[9] It is no surprise that the first reaction to the new approaches towards the history of eugenics in Latvia was somehow apologetic, assuming that Latvian eugenics were barely "racist" but merely intended to help "poor families".[10] This dispute is also apparent in this volume, through the different perspectives on eugenics in Latvia by Vladimir Kuznecovs and Björn Felder, as the latter questions the approach which labels Latvian eugenics as "liberal". It would seem that writing the history of eugenics and race in the Baltic states is still at the beginning stage. This volume seeks to contribute to this process.

1. Debates on Degeneration, Small Nations and Russian Eugenics

Eugenics-related ideas reached the Baltic provinces of the Tsarist Empire very early. They were associated with Darwinism, debates on degeneration, heredity, and the campaign against alcoholism, tuberculosis and sexual transmitted diseases.[11] In the St. Petersburg region, a vivid debate on eugenics began before First World War, especially involving professors and lecturers at the Military Medical Academy (*Voenno-medicinskaja akademija*).[12] Dorpat as a German University until russification in the 1880ies had attracted leading German medical students. The journal *Gigiena i Sanitarija*, published from 1910 to 1913 by Nicolaj Gamaleja, a well-known public health expert and lecturer at the Military Medical Academy and the University of Yur'yev' (Dorpat/Tartu), was a main journal exploring Western debates on eugenics. And, the Moscow-based journal *Priroda* frequently featured eugenics in the years leading up to World War One. Books by Western eugenicists were translated into Russian, including the works of Agnes Bluhm, Charles Davenport or Francis Galton.[13] Additionally, Russian researchers began publishing their first articles on the issue. The first eugenics monograph likely was the 1914 brochure *Basic principles and means of racial hygiene* by Evgenii Shepilevskii, a professor of bacteriology and hygiene at the University of Yur'yev'. Shepilevskii studied and worked at the Military Medical Academy in St. Petersburg and also studied at the Louis Pasteur Institute in Paris, and with Robert Koch in Berlin.[14] These eugenics debates, bio-medical and public health circles were the critical mass that produced Soviet eugenicists as Jurii Philipchenko and Nicolai Kolcov.[15]

The eugenics debates by Russian eugenicists at the Military Medical Academy in St. Petersburg and the University of Yur'yev' were crucial for the further development of Baltic eugenics. Looking at Baltic eugenics, one could imagine how Russian eugenics might have developed without the Bolshevik Revolution. Most Baltic eugenicists and physicians studied at one or both of these universities. Early debates in Russia were influenced more by German racial hygiene, and more by the genetics of Mendel and the deterministic genetic approach of Lombroso than by Lamarckist genetics.[16] This was especially the case at Yur'yev' with (speaking in eugenics terms) the radical proto-eugenics of psychiatrist Emil Kraepelin, who included eugenic ideas in his writings before the term "eugenics" was created in 1881 and the whole model of eugenics became known to a broader public. Kraepelin worked on the connection of alcoholism and mental diseases and was already interested in issues of heredity.[17]

There was also the political conservative and radical follower of Lombroso, psychiatrist Vladimir Chizh, who lectured at Yur'yev' and who followed Kraepelin as the chair of psychiatry. Gamaleja and Shepilevsky also lectured on eugenics. Shepilevsky even began to establish eugenics research at his institute, which was put to an end to by the onset of war in 1914. Concluding on the early Russian eugenics debate we have to question the assumption of Mark Adams, who in his nonetheless groundbreaking contribution to the history of Soviet eugenics (which recently was followed by Nikolai Krementsov) argued that eugenics was merely perceived by Russia as a means for "experimental biology". Looking at eugenics publications, it is clear that Russian scholars saw eugenics as a bio-political tool to shape the national body and solve health and social problems. For Shepilevsky, who was as enthusiastic for sterilisations as Gamaleja, racial hygiene was a new revolutionary means to "clean the race of criminals, alcoholics, mental ill and idiots – in other words, of all kind of degenerates".[18] Interestingly enough, both Gamaleja and Shepilevsky worked in the field of hygiene and saw eugenics in this context. It is not surprising that the early Soviet Union produced the first state agency dealing with public health, the Peoples Commissariat of Health, where social hygiene was declared to be the new national paradigm. Eugenics became an important tool of this new agenda.[19]

The fight against alcohol and sexually transmitted diseases, which were deemed to have mutagenic and teratogenic effects on offspring, constituted part of the debate on public health as a field for applied eugenics. In the late 1890s Russian physicians debated the most appropriate eugenics means in this field: For example at the meetings of the Livonian physicians in the 1880ies, the contributors decided to propose to the Russian government in St. Petersburg that it introduce mandatory examinations of students and prostitutes and the isolation of prostitutes and also discussed the ban on marriage for syphilitics to improve national heath and to limit the genetically harmful effects of sexually transmitted diseases.[20] There is a clear continuity from these late 19th century Russian eugenics debates to the Soviet ban on marriage of 1926 or the Latvian ban on marriage of 1937 both of which targeted mental and sexually transmitted diseases.[21] The eugenics agenda cannot simply be reduced to the general debate on genetics. It was part of a broader general debate on health and hygiene, and indeed this debate was to be found even in culture and the arts.

A significant coincidence lies with early Russian debates on eugenics and Baltic national movements. In the Baltic provinces of Tsarist Russia, social class corresponded to ethnicity. The urban German-Baltic middle class and aristocracy had the political and economic power, while Latvian and Estonians were mostly farmers, who were only slowly developing a middle class. With the beginning of Russification in the 1880s, the Tsarist government tried to limit German influence and wanted to foster a national consciousness toward a strategy of *divide et impera*. The outbreak of Baltic nationalism peaked in 1905. The revolt became violent due to social and ethnic tensions, and in a way had the character of an anti-German pogrom. Race and eugenics soon became part of the national debate. Estonian intellectuals saw "race" as a new means in their national struggle to symbolise the nation on a new and deeper level.[22] This was due to a general European shift toward the biologised nation in context of eugenics and race at that period, as Paul Weindling demonstrates in his chapter.

Eugenics was seen by physicians as a necessary strategy for the "small nation" of Estonia to secure its future. At the time, eugenics was considered in terms of public health what might be seen as bio-politics. The anti-alcohol movement had an important role in the pre-war Estonian national movement, and it also promoted eugenics ideas. [23] Estonian psychiatrist Juhan Luiga's 1904 dissertation at Yur'yev' addressed the problem of mental disease among the Baltic population.[24] Luiga connected the social situation of Estonian farmers to the problem of alcoholism and degeneration. Latvians did not embrace eugenics as much as Estonians, because in this different social structure there was already a strongly developed class of industrial workers; Latvian intellectuals discovered Marxism as a useful tool to articulate their demands for national emancipation. Genetics, degeneration and eugenics were also discussed among Latvian intellectuals, and were spread to a broader public.

In 1909, the medical student Hermanis Buduls published the first monograph on eugenics, which he called "racial improvement", in Latvian and the first in the entire Russian Empire. Buduls, who became the leading psychiatrist in independent Latvia and also an important figure of Latvian eugenics, referred mainly to the early German eugenicist Wilhelm Schallmayer and underlined the dangers of "degeneration".[25] Interestingly enough, the most important publications by Western eugenicists were received soon in the Baltics, including the influential and radical book by Geza von Hoffmann on racial hygiene in the United States. Shepilevsky cited Hoffmann in 1913, the year Hoffmann's book was published.[26]

This period of positivism and rising nationalism before First World War shaped the outlook or what Ludwik Fleck referred to as the *Denkstile* of the Baltic eugenicists. This generation of bio-medical scholars founded and organised their sciences after the war. They also assumed the leading role in the eugenics debate beginning in 1918 and in establishing national eugenics projects in the 1930s, the decade in which they became professors or head of institutes. This circle also educated a new generation of eugenicists at the notably nationalistic-oriented Baltic universities of Tartu, Riga, Vilnius and Kaunas. These newcomers looked to Germany, as their doctoral supervisors or doctor fathers did. An entire generation of Latvian psychiatrists and neurologists was taught by Ernst Rüdin at the Kaiser Wilhlem Insitute for Psychiatry at Munich.[27] Estonian eugenicists such as Hans Madissoon and Juhan Aul travelled to the Kaiser Wilhelm Institute for Anthropology, Human Genetics and Eugenics, in Berlin. Lithuanian racial anthropologist Jurgis Želinskas was taught by Rudolf Martin, the father of German racial anthropology, at the University of Munich in the early 1920s.[28] Volker Roelcke's contribution in this volume illustrates the importance of Rüdin for eugenics and genetics in Western Europe even after World War II. The transfer of eugenics knowledge and the influence of Germany's eugenics law of 1933 (*Gesetz zur Verhütung erbkranken Nachwuchses*) on Latvian legislation is explored in the chapter by Vladimirs Kuznetsovs.

Even if the eugenicists shared similar education and socialisation at the same alma mater, the particular eugenics debate and the path to the application of eugenics differed to some degree in each Baltic country. At least, the events in Estonia and Latvia were similar. As Ken Kalling illustrates, eugenics was quite popular after the First World War in the Estonian Republic. Established in 1924, the Estonian Eugenics Society demanded that a eugenics research institute be founded that was similar to the Institute for Racial Biology in Uppsala. Further, there was broad debate on eugenics in Estonia since the end of World War One, leading to the idea that eugenics was well known within all layers of Estonian society before the Second World War.

Debates on eugenics in Lithuania and Latvia, in a broader sense, began only in the start of the 1930s but intensified after 1934, when authoritarian rule was introduced in Latvia. Interestingly, the eugenics debate in the southern neighbour Lithuania, where the authoritarian rule of Antanas Smetona began in 1926, peaked in 1937, with a majority of Lithuanian bio-medical specialists making strong demands for sterilisations.[29] During the 1920s, mainly members of the Lithuanian bio-medical elite supported eugenics laws. These were newcomers and leading academic figures as Jouzas Blažys, who in 1935 became a professor of psychiatry at the University in Kaunas, as Björn Felder and Arūnas Germanivičius illustrate in their chapter. Actually, after writing about genetics and heredity in 1921, Blažys demanded the mass sterilisation of mentally ill and "feeble-minded" people already in 1926.[30]

In Lithuania, eugenics mainly was discussed by psychiatrists, though researchers in medicine and biology, public health specialists, politicians, lawyers also joined the debate. In contrast to both Estonia and Lithuania, Latvia had a small but growing community of plant geneticists who also took part in the eugenics debate in the early 1920s. Some of them not only taught genetics, which mainly was a subject for psychiatrists in terms of human genetics and pathology, but they were also very interested in eugenics and race, leading them to join the early debate in Latvia.[31]

While Baltic authors quite often referred to German racial hygienists such as Wilhelm Schallmayer and Alfred Ploetz, Soviet eugenics also had a presence in Latvia. Gustavs Reinhards, an ophthalmology lecturer at the University of Latvia, had Soviet abortion commissions in mind when he proposed controlled abortion in the context of eugenics in 1934.[32] He was the first university teacher who lectured on eugenics in mid-1930s Latvia. Actually, it was Reinhards' model that set the framework for Latvia's eugenics law of 1937. There are more examples of Soviet-Latvian contact. The genetic textbooks by Jurii Filipchenko and eugenicists from Leningrad was received in the Baltic countries. Filipchenko read Latvian biology journals,[33] perhaps because among his assistants at the Eugenics Bureau in Leningrad were Latvians such as Jānis Lūsis, Denis Lepiņ and A. Zuitiņš. Lūsis went to Latvia after Second World War and became a genetics professor in Riga. The Leningrad Bureau also had contact with the Estonian Eugenic Society. Another link is the Latvian surgeon Paul Stradiņš, one of the most important eugenics advocates in interwar Latvia, and who became an important figure in Latvian medicine after World War II. In the early 1920s he studied neuro-surgery in Leningrad and wrote his dissertation on this new field of psychiatry. He may have already been following a eugenics agenda at this time.[34] Just how deeply eugenics ideology influenced the political elite early in the Soviet Union is illustrated in Vsevold Bashkuev's chapter on the Soviet eugenics project that worked with German health specialists to fight syphilis in Buryatia.[35]

It is remarkable that before the 1930s, early Baltic eugenics enthusiasts took a radical approach – influenced by Germany – that in most cases favoured sterilisation and abortion. Therefore, it is not surprising that the Latvian abortion law of 1933, which legalised abortion in terms of "social" indications, permitted abortion based on eugenics indications. Abortion was legal in cases of children who would have been born with "severe mental and physical defects".[36] In her chapter, Ineta Lipša reveals that the abortion debate in Latvia was dominated by conservative opponents of liberal abortion, who argued about the "biological crisis" and pro-natalism for the sake of the nation. While eugenics was not discussed in this context, eugenic abortion seemed to be a broad agreement. This was actually the case in Catholic Lithuania, which also adopted an abortion law in 1935 that allowed abortions based on "medical indications" avoiding the term "eugenics"; therefore, eugenic abortions were also possible in Lithuania from 1935.[37]

With its 1933 abortion law, Latvia took the first step in the Baltics in the field of applied eugenics. This was not the first and only prohibition, as a marriage ban for mentally ill people was in force. This ban derived from Tsarist civil law was still in effect in all Baltic states until the 1930s. A civil law was introduced in Latvia in 1937 that maintained the marriage ban on mentally ill and people with sexually transmitted diseases.[38] In 1934 the first eugenics counselling bureau was opened in Lithuania. This type of bureau appeared in Estonia since 1927. In Latvia they were state funded only after 1938, but as a private initiative existed since 1929.[39] It was already under democracy, that the basis for "negative" eugenics was introduced in the Baltic states. But the more radical eugenic methods and broader national eugenic programmes were only possible under the later authoritarian rule.

2. Baltic Authoritarianism – Longing for the Racial State

It was part of the bio-political agendas of authoritarian regimes in Estonia under Konstantin Päts and Kārlis Ulmanis in Latvia that eugenic projects for national regeneration were established. It is difficult to call these authoritarian regimes "fascist". Even if these were strongly influenced by Italian fascism, one could hardly call them "totalitarian" in the classical definition of Hannah Arendt.[40] Baltic authoritarianism shared fascist features such as the chamber system in government, anti-Marxism, fascist aesthetics regarding uniforms, and the cult of personality. Latvia also had a governmental youth organisation. There were ruling parties in Estonia and Lithuania but not in Latvia. None of the regimes established a double administration of party and state, a central ideology or a mandatory manifesto as a type of "holy book". Most importantly, there was very little political violence and no physical elimination of "enemies". Actually, the far right in Estonia, Latvia and Lithuania was somehow more repressed than were Social Democrats. The regimes in Estonia and Latvia claimed that they had to prevent a political takeover by right-wing extremists such as the Estonian *Eesti Vabadussõjalaste Liit* (Estonian Union of Freedom Fighters) movement and the Latvian *Pērkonkrusts* (Thundercross). Ulmanis in Latvia banned all parties including his own *Farmers' Union*. His later ministers were drawn from several political directions.[41]

What all three dictatorships had in common was a form of agro-nationalism. The small farmer was praised and glorified as the basis of the nation and as the example for every citizen. In a way, Baltic authoritarianism was less futuristic than the regimes in Germany, Italy and the Soviet Union. This was due to the fact that two-thirds of the population was working and living in the countryside. And, the Baltic elites were farmers themselves, or at least children and grandchildren of farmers. Even statesmen such as Ulmanis, who studied agriculture in the US before the First World War, had their own private farms and vacationed during the hay harvesting in summer. Thus modernity and "modernisation" was one of the most important basics of the ideology. Therefore following the *Zeitgeist* in Central Europe, the regimes pursued national, biological regeneration closely linked to "modernisation".[42] The "improvement" of "quality" and "quantity" was especially important, as these were "small nations", at least according to the general self-perception. Bio-politics was implemented in the form of a national eugenics project in Latvia and Estonia. In 1936, the Latvian Council of Ministers decided to intensify research into race and eugenics, and to expand propaganda on both topics.[43] The national eugenics project had three main pillars: first, instruments of "positive" eugenics in the form of marriage counselling bureaus and pro-natalist measures; second, "negative" eugenics in the form of eugenic abortion and sterilisation; and third, eugenics propaganda. The pro-natalist measures were introduced even before the eugenics laws. In Estonia and Latvia, the programmes included financial support for families who moved to the countryside, government subsidies for children up to the age of 10, legal strengthening of women's rightsto claim alimony, credits to build housing for young couples in the countryside (similar to the German *Ehestandsdarlehen*), and health insurance for rural workers. All of these programmes were linked to rural workers and agro-nationalism: the small farmer was also seen as a biological and genetic resource to re-juvenate the nation - and this also had an ethnic agenda.[44] The Baltic agenda at the time followed the early eugenics agenda of "inner colonisation" in terms of improving population quality.

Laws concerning eugenic abortion and sterilisation were enacted in Estonia in 1936 (coming into force in April 1937) and in Latvia in 1937 (coming into force in January 1938). In Estonia the operations could be conducted against the will of the patient.[45] Estonia's law listed diagnoses of mental diseases, feeble-mindedness, epilepsy and physical handicaps as indications for sterilisation. Latvia's law did not mention any concrete diagnoses, which eugenicists labelled as a sign of progress and modernity, as eugenics practices could be adapted to the latest findings of genetic research.[46] The outcome in both countries was modest. In Estonia, there is official data for 41 sterilisations conducted from 1937 to 1939. We do not have complete numbers for abortions. It is clear that from summer 1939 to June 1940, the period of Soviet occupation that put an end to Estonian sterilisations, some 20 more sterilisations must have been conducted. There is no known evidence that the sterilisation programme, which was part of Estonia's national eugenics project, continued after the Nazi German occupation that began in 1941. However, most psychiatric patients were killed in the Baltics during German occupation – in Latvia by shooting and in Estonia and Lithuania by starvation. At least in the latter cases the local administration was involved in the Nazi "Euthanasia".[47]

In Latvia there were 63 documented cases of sterilisation during 1938 and 1939, but we do not have any numbers for 1940. There may have been some 20 more. Analysing the sterilisations in both countries, it is clear that the typical victim was "feeble-minded" and female.[48] About half of the victims were the so-called "feeble-minded". The diagnosis of "oligophrenia" was quite vague and included the subcategories of "idiocy", "imbecility" and "debility". Today, we would speak of a person being mentally handicapped, but also socially weak, under-educated or simply deviant. Therefore, "negative" eugenics was aimed at the under classes: this actually was a common policy in Europe at the time. The "feeble-minded" and "schizophrenics" were not only targeted in Nazi Germany but also in other countries with eugenic laws. In the case of Latvia, there are also hints that the diagnosis of "schizophrenia" often stood for "feeble-mindedness". Research on "feeble-minded" people was conducted at sites such as the Latvian

Eugenic Research Institute (*Tautas dzīvā spēka pētīšanas institūts*, Institute for Research on National Vitality), which covered the field of hereditary pathology. Hermanis Buduls, who was Latvia's leading psychiatrist, head of the Sarkankalns psychiatric hospital, head of the Institute for Psychiatry at the University of Latvia and dean of the medical faculty for several years, denounced the feeble-minded. In a 1936 article, he uttered his scepticism about eugenic sterilisation, arguing that genetic research did not yet support sterilisation for mental diseases. Still, he admitted that sterilisation would be a perfect method of eugenics for "feeble-minded" people, as this condition was clearly a hereditary disease, and the feeble-minded would have many more children than "normal" people.[49] Further, in the case of Latvia, there was a hunt for the "feeble-minded", as the administration and physicians in rural districts denounced "feeble-minded" people to the Riga Eugenics Commission. This did not occur by chance: the Latvian eugenics institute had been educating medical personnel and civil servants in the field of eugenics since 1938. The denounced victims, who were also candidates for abortions, often were described as poor alcoholics with many children and living in difficult social situations. There are also hints that a further argument in support of sterilising the under-class was to avoid the financial costs of social support.[50] Purging the Baltic under-class of its biologically and socially "inferior" parts accorded with the wider framework of the European context, and was similar to Nazi Germany and Sweden. This is illustrated in the chapter by Maija Runcis on sterilisation in Sweden. "Racial" purity, in the sense of "genetic purity" was meant to "reduce" the biological and social underclass: this was one aspect constituting the bio-political state.

Eugenics propaganda was another important part of national eugenics projects. It must be recognised that following the models of eugenics theorists, the ultimate aim was to develop the eugenics-conscious citizen. In all three Baltic states, eugenics propaganda spread during the mid-1930s. Even in Lithuania, which had no extensive eugenics agenda, bio-medical elites expanded the publication of articles on eugenics and sterilisation. In contrast to the politics of the Smetona regime, Lithuanian physicians demanded eugenic sterilisation at their 1937 congress in the capital Kaunas.[51]

In Latvia and Estonia, articles on eugenics even appeared in the daily press, especially after the introduction of sterilisation legacy. The rhetoric of strengthening the "national vitality" was a foundation of Baltic eugenics, a topos that frequently has found favoured by European eugenicists. "National vitality" meant the biological potential of a nation to survive the Darwinian struggle for existence. It was connected to both pro-natalism and eugenics. However, Latvian eugenicists emphasised that "quality" was more important than "quantity". Latvian eugenicist Verners Kraulis, who promoted pro-natalism and wanted to increase the birth rate, said there should be no support for "inferior families". Mothers of such families could be "only offered sterilisation".[52] Eugenics goals limited pro-natalist efforts.

Propaganda spread also in the form of public lectures and exhibitions. In spring 1939, an exhibition called "Work and Leisure" opened its gates in the centre of Riga. It informed visitors about statistics on industrial and rural work, production, occupational hygiene as well as demography, and eugenic and racial anthropology.[53]

In particular, race became an important issue among the authoritarian Baltic regimes during the 1930s. After the putsches in Estonia and Latvia, the discourse of national self-perception and national identity contained biological definitions of the nation that were much stronger than before. Race became a frequent topic in media and propaganda. While Päts and Ulmanis did not directly mention the racial state, they often referred to "blood" as a symbol of the nation. In general, the new nationalism that accompanied the new regimes was connected to the *völkisch* ideologies that were widespread in Europe at the time. This was also the case in Lithuania, where "purity of blood" at times was linked to elements of anti-Semitism.[54] In contrast to the emerging strong nationalism, anti-Semitism was officially banned by all three dictators, even in Lithuania, where anti-Judaism had its deepest roots. Still, this did not hinder anti-Semitic publications.

Even if there had been no direct anti-Semitism, each regime intended the homogenisation of a nation in cultural and biological terms. In Estonia and Latvia, the German minority was particularly targeted by a repressive language policy, expropriations and the reduction of cultural autonomy. In the Baltic economic sphere, where Jewish entrepreneurs played an important role, the Jewish minority was harmed.[55] There was also an unofficial policy that excluded non-autochthones such as German, Jews and Russians from the civil service. "Latvianisation" and "Estonisation" also affected the realm of bio-politics. The nation was seen as a biological entity, and its "purity" should be saved or restored. Latvians and Estonians were asked to procreate not only to increase the national population, but also not to marry partners from other ethnicities. This ideal of national purity was typical among Baltic Germans and to some extent among Jews in the sense of building a strong community. The racial state further meant to "improve" national health and "national vitality" by means of eugenics. It seems that in Latvia, and even more so in Estonia, eugenics measures such as abortion and sterilisation were performed more on autochthons than the minority population. The hunt for "feeble-minded" people and the policy of sterilising the socially weak and uneducated seem to have been features of the Baltic-style racial state. Baltic theorists of the racial state followed the paradigms of bio-politics: the new state should be united and homogeneous in terms of culture, ethnicity and race. For Latvian nationalist Ernests Brastiņš, race was an important topic of his state model. He had the idea that sooner or later the new nation would also form a monolithic union in terms of biology: through "breeding", Latvians would become "one race".[56]

In terms of Baltic authoritarianism, race and racial anthropology played important roles. In Latvia, Jēkabs Prīmanis was one of the leading lobbyists for eugenics, as Björn Felder illustrates in his chapter. Prīmanis spoke of a racial utopia that would again shape the Latvian nation into what it once was, as Prīmanis thought, a representative of a pure "Nordic" race. Therefore, mixed marriages with "inferior" races and ethnic minorities such as Jews, Russians and Poles should be avoided. Also limited should be the biological influence of Latgallians, the ethnic Latvian population of the eastern parts of the republic who Prīmanis also suspected of consisting of "East Baltic" and other "inferior" races.[57]

This connection of racial anthropology to bio-politics may be the point at which Baltic authoritarianism met with European fascism. Racial purity was a main feature of Baltic regeneration . Racial anthropologists were obsessed with the idea of finding the ancient root of the nation. They were not only engaged on writing bio-history out of bones, but also in creating a racial identity. One main task was to reconstruct the racial history of the nation as being part of the "Nordic race". As Estonians were described as "Asians" by Western European anthropologists, Estonian racial anthropologists eagerly tried to reverse this by promoting the image of a "Nordic" and European Estonian nation, as Leiu Heapost and Ken Kalling illustrate in their contribution to this volume. Maciej Górny examines in his chapter how scholars throughout Europe pursued the goal of writing racial history by praising one's own nation and condemning the other: "the old game", as Górny states.

The categories and hierarchies of racial anthropology that were important for eugenics programmes and defining the goals of human breeding for the sake of the nation. These categories were closely connected to the debate on the value of individuals, and the debate on "superiority" and "inferiority" that haunted European society at least since the imperialist age of the late 19th century. In medicine, patterns of thinking concerning the "inferior" were widespread, especially in fields that administered human pathology, such as psychiatry. Here, the debate on "inferiority" was most obvious, and the pitiable patients were often treated as such. It simply would be hard to bring a "living dead" – as a schizophrenic was labelled by a Baltic psychiatrist – back to life.[58] In psychiatry, the line between medical treatment and human experiment grew thin. Newly established "shock" therapies that used cardiazol, insulin and, later, electro-convulsion. Perhaps they only occasionally cost human lives, but they were extremely painful mentally and physically. Obviously, physicians at times may be tempted to give in to sadism or simply to academic fervour, overlooking compassion and empathy.[59] This kind of debate determined or at least influenced the discourse of "worthy" and "unworthy" life in the racial state, as this unfolded in Estonia, Latvia and Lithuania.

4. Conclusions

The implementation of practical eugenics in the Baltic states was moderate compared to Germany and Scandinavia. Even the compulsory sterilisations in Estonia produced fewer victims than in most other countries. In general, Baltic eugenics followed the European trend towards bio-politics and national rejuvenation, with its implications in the field of politics and the biologised image of nation. The strong influence of German racial hygiene, as well as the influence of Russian and Soviet debates and eugenic practices, was obvious. The forming of the racial state in the 1930s also differed from the examples in south-east European states and Nazi Germany. Race was a topos for a "pure" and "original" nation, free of "foreign" influences. Race, connected to ideas of a "Nordic race" current in anthropology, was an important tool of self-legitimising for the nation, creating both racial identity and national self-consciousness. The "Nordic" nation would have been "worthy" and could not be ignored by the old, established western European nations. Race also had a utopian side, as racial and national theorists planned to render the nation biologically homogeneous. The racial state should also become a pure ethnic state, and should eliminate the multi-ethnic communities in Latvia, Lithuania and Estonia. But again, the practical policies were modest. Ethnic cleansing took place only after World War Two began, and it was initiated by the occupation powers, starting with the "return" of the Baltic Germans from Latvia and Estonia in autumn 1939. Actually, this gave an example of a possible "peaceful" and non-violent ethnic cleansing that was welcomed by all sides – still it was forced upon the Baltic Germans by Nazi German. With the murder of Baltic Jews that followed in 1941, another, extremely brutal ethnic cleansing followed. Again, this genocide was initiated and carried out by Nazi German invaders, even if local anti-Soviet partisans and police officers were involved. Local "governments" under Nazi occupation had no direct influence on Nazi atrocities and genocide, but we have the impression that local actors may not welcome the violence but did so as regards the final outcome: after the Jews were killed, Russians and Poles were deported for slave labour. The Baltic nations came close to being an ethnically homogeneous nation and can be termed ethnic states. While anti-Soviet partisans partially conducted ethnic cleansing in the first weeks of the war, local actors even openly demanded ethnic cleansing.[60] In Estonia the new leadership favoured the deportations of Russians and dreamt of a geographically enlarged Estonian empire. In Lithuania the local health administration demanded the deportation of non-Lithuanian psychiatric patients in January 1942. In Latvia a first "National

Assembly" in July 1941 demanded in its agenda the ethnic cleansing of all non-Latvians.[61] It must be said that ethnic tensions were high in the summer of 1941 after the Soviet Terror and mass deportations, and due to the fact that because of selective Soviet ethnic mobilisation, the Stalinist regime was perceived by the locals as "Russian" and "Jewish" rule.[62] Still, the agenda of the ethnic state in 1941 was strongly connected to the pre-war racial patterns of thinking, eugenics and bio-politics in general. The racist and radical fascist movements – the *Eesti Vabadussõjalaste Liit* (Estonian union of freedom fighters) of Estonia, the *Pērkonkrusts* (thundercross) of Latvia, and the *Geležinis vilkas* (iron wolf) of Lithuania – were only in Estonia directly involved in the so-called "self-administrations" under Nazi occupation. In Lithuania and Latvia these groups were banned, even though their supporters joined military formations or became civil servants.[63] In general, the "self-administrations" had more in common with the pre-war regimes, and continued their bio-political agenda. In Latvia the national eugenics programme continued and was enlarged due to its pre-war programme.[64] The Nazi occupation made possible the implementation of the racial state in the Baltics through more radical methods.

The history of eugenics in the Baltics adds a further example to the diverse picture of Central and Eastern European eugenics movements. Characteristically, Baltic eugenics followed the pro-natalist approach due to small populations and references to agro-nationalism, stemming from the social structure. Despite similar history and culture, the eugenics debates and implementation differed among the three countries. Lithuania, with its very modest governmental eugenics policy, might be a special case, and could be compared to the Polish example. Even if eugenics as an ideology of national rejuvenation became a main agenda item in the region, this should not ignore the fact that there was a left-wing form of eugenics, at least in Estonia.[65] This is one of many open questions in the field of Baltic eugenics. The research presented here opens a new perspective in terms of politics, culture and science in the inter-war republics of Estonia, Latvia and Lithuania.

Notes

[1] See the letter of the IFEO to the Estonian Eugenic Association of March and May 1938: Eesti Riigiarhivi (Estonian State Archive – further ERA) 4855/1/6, 120, 149, 152.

[2] For Madissoon see his personal file at Tartu University: Eesti Ajalooarhiiv (Estonian Historic Archive – further EAA) 2100/2/596. About Latvian psychiatrists visiting Rüdin see: Felder, 2009.

[3] On the racial state see: Burleigh and Wippermann, 1991; Turda, 2010 and Weindling, 1991.
[4] Prīmanis, 1939, 6.
[5] Stranga, 2002, 209.
[6] Broberg and Roll-Hansen, 1996; Turda and Weindling, 2007.
[7] Some single works by Soviet authors condemning „bourgeois" eugenics in the 1950s: Zakas, 1959; Rikše, 1978. The first scholarly works were published only recently. For Estonia see: Kalling, 1999; Kalling, 2007a, for Latvia see: Felder, 2005; Felder, 2009a. For Lithuania see recently: Germanavičius, 2012.
[8] The first narratives of Soviet genetics (eugenics) were written by students of the early Soviet eugenicists - eugenicists themselves: See f. e.: Lusis, 1973; Medwedjew, 1971.
[9] An example of the conference might be: Lindberga, 2003. On Backman see the contribution by Björn Felder in this volume, see also: Kott, 2009.
[10] Zelče, 2006.
[11] On the reception of Darwin in Estonia see: Kalling and Tammiksaar, 2009. On the reception in Russia: Kolčinskij, 2011; Vucinich, 1988.
[12] For early eugenic debate in Russia before World War I see: Felder, 2011; Felder, 2012.
[13] On Gamaleja see: Felder, 2011. On Priroda see also: Krementsov, 2011.
[14] Šepilevskij, 1914; for Shepilevskij see: Felder, 2011.
[15] On the history of Soviet eugenics see: Graham, 1977; Adams, 1990; Krementsov, 2011.
[16] See: Beer, 2008; also Felder, 2011.
[17] On Kraepelin and degeneration see: Weber, Burgmair and Engstrom, 2006, A 2685-2690; Engstrom, 2007, 389–398. In general see: Steinberg, 2001; Steinberg and Angermeyer, 2001, 297-327.
[18] Shepilevskii, 1914, 56.
[19] Semashko, 1923.
[20] *Protokolle des vierten Livländischen Ärztetages in Wenden vom 14. bis 16. September 1896*, St. Petersburg, 1897, 41-44.
[21] On the Soviet Union see: Bernstein, 2010, 102. On Latvia see: *Civillikums* [Civil law] - Kodifikācijas nodaļas 1937 gada izdevums. 3. iespediums, 2. Nachdruck der Ausgabe Riga, 1938, Hamburg: Loeber, 1990[1938].
[22] Ridala, 1913; On the racial debate in Estonia see: Karjahäm, 1993.
[23] See the chapter by Ken Kalling on the Estonian eugenics in this volume.
[24] Lujga, 1904.
[25] Buduls, 1909; On Schallmayer's eugenic agenda see: Weiss, 1987.
[26] Shepilevskii, 1914, 72.
[27] Felder, 2009, 275-285.

[28] For Zhelinskas see his personal file at the Kaunas University: Lietuvos Centrinis Valstybes Archyvas (Lithuanian Central State Archive – further LCVA), 631-3-857.
[29] Blažys, 1939.
[30] Blažys, 1921; Blažys, 1926, 653-77.
[31] Ābele, 1929; Lūke, 1939.
[32] Reinhards, 1934.
[33] See his manuscripts at the Russian National Library in St. Petersburg: 813/1293.
[34] See also his personal file of the Latvian University: Latvijas Valsts Vēstures Archīvs (Latvian State Historical Archive – further: LVVA), 7427/13/1639.
[35] Solomon, 1993, 204-232. On the German-Soviet cooperation on the field of medicine and hygiene in general see: Solomon, 2006; Weindling, 2003.
[36] See the chapter of Ineta Lipša in this volume. On the abortion law see: *Likumi un Ministru Kabinetu noteikumu Krājums*. Riga, 1932, 311.
[37] See the contribution by Björn Felder and Aruna Germanivičius in this volume.
[38] Civillikums, 1990[1938].
[39] Blažys, 1939; On Latvia see the chapter by Björn Felder in this volume, on Estonia see the chapter on Estonian eugenics by Ken Kalling in this volume.
[40] In the broad definition of Roger Griffin they could be called „fascist", but not totalitarian following H. Arendt: Arendt, 1986. For the fascist theory by Griffin see: Griffin, 2006; Griffin, 2007.
[41] There are still only few works done on the Baltic dictatorships. On Latvia see: Butulis, 2001; On Estonia: Pajur, 2001; Mertelsmann, 2012; And on Lithuania: Lopata, 2002 and Richter, 2010. A general picture gave: Misiunas and Taagepera, 1983; Rauch, 1977. On the rightwing movements on Estonia see: Kasekamp, 2000. On Latvia see: Ezergailis, 1996; Felder, 2004 and Felder, 2009a.
[42] Turda, 2010.
[43] "Ministru cabinets" *Valdības vēstnesis*, 15. July 1936, 1. See also: Arājs, 2005, 148.
[44] On the pro-natalist instruments in Latvia see: Starcs, 1939.
[45] About eugenics legislation in Estonia see the chapter by Ken Kalling in this volume. For Latvia see Felder, 2009a, 275-285; and the chapter by Björn Felder in this volume.
[46] See the comment of the Estonian eugenicists Hans Madissoon: Madissoon, 1938.
[47] On Nazi „Euthansia" in Estonia see: Kalling, 2007b, 95-104. On Lithuania see: Felder, 2013a; and on Latvia see: Vīksne, 2003, 324–350; Felder, 2009a, 206, 287, 294-296. In general see also: Felder, 2013b.

[48] The Latvian sterilisation files can be found in the documents of the Latvian Eugenics Commission: LVVA, 4578/1/202, 4578/1/203, 4578/1/205, 4578/4/639, P-1023/1/40. For Estonia no documents of the Eugenics commissions are preserved. The sterilizations could only be partially reconstructed by analyzing patient files and operation books of several clinics as the First Women's Clinic in Tartu (EAA, 2451/2/184, 2451/2/186 and 2451/2/ /188). The First Wound Clinic in Tartu (*Tartu Ülikooli Kliinikum arhiivi* – Archive of the University Clinic Tartu), and the Tallinn City Hospital (*Tallinna Linnaarhiiv* – Tallinn Town Archive, 369/1/175, 369/1/178 and 369/1/179).
[49] Buduls, 1936.
[50] See the chapter by Ken Kalling on Estonian Eugenics in this volume.
[51] Blažys, 1939.
[52] Kraulis, 1939, here 167.
[53] Arājs, 2005.
[54] See: Vareikis, 2004.
[55] See: Rauch, 1977.
[56] See the chapter by Björn Felder on race and eugenics in this volume.
[57] Ibid.
[58] The Latvian psychiatrist Verners Kraulis described schizophrenic patients as "living dead": Kraulis, 1939, here 49.
[59] For human experiments in Lithuanian psychiatry see: Felder, 2013a.
[60] See the report of the district physician of Dünaburg to the Generalkommissar in Riga, 6 June 1942, on ethnic cleansing in Eastern Latvia: *Bundesarchiv Berlin* – Federal Archive Berlin, R-92/541.
[61] For Estonia see the Einsatzgruppen report of September 1941: *Bundesarchiv Freiburg – Militärarchiv* (Federal Archive Freiburg– Military Archive) 19III/470, 51-61. For Lithuania see: Letter from Matulionis to the Generalkommissar, 23 January 1942: LCVA, R-627/1/46, 63; For Latvia: Felder, 2009a, 253-265.
[62] Levin, 1995; Felder, 2009b.
[63] For Estonia see: Kasekamp, 2003. On Latvia see: Felder, 2004 and Felder, 2009a, 235-245, 309-311; On Lithuania see: Rauch, 1977 and Misiunas and Taagepera, 1934. On the local administration see for Estonia: Kasekamp, 2000; Weiss-Wendt, 2009; for Lithuania: Rauch, 1977 and Misiunas and Taagepera, 1934.
[64] See: Felder, 2013c (forthcoming).
[65] See the chapter on Estonian eugenics by Ken Kalling in this volume.

References

Ābele, K. (1929), "Rašu Higiēna (Eugenika)" *Daugava Literātūras, mākslas un zinātnes mēnešraksts*, vol. 3, 334-342.

Adams, M. B. (1990), "Eugenics in Russia" in: Adams, M. B. (ed.) *The Wellborn Science. Eugenics in Germany, France, Brazil, and Russia*. New York, Oxford: Oxford University Press, 153-216.

Arājs, K. (2005), *Latviešu kauli...Atmiņas par anatomikunu un antropoloģijas ziedu laikiem Latviā* [Latvian Bones...Memories on the the blooming Years of Anatomy and Anthropology in Latvia]. Riga: Rigas Stradiņa Universitāte.

Arendt, H. (1986), *Elemente und Ursprünge totaler Herrschaft*. München: Piper.

Beer, D. (2008), *Renovating Russia. The Human Sciences and the Fate of Liberal Modernity 1880-1930*. Ithaca, London: Cornell University Press.

Bernstein, L. F. (2010), "Behind the Closes Door - VD and Medical Secrecy in Early Soviet Medicine" in: Bernstein, L. F.; Burton, C. and Healey, D. (eds.), *Soviet Medicine. Culture, Practice, and Science*. Dekalb: Northern Illinois University Press, 92-110.

Blažys, J. (1921), "Paveldėjimo klausimas ir mendelizmas [The Question of Heredity and Mendelism]" *Medicina*, vol. 2, no. 3, 65-73.

Blažys, J. (1926), "Išsigimimo ir jo įspėjimo klausimas bendrai ir ypatingai Lietuvoje [The Question of Degeneration and its Prevention in general and especially in Lithuania]" *Medicina*, vol. 7, no. 12, 653-677.

Blažys, J. (1939), "Eugenika" in: *Lietuviškosios Enciklopedijos*, vol. 7., Kaunas:Varpas, 1250-1252.

Broberg, G. and Roll-Hansen, N. (eds.) (1996), *Eugenics and the Welfare State. Sterilization Policy in Denmark, Sweden, Norway, and Finland*. Uppsala Studies in History of Science, vol. 21. Michigan: Michigan State University Press.

Buduls, H. (1909), *Lauliba un zilweka dsihwes mehrķis. Bioloģisks un etisks apzerejums* [Marriage and the Goal of Human Life. Biological and Ethical Essay]. Riga: Ed.Sirgela.

Buduls, H. (1936), "Rases labidzimtība [Eugenics of Race]" *Jaunākās Ziņas*, 6 June, 4.

Burleigh, M. and Wippermann, W. (1991), *The Racial State: Germany 1933-1945*. Cambridge: Cambridge University Press.

Butulis, I. (2001), "Autoritäre Ideologie und Praxis des Ulmanis-Regimes in Lettland 1934-1940" in: Oberländer, E. (ed.), *Autoritäre Regime in Ostmittel- und Südosteuropa 1919-1944*. Paderborn [u.a.]: Schöningh, 249-298.

Engstrom, E. J. (2007), "On the Question of Degeneration by Emil Kraepelin (1908)" in: *History of Psychiatry*, vol. 18, no.3, 389-398.

Ezergailis, A. (1996), *The Holocaust in Latvia 1941-1944. The Missing Center*. Riga: Historical Inst. of Latvia.

Felder, B. (2004), ""Die Spreu vom Weizen trennen... " Die Lettische Kartei - Pērkonkrusts im SD Lettland 1941-43" in: Nollendorfs, V. (ed.), *Sphere of Influence*. Riga: OMF, 47-68.

Felder, B. (2005), "Mazvērtīgo samazināšana- eigēnika Latvijā [Reducing the Inferior - Eugenics in Latvia]" *Kultūras Diena*, 23 April, 16-17.

Felder, B. (2009a), *Lettland im Zweiten Weltkrieg. Zwischen sowjetischen und deutschen Besatzern 1940-46*. Paderborn: Schöningh.

Felder, B. (2009b), "Stalinismus als "russisch-jüdische Herrschaft". Sowjetische Besatzung und ethnische Mobilisierung im Baltikum 1940 bis 1941" *Zeitschrift für Geschichtswissenschaft*, vol. 57, no. 1, 5-25.

Felder, B. (2011), "Rassenhygiene in Russland. Der Hygieniker und Bakteriologe Evgenij A. Shepilevskij (1857-1920) und die Anfänge der Eugenik im Russischen Zarenreich. The Hygienist and Bacteriologist Evgenii A. Shepilevsky and the Beginning of Eugenics in Russia" in: Riha, O. and Fischer, M. (eds.), *Naturwissenschaft als Kommunikationsraum zwischen Deutschland und Russland im 19. Jahrhundert. Internationale Tagung. Leipzig, 29.2. - 1.10.2010*. Aachen:Shaker, 335-365.

Felder, B. (2012), "Rasovaja gigiena v Rossii. Evgenij Alkeceevich Shepilevskij (1857-1920) i zarozhdenie evgeniki v Rossijskoj imperii" *Istoriko-biologicheskie Issledovanija*, vol. 4, no. 2, 39-60.

Felder, B. (2013a), "National-Socialist Euthanasia, Human Experiments and Psychiatry in Nazi-Occupied Lithuania 1941-1944" *Holocaust and Genocide Studies* 2013 (forthcoming).

Felder, B. (2013b), "Nationalsozialistische Krankenmorde in den Baltischen Staaten und die baltische Eugenik der Zwischenkriegszeit 1918-1945" in: Friedmann, A. and Hudemann, R. (eds.), *Krankenmorde in Weißrussland*, (forthcoming).

Felder, B. (2013c), "In Pursuit of Biological Purity. Eugenics and Racial Paradigms in Nazi-Occupied Latvia, 1941-45" in: Weiss-Wendt, A. and Yeomas, R. (eds.), *Racial Science in Hitler's new Europe, 1938-1945* (forthcoming).

Germanavičius, A. (2012), "Lietuvos akademinės psichiatrijos raida 1918-1940 metais [Development of academic psychiatry in Lithuania during 1918-1940]" *Nervų ir Psichikos Ligos*, vol. 56, no.1, 12-14.

Graham, L. R. (1977), "Science and Values: The Eugenics Movement in Germany and Russia in the 1920s" *The American Historical Review*, vol. 82, no. 5, 1133-1164.

Griffin, R. (2006), *Fascism, Totalitarism, and Political Religion*. New York, London: Routledge.

Griffin, R. (2007), *Modernism and Fascism: The Sense of a beginning under Mussolini and Hitler*. Houndmill: Palgrave.

Kalling, K. (1999), "Introduction to the history of Estonian eugenics" *Annual Report 1998, Tartu University History Museum*, no. 3, 31-42.

Kalling, K. (2007a), "The Self-Perception of a Small Nation: The Reception of Eugenics in Interwar Estonia" in: Weindling, P. and Turda, M. (eds.) *Blood and Homeland: Eugenics and Racial Nationalism in Central and South-East Europe 1900-1940*. Budapest:Central European University Press, 253-262.

Kalling, K. (2007b), "Estonian Psychiatric Hospitals during the German Occupation (1941-1944)" *International Journal of Mental Health*, vol. 36, no. 1, 95-104.

Kalling, K. and Tammiksaar, E. (2009), "Descent versus extinctions: The Reception of Darwinism in Estonia" in: Engels, E.-M. and Glick, T. F. (eds.) *The reception of Charles Darwin in Europe*. London [u.a.]: Continuum.

Kasekamp, A. (2000), *The Radical Right in interwar Estonia*. Basingstoke [u.a.]: Macmillan.

Kasekamp, A. (2003), "The Ideological Roots of Estonian Collaboration during the Nazi Occupation" in: Koll, A. M. (ed.), *The Baltic Countries under Occupation. Soviet and Nazi Rule.1939-1991*. Stockholm: Dep. of History, 85-96.

Kolchinskij, E. I. (2011), "A German Trace in the Russian Evolutionism of the 19th Century" in: Riha, O. and Fischer, M. (eds.), *Naturwissenschaft als Kommunikationsraum zwischen Deutschland und Russland im 19. Jahrhundert. Internationale Tagung. Leipzig, 29.2. - 1.10.2010*. Aachen: Shaker, 406-421.

Kott, M. (2009), "Anthropologen Gaston Backman och den uppsaliensiska rasbiologins spridning i tid och rum" in: Müssener, H. and Jegebäck, P. (eds.) *Rasen och vetenskapen*. Uppsala: Centrum för multietnisk forskning. Programmet för studier kring Förintelsen och folkmord, 59-82.

Kraulis, V. (1939), "Eigenikas ceļi Latvijā [The Way of Eugenics in Latvia]" *Tautas Labklājības Minstrijas Mēnešraksts*, vol. 3, 161-168.

Kraulis, V. (1939), "Über die Behandlung der Schizophrenie mit protrahiertem Insulinchock" *Zeitschrift für die gesamte Neurologie und Psychiatrie*, vol. 164, 36-49.

Krementsov, N. (2011), "From "Beastly Philosophy" to Medical Genetics: Eugenics in Russia and the Soviet Union" *Annals of Science*, vol. 68, no. 1, 61-92.

Levin, D. (1995), *The Lesser of Two Evils. Eastern European Jewry under Soviet rule, 1939 - 1941*. Jerusalem, Philadelphia: Jewish Publication Society.

Lindberga, R. (2003), "Medicīnas Doktors, Professors Gastons, Viktors Bakmanis - Leģendāra personība Eiropas un Latvijas zinātnē [Medical Doktor, Professor Gaston Viktor Backman - Legendary Person of European and Latvian Science]" *The Professor Gaston Victor Backmann scientific conference*. Riga: Anatomicum, 42-44.

Lopata, R. (2001), "Die Entstehung des autoritären Regimes in Litauen 1926 - Umstände, Legitimation, Konzeption" in: Oberländer, E. (ed.), *Autoritäre*

Regime in Ostmittel- und Südosteuropa 1919-1944. Paderborn: Schöningh, 95-142.

Lujga, I. (1904), *Prizrenie duševnobol'niyh v Pribaltijskom krae* [The Medical care for Mental Illnesses in the Baltic Provinces]. Dorpat: Universität Dorpat.

Lūke, J. (1939), *Rāsu Lietā* [About Race]. Riga: Autora Izdevums.

Lusis, J. (1973), *Problemy genetiki i evoljucii* [Problems of Genetics and Evolution]. Riga.

Madissoon, H. (1938), "Lettlands Sterilisations- und Schwangerschafts-unterbrechungsgesetz" *Pro Juventute Baltica*, vol. 2, no.1, 1-9.

Medwedjew, S. A. (1971), *Der Fall Lyssenko. Eine Wissenschaft kapituliert*. Hamburg: Hoffmann & Campe.

Mertelsmann, O. (2010), "Führer im Ausverkauf. Estland 1939-1945" in: Ennker, B. and Hein-Kricher, H. (eds.), *Der Führer im Europa des 20. Jahrhunderts*. Marburg: Herder-Institut, 234-252.

Misiunas, R. and Taagepera, R. (1983), *The Baltic States. The Years of Dependence 1940-1980*. Berkeley: Univ. of California Press.

Pajur, A. (2001), "Die "Legitimierung" der Diktatur des Präsidenten Päts und die öffentliche Meinung in Estland" in: Oberländer, E. (ed.) *Autoritäre Regime in Ostmittel- und Südosteuropa 1919-1944*. Paderborn: Schöningh, 163-214.

Prīmanis, J. (1939), "Tautas dzīvā spēka pētīšanas uzdevumi [Tasks of Research into the National Vitality]" *Tautas dzīvā spēka pētīšanas institūta raksti*, vol. 1, 3-9.

Rauch, G. v. (1977), *Geschichte der baltischen Staaten*. München: DTV.

Reinhards, G. (1934), "Kā pasargāt tautas no bojās ejas [How to save the Nation from Decay]" *Universitas*, vol. 9, 134-137.

Richter, K. (2010), "Der Kult um Antanas Smetona in Litauen (1926-1940)" in: Ennker, B. and Hein-Kricher, H. (eds.), *Der Führer im Europa des 20. Jahrhunderts*. Marburg: Herder-Institut, 111-136.

Ridala, V. (1915), *Tõu küsimus* [The Racial Question]. Tartu: Postimees.

Rikshe, V. J. (1978), "“Rassovaja gigiena” na sluzhbe latishskoy nacionalisticheskoj burzhuazii [“Racial hygiene” in service of Latvian nationalist Bourgeoisie]" *Medico-socialnye izsledovanija*, 139-142.

Semashko, N. (1923), “Ochenednye zadachi social'noj gigieny v Rossii” *Socialnaja gigiena*, vol. 2, 1-8.

Shepilevskii, E. A. (1914), *Osnovy i sredstva rasovoj gigieny (gigiena razmnozhenija)*[Basic Principles and Means of Racial Hygiene (hygiene of procreation)]. Jur'ev': Mattisen.

Starcs, P. (1939), “Kas darīts, kas vēl jādara tautas dzīvā spēka pieauguma veicināšanai [What is done and what have to be done for the Advancement of the Growth of the National Vitality]” *Tautas Labklājības Minstrijas Mēnešraksts*, vol. 5, 351-362.

Steinberg, H. (2001), *Kraepelin in Leipzig. Eine Begegnung von Psychiatrie und Psychologie.* Bonn: Ed. Das Narrenschiff.

Steinberg, H. and Angermeyer, M. C. (2001), “Emil Kraepelin’s years at Dorpat as professor of psychiatry in nineteenth-century Russia” *History of Psychiatry*, vol. 12, 297-327.

Stranga, A. (2002), Ebreij un diktatūras Baltijā 1926-1940 [The Jews and the Dictatorships in the Baltics] Riga: Latvijas Universitāte.

Solomon, S. G. (1993), “The Soviet-German Syphilis Expedition to Buriat-Mongolia, 1928: Scientific Research on National Minorities” *Slavic Review*, vol. 52, no. 2, 204-232.

Solomon, S. G. (ed.) (2006), *Doing Medicine together: Germany and Russia between the wars.* Toronto: University of Toronto Press.

Karjahärm, T. (1993), “Tõuküsimus Eestis iseseisvuse eel. [Racial Issue in Estonia before the Era of Independence]” *Akadeemia*, no. 7, 1347-1365.

Turda, M. (2010), *Modernism and Eugenics*. Houndsmill: Palgrave Macmillan.

Turda, M. and Weindling, P. (eds.) (2007), *"Blood and Homeland": Eugenics and Racial Nationalism in Central and Southeast Europe, 1900-1940*. Budapest: Central European University Press.

Vareikis, V. (2004), "Anti-Semitism in Lithuania (Second half of 19th - First half of 20th century)" in: Truska, L. and Vareikis, V. (eds.), *The Preconditions for the Holocaust. Anti-Semitism in Lithuania (Second half of the 19th Century - June 1941)*. Vilnius: Margi raštai, 119-172.

Vucinich, A. S. (1988), *Darwin in Russia*. Berkely: University of Berkely Press.

Weber, M. M.; Burgmair, W. and Engstrom, E. J. (2006), "Zwischen klinischen Krankheitsbildern und psychischer Volkshygiene: Emil Kraepelin 1856–1926" *Deutsches Ärzteblatt*, 13 October 2006, 2685-2690.

Weindling, P. (1991), *Health, Race and German Politics between National Unification and Nazism, 1870 - 1945*. Cambridge: Cambridge Univ. Press.
Weindling, P. (2003), *Epidemics and Genocide in Eastern Europe 1890-1945*, Oxford: Oxford Univ. Press.

Weiss, S. F. (1987), *Race Hygiene and National Efficiency: the Eugenics of Wilhelm Schallmayer*. Berkeley: University of California Press.

Vīksne, R. (2003), "Garīgi slimo iznīcināšana Latvijā vācu okupācijas laikā" in: Caune, A. (ed.), *The Issues of the Holocaust Research in Latvia. Reports of an International Seminar 29 November 2001, Riga and the Holocaust Studies in Latvia 2001–2002*, Riga: Latvijas Vēstures Institūta Apgāds, 324–350.

Weiss-Wendt, A. (2009), *Murder without hatred. Estonians and the Holocaust*. Syracuse, New York: Syracuse University Press.

Zakas, I. (1959), *Rasizmas ir Eugenika buržuazinėje Lituvoje* [Racism and Eugenics in Bourgeois Lithuania]. Vilnius: Valstybinė politinės ir mokslinės literatūros leidykla.

Zelče, V. (2006), "Vara, zinātne, veslība, un cilvēci. Eigēnika Latvijā [Power, Science, Health and Men. Eugenics in Latvia]" *Latvijas Arhīvi*, vol. 3, 94-137.

Part 1

Eugenics in the Baltics

Race, Eugenics and National Identity in the Eastern Baltic: From Racial Surveys to Racial States

Paul J. Weindling

Abstract

This chapter outlines the emergence of the idea of an East Baltic Race against the background of a shift from multi-ethnic dynastic Imperial states to the states based on the interwar idea of national self-determination. Key issues were ethnic composition and social welfare. Here the issues of eugenics and ideas of race assumed increasing importance. The Austrian racial anthropologist Rudolf Pöch exercised considerable influence. His ideas of an "East Race" were taken up and developed by Michael Hesch, whose the "East Baltic" race thus marked a transitional point between earlier physical concepts of race and the nazified fusion of race and psychological attributes.

1. Nation and Race

The fascination of considering eugenics in the Eastern Baltic region is that fundamental historical processes of scientific innovation, social stratification and state formation intersect. Eugenics provides a historical bridge between contrasting forms of the state in European history. In broad terms, we are confronted by a shift from the great multi-ethnic European dynastic and imperial states that finally expired in the conflagration of the First World War, to the post-1918 birth of democracies based on the principle of national self determination (or, as in Bolshevik propaganda, "socialism in one country"). Eugenics was a key factor in the transformative process, which involved welfare provision on a selective basis, population policies and transfers, and migration controls. Eugenics and related forms of biologically rationalised social policies offer an alternative to interpreting the shift in terms of democratisation and national self-determination in the wake of the social upheavals in 1917-19. At the same time eugenics deepens the understanding of the emergence of nationalist movements by stressing biological and ethnic/racial identities in national state systems.

During the later nineteenth century tensions increased concerning the great powers blocs of Germany and Russia around the Eastern Baltic with the clash of policies of Germanisation and Russification, while at the same time the Scandinavian/ Nordic states exerted strong influence. As elsewhere in Europe tensions increased between population groups of varying cultural and ethnic background who had long co-existed. The Baltic region can be seen as a flashpoint in a transformation, culminating in the political independence and ethnic composition of the newly independent Baltic states. Yet the situation was complex, both because of the ethnic diversity of each of these states within a wider Eastern Baltic region, and because of long-running debates on the origins of the ethnic groups within a broader demarcation between "the European" and "Asiatic", often linked to a putative dichotomy between the civilised and the primitive. The Baltic region was thus an ethnological flashpoint, as well as an ideological bulwark for various ethnicities.

The later nineteenth century academic landscape across Europe saw a new feature with the founding of anthropological societies (notably the *Deutsche Gesellschaft für Anthropologie* and the *Anthropologische Gesellschaft* in Vienna, both in 1870).[1] These represented combined academic, state, and public and civic interest in anthropology, as a type of stock-taking and definition of the characteristics of imperial populations. The variations in human types were recognised but in an inclusive way as imperial administrations sought legitimacy that was increasingly challenged by nationalist movements. New methods of anthropological investigation were constantly being forged. The trend was from measurement of physique to biological methods such as serology, and analysis of genealogical patterns to correlate a range of traits. As a consequence surveys of racial types were undertaken with greater frequency focusing on populations touched by state and military administrations – notably school children, military recruits, and then prisoners. These revealed the variety of the peoples of each of the dynastic Empires. Museums became repositories of such survey material. The anthropological-ethnological department at the *Naturhistorisches Museum*, Vienna, founded in 1876, carried out what amounted to a national stock taking exercise. The anthropologists aligned with the Habsburg state wished to survey the ethnic and physical types of the heterogeneous Empire. Ideally, this should show common physical types, in order to establish unity, in contrast to the evident linguistic and political tensions. By the 1880s there were a number of surveys in Germany and Austria-Hungary – whether of school children as by the pathologist and liberal politician Rudolf Virchow in Germany, and of Habsburg military recruits.[2] In this liberal phase of anthropology, the aim seems to be a stock-taking of citizens' bodies and physique. The anthropological surveys coincided with social surveys of poverty and poor housing; the primary aim was to map the physical types of citizens.

The agenda of imperial eugenics between 1890 and 1914 became more complex, as social, gender-related and ethnic problems forced themselves onto the agendas of imperial states to administer welfare, promote fitness, and to exercise social control. The emergence of school doctors from the 1880s provided an opportunity to classify children on an immense scale. Physical characteristics such as height and weight, eyesight and nutritional status became the order of the day. From the turn of the century mental ability became a concern. Similarly, police authorities became interested in criminal types, developing ideas of the Italian anthropologist Cesare Lombroso. There was an interest not only in the normative but also in the pathological in the context of composite surveys. Motives were a fit and healthy population, not least as a reservoir for military recruits. This liberal phase allowed for a plurality of physical types, and the improving effect of liberal cultural values.

Attached to these descriptive surveys was the question of the historic origins of the physical types and the reconstruction of the migration patterns in the past. National identity became attached to the idea of geographical origins. Nations could be viewed as composites of racial stocks - Great Britain being a prime example of the Celtic, Nordic, and Germanic. Here, the hope was that emergent Irish nationalism associated with a Celtic revival could be subsumed within an overarching British national identity. Multi-ethnic imperial states found anthropology to be two-edged. It might show that there were distinctive composite groups, on a geographical basis – for example, an Alpine type, and yet while the hope was for different ethnicities and physical types to be encompassed in an over-arching imperial identity, the mapping of difference could result in polarities. Nationalists increasingly argued for race as a basis of smaller political entities. The German Reich, Austria-Hungary and Imperial Russia were composites of a wide plurality of ethnic, linguistic, religious and physical types. Rulers hoped to provide a focus for a variety of ethnic groups, using notions of cultural homogeneity. Nationalists strongly disagreed, and nationalism on a racial basis developed from the 1880s. Again, there were challenges to the metropolitan centres of power. Polish nationalists realised that they needed to combine territories from three imperial power blocs, while a Jewish nation state seemed to the nascent Zionist movement only possible outside Europe. National homogeneity thus posed new problems. Yet – as eugenics in Imperial Austria showed – it might be attached less to ideas of racial purity and more to ideas of welfare and a “human economy”.[3]

2. The Eastern Baltics

The Baltics were poised between two imperial powers, Russia and Germany. Imperial Russia promoted russification of cultural institutions like the German university of Dorpat (Yur’yev’/Tartu). Here was a location where the German anatomist and renowned embryologist, Carl Ernst von Baer (a Baltic German whose career was at the Prussian university of Königsberg and then at Dorpat University) developed anthropological surveys.[4] At the same time Baltic peoples fostered national movements with demands for national self-determination free from Russia. German ethnic groups saw a growing divide from others, and Jews, subject to increasing persecution prior to the First World War, saw a range of options from a chimeral Zionist state to developing a range of Jewish institutions within changed political frameworks. The YIVO Institute for Jewish cultural research was thus founded at Vilna (then in Eastern Poland) in 1925. That Professor Jules Brutzkus (a sero-anthropologist from Kaunas) was involved, indicates a biological strand in interwar Jewish cultural studies, that also manifested itself in a self-consciously Jewish eugenics, that was oriented towards the building of a future Jewish state. Brutzkus took part in the building of an international academic coalition against Nazi racial anthropology.[5]

The shift from the craniological and body measurement approach to defining race to one based on biology and heredity had profound institutional implications. Anthropology grew out of anatomical institutes and museum collections into institutions that sought involvement in social policy, not least on questions of criminality and paternity. State anthropological institutes as at Uppsala in Sweden, founded in 1922, or the Kaiser Wilhelm Institute for Anthropology, Human Heredity and Eugenics, established 1927 in Berlin under Eugen Fischer, are seen in the literature as representing a state interventive approach.[6] Here anthropological surveys shifted to such biological markers as blood group and marked a preliminary for coercive sterilisation, and in the German case the sifting of populations for their racial value for a Greater Germany. The Vienna Museum had a Racial Biological Museum, opened in 1936, although here the intention (in an era of Austro-fascism) was still to prove a distinct Alpine race, essentially different from Germany. Things went in the state-biological direction, but it was only one direction.

Set against the scientific endeavours to establish a biology of race was the persistence of the Virchowian tradition with the anthropologist Franz Boas, who was Virchow's co-worker from 1885. Boas joined also with the geographer Adolf Bastian in supporting the ideal of the psychic unity of mankind. He pointed out that physical attributes could be shaped by cultural adaptation. Boas was particularly interested in migration patterns over time, as explaining similarities between distinct population groups. It was Boas who was a consistent opponent of racial discrimination as against African Americans, and he became the rallying point of an anti-Nazi ethnology in the 1930s.[7]

The biological approach dominated the discussion of the Mongol issue. Leiu Heapost and Ken Kalling discuss in this volume the Mongol problematic with regard to racial studies of Estonians. By way of contrast, the German evolutionary biologist and crusader for Darwinism, Ernst Haeckel saw the Mongols as one of the twelve primal human species. The dark, mongoloid, round headed Finno-ungaric race was ranked by Haeckel as "too primitive" to form a state (state formation being a mark of an evolutionary high level of development). The Finns were progressive, but less evolved than the long headed, Germanic and Nordic Scandinavian. A state then was a higher evolutionary form. Haeckel's views thus supported the continuing of Imperial conglomerate states, directed by "higher" racial elites, but containing "lower" racial types who did not have the capacity to form their own states.[8]

The idea of an "Eastern Baltic" race was devised by a Swedish commentator of Finnish origin Rolf Nordenstreng (1878-1964) in his treatise on European races, which was published in several gradually amended editions between 1917and 1926.[9] The popular author Hans F.K. Günther incorporated the idea of a distinct Eastern Baltic race as part of his study of the German racial types.[10] Günther took up the idea of an Eastern Baltic race, which he described as a Nordic-mongol hybrid: "short, short-headed, broad-faced, with heavy, massive under-jaw, chin not prominent, flat, rather broad, short nose with low bridge; stiff, light (ash-blond) hair; light (grey or whitish blue) eyes, standing out; light skin with a grey undertone." He characterised this race as highly physical, and thus having the capacity of tenacity and endurance. Günther saw this type as belonging to a wider race called the "Ostic", thereby drawing a contrast to the Nordic.[11]

Günther's popularisation occurred in the shadow of the idea of a racial hygiene, and ideas of an "east race" or *race orientale*. The concept of an Eastern primary race was originally advocated by the French anthropologist Joseph Deniker in 1900 and then adopted by the mercurial Vienna anthropologist, Rudolf Pöch, who sought to develop racial studies on a combined biological and physical basis. The influential term, "racial hygiene" was introduced by Alfred Ploetz in 1895 as a composite between biological anthropology and hygiene as an experimental science.[12] Both elements of this hybrid need to be considered, that is to say race on the basis of reproductive biology, and hygiene as a medical discipline concerned with the experimental approach to population health. Ploetz can be seen as bringing together a biological programme of racial studies. He brought about contacts between figures like Eugen Fischer and the genetic statistician Fritz Lenz.[13]

Eugenics emerged, with middle class elites, often professionals in medicine as well as university academics, deploying biology to frame social problems. The idea was that the population groups could be sifted and refined to maintain desired elite qualities, and if not to eradicate then to prevent the reproduction of the undesirable. This meant a redefining of the role of the state as a eugenic agency, and a shift to technocratic elites to "solve" the problems of disease, poverty and crime. Instead of history being applied to nature, we find the reverse: nature – particularly the forces of heredity - became the determinant of history and of how society should be constructed.

How these eugenic studies emerged and impacted on politics and administration in different states, cities and communities presents historians with a complex and challenging set of research agendas. The history of eugenics as it emerged in the 1970s and continued to expand has had no sense that this area of historical study is somehow complete. Recent research has achieved much in identifying eugenics and public health in virtually unknown contexts in eastern and south east Europe. Just as there was in the1980s surprise – and indeed scepticism on the part of historians of national politics - that there was actually Weimar or Soviet eugenics, the reconstruction of Polish, Czech, Hungarian or indeed Greek movements have been greeted with similar sense of scepticism that these were at most marginal to mainstream politics.[14] Yet we have a similar set of ingredients – new national states, emergent professionals, intense cultural interest in nature, and immense medical and social problems. In this sense, north-eastern, central and south-eastern Europe have been no different to Britain, France or Germany in Western Europe. Yet the very different demographic, political and cultural circumstances mean that while certain elements, not least the constituting of eugenic societies and lobbies may be familiar, the significance and impact takes on its own distinctive narrative evolution.

Placing these movements within a broader international context, it is clear that we have far more than a "backward" (in terms of political evolution) central Europe. Central Europeans like the Hungarian Géza von Hoffmann, or – in public health - the Croatian Andrija Stampar were pace-setting social visionaries when considered internationally.[15] These exemplars of influence from Central Europe to the West indicate how models of influence from west to east have to be revised. One such model is that of sterilisation from California to Nazi Germany. It was a more complex process with the sterilisation legislation of 1933 as a distillation from a variety of sources. To the Californian and North American influences, we must add Danish legislation on sterilisation from 1929 and currents from within Germany dating from the turn of the century. This in turn means that sterilisation proposals in the Baltic region should be seen to have complex roots.

3. Race

From the turn of the century racial maps began to be superimposed on political maps, indicating the need to redraw European frontiers. Joseph Deniker (1852-1918), a French naturalist and anthropologist, systemised racial descriptions in his work, *The Races of Europe*.[16] Deniker considered that there were ten basic European racial types. One of these was new - the Nordic racial type. The term was descriptive of physical characteristics. Deniker also spoke of an Eastern race. Parallel to Deniker, William Z. Ripley, who was a racial cartographer, published his *Races of Europe*.[17] This was based on a simpler system of just three races, Teutonic, Mediterranean and Alpine. Like Deniker, Ripley's work was more concerned with physical characteristics than cultural and mental characters.

According to the American anthropologist, Madison Grant, the Nordic type was superior but by the time of the First World War under immense threat. Grant represented a new racist turn in the discourse, coinciding with the War. In his influential study, *The Passing of the Great Race*, he emerged as an ideologist of Nordic racial superiority.[18] He portrayed the Baltic as a racial pressure point. He showed that around the year 1100, a dynamic Nordic expansion took place, affecting the whole of Europe. A consequence was that the Baltic area became an area of confrontation between the Nordic and Asiatic.

The First World War which ushered in a fundamental set of shifts as regards the state espousal of population policy, and as well as a final fragmenting of the imperial powers. The war also saw a fundamental transformation from liberal anthropology concerned with measurement to a more interventive biological determinism. In the context of war, culture became an extension of the biological, and welfare was subordinated to the struggle for existence between nations. Although continuously criticized and challenged, biologised thinking gained great influence, not least in central, eastern and south-eastern Europe. International organizations like the magnanimous Rockefeller Foundation did not see the rise of eugenics as problematic, but rather as part of a modern health administration in the newly independent states.[19]

The new health structures offered prospects of national cohesion and stabilisation, as well as remedying problems of persistently high mortality. But interwar states based on national self-determination meant minorities – religious, national, ethnic/ racial – were inherently vulnerable. This observation was made by the Polish lawyer, Raphael Lemkin who saw how biology might become part of a genocidal arsenal.[20] There was a lack of safeguards to protect minorities, despite the work of Fridtjof Nansen for displaced Greeks and Turks in the engineering of massive population transfers and feeding programmes and clinics by other welfare organisations, not least to counter famine and epidemics in Russia. Ethnic minority communities within states were vulnerable, state borders were themselves fragile, and people might opt for whatever nationality was most advantageous. Around the Baltic we see Danzig/Gdansk, Memel/Klapeida, and the multi-ethnic Baltic states as all posing intractable and highly charged ethnic and political questions.

The American Nordic ideologue, Lothrop Stoddard saw the "New Poland" as doomed - as racially and territorially unstable with its sizeable German, Russian and Jewish minorities.[21] Stoddard represented a new viewpoint that state, nation and race should be identical. This gave a racial interpretation to the post-Versailles ideology of national self-determination.

Race and territory became contentious questions. Racial minorities (as the Hungarians in Romania, and German ethnic minorities in the Baltic states) found protection from outside national borders as coming from their mother state. The unresolved problem concerned those scattered ethnicities, notably Jews and Roma, who had no protecting state to take up their cause.

In the wake of the cataclysmic mass slaughter, hunger and epidemic typhus and influenza of the First World War, there was an increasing sense that the German race under severe threat.[22] In response, Germanic racial anthropology had a dual role: first, to regenerate the nation saving German populations from an imagined extermination imposed by the Germany's enemies; second, to sustain ethnic German communities outside Germany's borders – for example, the Siebenbürgen Sachsen in Romania, the Volga Germans in the Soviet Union, and the Baltic Germans within a new political framework. The hidden agenda was to reinforce these communities as outposts for an eventual restoration of a Greater German Reich, recovering Lebensraum and uniting the ethnic Germans.

Immigration quotas and controls were the corollary to eugenic welfare. Stoddard defended a Nordic and racially white North America, protected by draconian immigration restriction laws. He criticised both nationalism and internationalism – arguing for a racial internationalism that inevitably would lead to an overthrow of the post-Versailles order of national self-determination. In fact, one could see racial power blocs emerging with a Nordic-German allied with the fascist Mediterranean to roll back the Alpine-Slav – a quite horrific prospect of a racial Armageddon. Certainly, the US immigration restriction laws exacerbated the containment of the persecuted in German-occupied Europe. Immigration restrictions were a major factor in precipitating the Holocaust, providing the Nazi invaders with highly vulnerable populations.

The German political geographer Karl Haushofer, later to be highly favoured under the Nazis, saw Estonia and Latvia as intermediate areas – on the one hand, subject to German and Nordic culture, but on the other part of the Slavonic land mass. Consequently they were aptly termed "Zerrlandschaften" – landscapes subject to rupture. Some commentators saw them as part of a Germanic-Scandinavian sphere, but even so a peripheral region.[23] This situation adds considerably to the complexity of the history of eugenics. For a long time a contrast was drawn between Latin eugenics that was pro-natalist and welfare oriented, and a Nordic eugenics that was selectionist and increasingly focused on sterilisation. In central and south east Europe this presumed dichotomy overlooks a spectrum of eugenic lobbies within each national context. Again, the Baltic context adds considerably to this complexity as we find a spectrum between Lithuania, that lacked a eugenics society, and the more Nordic Estonia that had a range of eugenic institutions as well as sterilisation legislation. Sterilisation policies well illustrate this complexity in the Baltic states as where there was legislation (as Estonia 1936 and Latvia in 1937) it came late and appears to have been limited in impact.

Nazi racial policy aimed for a "rejuvenated" and "purified" German race. In 1938 the Nazis sponsored a policy of return, known as *Heim ins Reich*. The hidden agenda was one of re-settlement in conquered Eastern Territories. Baltic Germans, as a notable cultural and economic elite, emerged as stalwarts of the German scientific and racial research, as well as bitterly anti-Bolshevik. The Nazi ideologist and advocate of Lebensraum, Alfred Rosenberg, born in Tallinn/Reval left Latvia in 1918 with the failure of anti-Bolshevik resistance. The pioneering racial hygienist, Lothar Stengel-von Rutkowski was of Baltic stock, and had parents killed by Bolsheviks. By 1930 he was a Nazi and SS member, and worked energetically at the Thuringian Office for Racial Hygiene at Jena and then Prague building up the Nazi racial juggernaut. During the war von Rutkowski was involved in the German occupation of Southeast Europe before finally being taken a Soviet prisoner of war.[24] The Baltic German Jakob von Uexkuell pioneered *Umweltforschung* – environmental research and animal behaviour studies, as well as philosophical conceptualisation of biology - after war wiped out his family fortune and he moved to a position at Hamburg University. Racial hygiene was one point on a spectrum from innovative biology (Uexkuell) to racial ideology (Rosenberg).

Race became increasingly dependent on hereditary biology, and a deterministic psychology as can be seen with Hans Günther's racial typologies.

The East Baltic emerged as a frontier between the Slavic roundheads and Nordic longheads. There was a need to determine the original physical type – the *Ur-Typ* – as opening the way to regeneration. Ethnic German communities were deemed outposts of a beleaguered Germanic race and nation. The pioneering racial hygienist, Fritz Lenz, had been in Eastern Europe during the First World War, where he saw German populations in the Ukraine as a threatened species. Lenz was also a firm advocate of the superiority of what were presumed to be Nordic racial qualities. Eugen Fischer and Lenz provided support in their classic textbook on human heredity and racial hygiene, a textbook which was pioneering in placing eugenics on the basis of Mendelian genetics during the interwar period.[25]

Lenz had a historical view of migration – the earlier migration had taken place, the more dynamic the qualities of the population. The Baltics were settled before the Germans settled on the Volga, so the stock reaching the Baltics was more dynamic. Lenz came to the view that German migration produced an outstanding racial stock but one that was in danger of being subsumed.[26]

The critical study *We Europeans* (1st edition 1936, Penguin edition 1939), written by the Cambridge anthropologist A.C. Haddon in conjunction with the mercurial biologist (and eugenicist) Julian Huxley, accepted the physical observations, but kept clear of the racial genealogies and the link of psychological to physical characteristics. The book attacked racial mythologizing, and launched a critique of the Nazi politics and ideology. The account included characterisations of the ethnic composition of Estonia, as well as a characterization of "Letto-Lithuanians".[27] Here was in physical terms a patch work of human types but psychological characteristics were not imputed.

The idea of a blond variant of the *Ostrasse* or East Race developed rapidly from 1921. In 1933 the concept of a distinctively blond skin and blond haired Baltic race – as opposed to a dark East Race - was given further scientific underpinnings during the 1920s. This drew on studies by Pöch, the Swedish eugenicist Hermann Lundborg, and studies on Livonians by the Latvian anthropologist Jānis Vilde in 1924. Michael Hesch was an anthropologist from the Transylvanian Saxon community, known as the Siebenbürgen Sachsen. He was originally a Hungarian national, and his studies had taken him to the University of Vienna. He studied under Rudolf Pöch, for whom he conducted skull measurements on prisoners of war in 1915, when a pioneering skull index was developed on Russian prisoners.[28] Anthropologists at the Museum research on blood groups in the Romanian Banat town of Marienfeld.[29] In 1933 Hesch published materials left by Pöch at the Vienna Natural History Museum. Using the papers, dealing with studies of prisoners of war in 1914-18 and first written up in 1921, he drew attention to the concept of the "East Baltic" race as a contribution to the debate on the existence of an "East Race", as postulated first by Pöch – indeed his book appeared as a volume in a series deriving from the papers of Pöch. The prisoners – deriving from western Russia as well as the Baltic area – appeared to represent a distinct racial type.[30] Extensive measurements showed a correlation of the features of white Russians, Latvians and Lithuanians – enough to indicate the existence of a composite East Baltic race. Hesch argued that the region constituted a cultural unity, not least because of pressure from the Germans in the west, and the Russians in the east since the thirteenth century.[31]

Hesch's career manifested the racialisation of anthropology. Having been assistant to the racial anthropologist Otto Reche when appointed to a chair in Vienna in 1924, he moved with Reche from Vienna to Leipzig.[32] Hesch joined the Nazi Party in 1933 (by now as a German citizen), and the SS in 1935, and then he went on to take a successive posts in developing the racial administrative machinery of the SS and the Nazi state. His study of the "East Baltic" race thus marked a transitional point between earlier physical concepts of race and the nazified fusion of race and psychological attributes.

4. Arising Issues

Biology had a transformative impact on social policies along with biologised models of social change on the state. The new demand was that each state should have a national institute for eugenics and racial biology. Here, Sweden offered a model with a major national institute for racial biology established by Lundborg in 1922 at Uppsala. This represented a shift from the museum as a repository and survey centre, to an institute which should take a role in actively managing populations, backed up by state legislation. While eugenics and race became central to a new model of a racial state, it was one in which professional elites – as opposed to populist mass movements of the ultra-right - were intended to take a central role in administering state social policies. This was more a racial state in the cultural sense of an "état raciste" as articulated by French academic opponents of Nazi racism, than the modern concept of an intrusive state.[33]

Despite nazification, race remained a contested concept, none more so than that of a Germanic or Aryan or Nordic race. This remained the case under National Socialism. We see this in a book compiled by Charlotte Kohn-Behrens, *Was ist Rasse?*[34] This compilation by scientists and party ideologists indicates that there was no orthodoxy on race in Nazi Germany, when it was always contested and disputed. The Nazi state saw a range of competing authorities which radicalized ideas of nation and race into those of racial rebirth through mass destruction of supposedly degenerate populations.

The Baltic case shows the ambiguities of race. The eastern Baltic region was a racial patchwork, intrinsically highly complex, multi-ethnic and with political and cultural divisions. At the same time, the mix of ethnic cultures that found a physical expression in the idea of distinct "races" meant the Baltics were regarded as both a crucial frontier and a distinctive region between what was dangerously represented as the "Germanic" and "Asiatic". This ideological fault line had cataclysmically seismic repercussions in the context of the German onslaught on Russia in 1941, given the German sense of a racial war for survival.

Notes

---

[1] Pusman, 2008; Ranzmaier, 2011, 1-22.
[2] Penny and Bunzl, 2003; Zimmerman, 2001; Zimmerman, 1999, 409-429.
[3] Weindling, 2009, 81-113.
[4] Oppenheimer, 1990, 75-82.
[5] *Brutzkus*,1940; For context see: *Cantor and Swetlitz, 200.*
[6] Kühl, 1997.

[7] Weindling, 2006, 263-80. Weindling, 2010, 49-64. Revised English translation as: "Critics and Opponents of Eugenics" *East Central Europe*, vol. 38, no.1, 2011, 79-96.
[8] Weindling, 1989.
[9] Kemiläinen, 1993, 237-242.
[10] Günther, 1928.
[11] Günther, 1922.
[12] Weindling, 1989, 123-131; Ploetz, 1895.
[13] For details see: Weindling, 1989.
[14] For landmark studies see: Turda and Weindling, 2006.
[15] See: Turda, (forthcoming).
[16] Deniker, 1900.
[17] Ripley, 1899.
[18] Grant, 1916.
[19] Weindling, *2011, 27-54; Weindling,* 1993, 253-267.
[20] Lemkin, 1944.
[21] Stoddard, 1924.
[22] Weindling, 1989.
[23] Sievert, 1938.
[24] Stengel von Rutkowski, 1934; Hoßfeld and Š imůnek, 2008; Weindling, 2003, 1013-1026.
[25] Baur, Fischer and Lenz, 1921.
[26] Baur, Fischer and Lenz, 1921.
[27] Haddon and Huxley, 1936, 151(with a chapter on Europe overseas by Carr Saunders).
[28] For the wider context of prisoner of war studies: *Evans*, 2010.
[29] Teschler-Nicola, 2006.
[30] Hesch, 1933, 58-62.
[31] Ibd.
[32] Weindling, 2009, 81-113.
[33] Lichtenberger, 1936.
[34] Köhn-Behrens, 1933.

References

Baur, E.; Fischer, E. and Lenz, F. (1921), *Grundriss der menschlichen Erblichkeitslehre und Rassenhygiene*. Munich: J.F. Lehmann.

Brutzkus, J. (1940), *Les groupes sanguins parmi les populations juives*. Paris: Impr. centrale commerciale.

Cantor, G. and Swetlitz, M. (eds.) (2006), *Jewish Tradition and the Challenge of Darwinism*. Chicago: Chicago UP.

Deniker, J. (1900), *The races of man: an outline of anthropology and ethnography*. London: Scott.

Evans, A. D. (2010), *Anthropology at War: World War I and the Science of Race in Germany*. Chicago: University of Chicago Press.

Grant, M. (1916), *The Passing of the Great Race*. New York: Charles Scribner's Sons.

Günther, H. F. K. (1922), *Rassenkunde des deutschen Volkes*. Munich: J. F. Lehmann.

Günther, H. F. K. (1928), *Kleine Rassenkunde des deutschen Volkes*. Munich: J. F. Lehmann.

Haddon, A. C. and Huxley, J. (1936), *We Europeans*. London: Harper

Michael Hesch (1933), *Letten, Litauer, Weissrussen. Ein Beitrag zur Anthropologie des Ostbaltikums mit Berücksichtigung der siedlungs und stammesgeschichtlichen Grundlagen, (Rudolf Pöchs Nachlass. Serie A: Physische Anthropologie,* vol. 3), Wien: Verlag der anthropologischen Gesellschaft in Wien.

Hoßfeld, U. and Šimůnek, M. (2008), *Die Kooperation der Friedrich-Schiller-Universität Jena und [der] Deutschen Karls-Universität Prag im Bereich der „Rassenlehre" 1933-1945*. Erfurt: Landeszentrale für Politische Bildung Thüringen.

Kemiläinen, A. (1993), *Suomalaiset, outo Pohjolan kansa* [Finns, a Weird Northern Nation]. Helsinki: Suomen Historioallinen Seura.

Köhn-Behrens, C. (1933), *Was ist Rasse? Gespräche mit den grössten Deutschen Forschern der Gegenwart*. Munich: Eher-Verlag.

Kühl, S. (1997), *Die Internationale der Rassisten. Aufstieg und Niedergang der internationalen Bewegung für Eugenik und Rassenhygiene im 20. Jahrhundert*. Frankfurt/ M.: Campus.

Lemkin, R. (1944), *Axis Rule in Occupied Europe: Laws of Occupation, Analysis of Government, Proposals for Redress*. Washington, D.C.: Carnegie Endowment for International Peace.

Lichtenberger, H. (1936), *L'Allemagne nouvelle*. Paris: Flammarion.

Oppenheimer, J. (1990), "Science and Nationality: The Case of Karl Ernst von Baer" *Proceedings of the American Philosophical Society*, vol. 134, no. 2, 75-82.

Penny, G. H. and Bunzl, M. (eds.) (2003), *Worldly Provincialism. German Anthropology in the Age of Empire*. Ann Arbor: University of Michigan Press.

Ploetz, A. (1895), *Grundlinien einer Rassen-Hygiene, Vol. 1, Die Tüchtigkeit unsrer Rasse und der Schutz der Schwachen: ein Versuch über Rassenhygiene und ihr Verhältniss zu den humanen Idealen besonders zum Socialismus*. Berlin: S. Fischer.

Pusman, K. (2008), *Die „Wissenschaften vom Menschen" auf Wiener Boden (1870-1959). Die Anthropologische Gesellschaft in Wien und die anthropologischen Disziplinen im Fokus von Wissenschaftsgeschichte, Wissenschafts- und Verdrängungspolitik*. Vienna: LIT Verlag.

Ranzmaier, I. (2011), "The Anthropological Society in Vienna and the Academic Establishment of Anthropology in Austria, 1870-1930" *Histories of Anthropology Annual*, vol. 7, 1-22.

Ripley, W. Z. (1899), *Races of Europe: A Sociological Study*. New York: Appleton.

Sievert, W. (1938), *Der Ostseeraum*. Leipzig: Teubner.

Stengel von Rutkowski, L. (1934), *Grundzüge der Erbkunde und Rassenpflege*. Berlin-Lichtenfelde: Langewort.

Stoddard, L. (1924), *Racial Realities in Europe*. New York: Charles Scribner's Sons.

Teschler-Nicola, M. (2006), "Volksdeutsche and Racial Anthropology in Interwar Vienna: "The Marienfeld Project"" in: Turda, M. and Weindling, P. (eds.), *Blood and Homeland*, 55-82.

Turda, M. and Weindling, P. (eds.) (2006), *Blood and Homeland: Eugenics in Central Europe 1900-1940*. Budapest: Central European University Press.

Turda, M. (2013), "In Pursuit of Greater Hungary: Eugenic Ideas of Social and Biological Improvement, 1940-1941" *Journal of Modern History*, vol. 48, no. 1, (forthcoming).

Weindling, P. (1989), *Health, Race and German Politics between National Unification and Nazism, 1870-1945*. Cambridge: Cambridge University Press.

Weindling, P. (1993), "Public Health and Political Stabilisation: Rockefeller Funding in Interwar Central/Eastern Europe" *Minerva*, vol. 31, 253-267.

Weindling, P. (2003), "'Mustergau' Thüringen. Rassenhygiene zwischen Ideologie und Machtpolitik" in: Hossfeld, U. et al. (eds.), *"Kämpferische Wissenschaft". Studien zur Universität Jena im Nationalsozialismus*. Vienna: Böhlau, 1013-1026.

Weindling, P. (2006), "Central Europe Confronts German Racial Hygiene: Friedrich Hertz, Hugo Iltis and Ignaz Zollschan as Critics of German Racial Hygiene" in: Turda, M. and Weindling, P. (eds.), *Blood and Homeland*. Budapest: Central European University Press, 263-280.

Weindling, P. (2009), "A City Regenerated: Eugenics, Race and Welfare in Interwar Vienna" in: Holmes, D. and Silverman, L. (eds.), *Interwar Vienna: Culture between Tradition and Modernity*. New York: Camden House, 81-113.

Weindling, P. (2010), "Krytycy, komentatorzy i przeciwnicy eugeniki 1880-1950" in: Gawin, M. and Uzarczyk, K. (eds.), *Eugenika – biopolityka – państwo. Z historii europejskich ruchów eugeniczyych w pierwszej połowie XX w.* Warsaw: Wydawnictwo Neriton, Instytut Historii PAN, 49-64.

Weindling, P. (2011),"Racial Expertise and German Eugenic Strategies for Southeastern Europe" in: Promitzer, C.; Turda, M. and Trubeta, S. (eds.), *Health, Hygiene and Eugenics in Southeastern Europe to 1945*. Budapest: CEU Press, 27-54.

Zimmerman, A. (1999), "Anti-Semitism as Skill: Rudolf Virchow's Schulstatistik and the Racial Composition of Germany" *Central European History*, vol. 32, 409-429.

Zimmerman, A. (2001), *Anthropology and Antihumanism in Imperial Germany*. Chicago: University of Chicago Press.

The Application of Eugenics in Estonia 1918-1940

Ken Kalling

Abstract
From the 19th century onwards, the biologisation of social discourse concerning Estonian nation-building was prompted by the supposed threat of degeneration and demographic concerns. The Estonian eugenicists constituted mainly a pro-natalist movement which stood behind several prominent activities aimed at the improvement of demographic situation. The eugenic movement was strong: in 1924 a eugenics society was founded. Eugenics was well perceived by the authoritarian political system in Estonia – the latter, besides attempting to biologise the national concept through the term of national entity, was aiming towards a corporate system of state-building, and fostered applied eugenics.

1. Introduction

The ideology of eugenics has been more widespread than its application. The latter is mostly connected to eugenic legislation, notably the so-called sterilisation laws. Eugenic legislation in Europe was imposed in all Scandinavian countries, a canton in Switzerland and Nazi-Germany.[1] Lesser known are the cases of Latvia and Estonia, both of which introduced legislation in the second half of 1930s that permitted eugenic sterilisations and abortions. These two Baltic states, which turned towards authoritarianism in 1934 united the emerging welfare societies in Scandinavia on one hand with Nazi-German totalitarianism on the other.

This chapter aims to broaden the knowledge of the eugenics movement in Estonia. The study sheds light on the reception and application of an ideology based on scientific ideas by a society still in its formation.

The historiography of this particular topic is not extensive, consisting of an early essay by Eduard Laaman (1888-1941),[2] and more recent studies by Toomas Karjahärm,[3] Leena-Kurvet Käosaar[4] and by the undersigned.[5] The sources for the present article are partly archival, but in most cases it relies on Estonian-language literature, both popular and scientific.

2. The Fear of Degeneration

The ideology of eugenics appeared in Estonia, then under strong German cultural dominance at the turn of the 20th century. At this time, Estonians constituted a peasant population (so-called *Naturvolk*) with seemingly “no history”, who thus supposedly were also doomed to have no future as a political nation (*Kulturvolk*). The latter privilege for the territory – now comprised of Estonia and Latvia, and then part of Russia – belonged to local, German-speaking elites who dominated the Baltic economic and political realm until World War I. In the final decades of the Tsarist era, German dominance was challenged by growing russification. A challenge also came in the form of a national emancipation movement by local, indigenous Estonians and Latvians.

The Estonian national awakening that emerged in the middle of the 19th century, and which continued to gain strength, did not accept a fate of not achieving nationhood. Instead of political theories, the "laws of nature" were to justify emancipation: if humans evolved from apes, why then should Estonians not escape the limits of a peasant class and become a complex society?[6] Accordingly, the basis for a nationality should be built upon natural laws rather than traditional historical theories.[7] The concept of "nation" – in Estonia, mainly a linguistic phenomenon – was enhanced in such a way by a strong biological emphasis.

Eugenic discussions held in the Estonian language emerged at the turn of the century. Eugenics became important because of not only demographic but also racial concerns. The Estonian word for "eugenics" became *tõutervishoid* – "the stock hygiene". The word *tõug* (animal/cattle stock), in the particular historical context, united linguistic, mental, cultural and biological categories of ethnicity.[8] This was also the case with the word for "race" – *rass* – primarily meaning strictly human biological categories but which was often understood to possess a "deeper" meaning.[9] Thus, phrases such as "racial hygiene" and "stock hygiene" in the Estonian language could be understood as being synonymous with complex methods that focused on both biomedical as well as socio-political goals. On the other hand, the word *eugeenika*, which was also used, could be explained mainly on the basis of medical and biological criteria.[10]

In the self-definition of Estonians, the central category had become the so-called self-perception of a small nation.[11] The fear of becoming extinct (the loss of a language) was a realistic prospect at the turn of the 20th century, as many educated Estonians became Germanised or emigrated to Russia. The loss of elites must have been a serious concern, particularly in the light of the so-called "categoric imperative" of Estonian nationalism, which declared that even without becoming great in number, Estonians should try to achieve greatness through their mental powers.[12]

The main fear of a small nation concerned "degeneration" – a deteriorating biological process due to factors such as alcohol consumption and sexually transmitted diseases which threatened both the quantitative and qualitative traits of the population. In a characteristic statement, Estonian psychiatrist and leading eugenics propagandist Juhan Luiga (1873-1927) warned that only degenerated nations could become extinct.[13] For Luiga, the main pathogenic agents that threatened Estonians were unfavourable socio-political conditions and alcohol.

Luiga's thesis (1904) dealt with the distribution of mental illness in Baltic provinces. His view was that psychiatric problems among Estonian peasants were mainly caused by economic and political oppression.[14] This argument placed him in opposition to local German elites, who blamed him for introducing politics into the academic discourse.

Socio-economic factors were mainly stressed for outsiders; among fellow Estonians, however, Luiga seems to have acknowledged the presence of biological deterioration. In 1909 he diagnosed his nation as still being in a state of degeneration, but already showing hints of recovery.[15] Luiga's ideas were shared by many. Peeter Hellat (1857-1912), an outstanding Estonian physician and populariser of science who worked as an otorhinologist in St. Petersburg, was a highly respected authority among most of his Estonian colleagues. Hellat fully supported a statement in Luiga's thesis by which "the fact of degeneration has to be opposed by the fact of regeneration".[16]

Hellat was convinced he had diagnosed a particular mental pathology, which he named *Hysteria estonica*. It seemed to him that Estonians were often also neurasthenic (only the mental pathologies of Latvians could be compared to them).[17] He assumed that emancipation increased mental stress; for example, there seemed to be more mental disorders among educated Estonians than peasants and local Germans.[18] Luiga also warned about the threats posed by cultural developments such as urbanisation and female emancipation.[19]

The main opening through which certain Estonian public figures at the beginning of the 20th century proclaimed the threat of degeneration and developed a perspective for eugenics was the anti-alcohol movement.[20] By 1885, four years before the first Estonian abstinence society met, an idea was proposed that the anti-alcohol movement should be used to support Estonian national emancipation.[21]

Anti-alcohol societies in Estonia emerged sporadically due to local initiatives, as they were among the few forms of public initiative accepted by Tsarist officials. There were 61 abstinence societies in the movement's peak year of 1903. Anti-alcohol activist Villem Ernits (1891-1982) writes that abstinence societies in fact were the first to carry out broad popular educational work among the general public.[22] Villem Reiman (1861-1917), a pastor and leader of the temperance movement in Estonia, described in 1890 alcohol as a danger to children of drinkers.[23] Jaan Tõnisson (1868-1941?), a leader of the Estonian national movement, eagerly spread the ideas of the Swiss anti-alcohol campaigner (and zoologist) Auguste Forel (1848-1931) after participating at the International Anti-Alcohol Congress in Vienna in 1901. Tõnisson declared in 1907 that even if people did not share anti-alcohol sentiments, they should support the movement because of its eugenic agenda.[24]

Following the Russian Revolution of 1905, the popularity of abstinence work diminished. The revolution created new conditions for local political life: the first Estonian political party emerged, and many anti-alcohol societies became inactive during the revolution because their leaders were prosecuted. Another reason for the movement's decline was the leadership's alienation from the practical interests of ordinary members. This was caused by an umbrella institution, the Central Committee of Estonian Abstinence Societies (*Eesti Karskusseltside Kesktoimkond*, established in 1906) which came to be controlled by leaders of the national movement who declared abstinence work to be simply a means and no longer a goal of their work. The everyday local routine had become unimportant in light of the broad strategic aims.[25]

3. The Estonian Eugenics Society

It was only after Estonia successfully fought for independence in the aftermath of the Russian Revolution and Civil War (peace with Soviet Russia was signed in 1920) that the ideology, which was already relatively well-known and accepted within educated circles, was institutionalised. The Estonian Eugenics Society "Stock Health" (*Eesti Eugeenikaselts "Tõutervis"*) was formed in 1924. It continued to act in close cooperation with the Estonian Abstinence Union (*Eesti Karskusliit*). The journal *Tulev Eesti* (Future Estonia) was jointly issued, strategic goals were declared to be common, and so on.

The first chairman of the eugenics society, anti-alcohol activist Alexander Rammul (1875-1949), was a professor of hygiene at the University of Tartu. Members of the society's first board were paediatrics professor Aadu Lüüs (1878-1967), the society's chairman in 1925-37; biologist Alexander Audova (1892-1932); Dr. Juhan Vilms (1893-1952), who studied to be a physical education professor but later opened a private practice in venereology in Tallinn; professor of theology and later rector of the University of Tartu and the Archbishop of Estonia in exile, Johan Kõpp (1874-1970); socialist feminist activist Minni Kurs-Olesk (1879-1940); and leftist historian and later professor Peeter Treiberg (1894-1953).

Members of the society included famous people, medical doctors and university professors. An extant list includes 94 names from 1938, by which time the organisation had been renamed (in 1931) the Estonian Society of Eugenics and Genealogy (*Eesti Eugeenika ja Genealoogia Selts*), with history Professor Juhan Vasar (1905-1972) heading its genealogy board. (Genealogy was to add a scientific tone to the eugenics propaganda, as it was considered an important element in the study of medical genetics.)[26]

The aims of the eugenics society were broad, as were its planned means to achieve them. The 43 watchwords of the society carried mostly pro-natalist ideas, but also stressed the topic of negative eugenics (i.e. sterilisation).[27]

Some of the eugenicists' activities were carried out through professional medical organisations, such as the annual Estonian Doctors' Days, which acted as the "parliament" of doctors. The first major public achievement of local eugenicists came in 1927: the foundation of the first Baltic-area marriage-advice centre in Tartu (headed by Hans Madissoon, 1887-1956).[28] In Tallinn, a similar institution (headed by Vilms) was opened the following year. Also in 1927, a section of the eugenics society was established in Tallinn, mainly with military doctors as members.[29]

Beginning in 1926, the society supported student research at the University of Tartu, a practice taken over from the abstinence movement. Through their activities directed toward students, eugenicists managed to identify and educate a younger generation of eugenic enthusiasts. (It must be stated here that scientific research in the field in Estonia remained modest, with the work conducted most often being compilations rather than original research.)

A eugenics course at the University of Tartu was first taught in 1928 (since 1921 the issue had been addressed within the field of hygiene).[30] The university opened its Institute of Eugenics in 1939. A chair of eugenics at the medical faculty was established a year earlier, with Madissoon heading this as well as the institute. The idea for the institute dated from 1924, when the eugenics society proposed an institution in Estonia similar to that operating in Sweden (*Svensk Sällskap for Rashygien*). In March 1940 the institute was complemented by a eugenics clinic.[31] Eugenicists prepared several (text)books on the topic;[32] chapters of eugenics were added to high school textbooks; and many writings emerged in the academic and general press.

Eugenicists were active in offering solutions to various public concerns. During the 1920s and 1930s, their voice was well heard in discussions about topics such as abortion, marriage legislation and social support. Eugenicists also organised the two Congresses of National Education in 1927 and 1935 (see below).

In 1928 the Estonian Eugenics Society became a co-operating member in the International Federation of Eugenics Organisations.[33] In spring 1938 Estonia was proposed to host the Assembly of the Federation, as Hungary had dropped out as an organiser. This was a politically complicated era during which the eugenics movement split according to lines that later divided nations. Estonia, supposedly neutral, seemed to have been well-suited for the event, and local eugenicists were willing to meet the participants. Due to delayed arrangements, however, the assembly was not held.[34]

4. Soviet and Nazi Occupations

In autumn 1940, when the occupying Soviets banned most public organisations, the eugenics society was on this list. It handed its property over to the Estonian Naturalists' Society (*Eesti Looduseuurijate Selts*), which proposed opening a section of medical biology and genetics (which did not in fact happen).[35] In February 1941 the Institute of Eugenics became the Laboratory of General Biology of the University of Tartu. The Law of Sterilisation was revoked in October 1940, as it contradicted the "humanistic principles of the moral norms of socialism".[36] Estonia, possessing a similar fate as Latvia, became the first nation to have its sterilisation policy abolished and its eugenics movement put to an end.

Scholars who were active in the movement at the end of the 1930s survived the first year of Soviet repressions, in 1940-41. Accordingly, the institute was re-established when the German occupation began in 1941,[37] as was the eugenics society in 1944.[38] Both institutions were closed again when the Soviets returned in autumn 1944. The second Soviet occupation was much more devastating for the community of eugenicists.

Vilms and Lüüs emigrated before the beginning of the second Soviet occupation. Among the once-active eugenicist remaining in Estonia, Madissoon was imprisoned in 1945 (as a forensic specialist, he had been involved in the work of commissions organised by the Nazis to investigate Soviet massacres).[39] Several others were hindered in their academic work, especially during a wave of repression directed at local academic elites in 1947-50. Such a fate befell zoologist and physical anthropologist Juhan Aul (1897-1994), who had to give up his position at the University of Tartu. Aul himself felt that the main problem with his academic track record could have been his work at the Institute of Racial Studies at the University of Tartu in 1943.[40] But it also could have been eugenics – more precisely, genetics – that caused problems for him during the era of Lysenkoism.[41]

5. The Political Background of Leading Estonian Eugenicists

Eugenics ideas were well known within the wide spectrum of Estonian public thought. Several leading Estonian eugenicists belonged to the Labour Party (*Tööerakond*), which often participated in centrist coalitions. The party's programme included eugenics demands (without mentioning sterilisations).[42] Madissoon and Luiga were members of the Labour Party, as was Vilms, during the 1920s. Luiga belonged to the Estonian Students' Society (*Eesti Üliõpilaste Selts*), Vilms and Madissoon were members of the student society *Raimla*. Members of *Raimla* included such outstanding eugenics-minded scholars as Aul, Dr. Voldemar Üprus (1902-1956), psychiatrist Karl Toomingas (1892-1969) and geographer Edgar Kant (1902-1978).

Student fraternities and societies played an important role in Estonian political life – especially during periods of diminished democracy, such as the pre-independence Tsarist era and the post-1934 autocratic years – and they can be used to explain political sympathies among elites. The above-mentioned academic organisations are important for illustrating the history of Estonian eugenics.

Raimla's roots were in the radical Estonian student movement that emerged at the turn of the century. In 1907 it evolved into a student society *Ühendus* (Alliance), whose ideological goal seems to have been a "third way" between solutions proposed on one hand by the moderate liberals, who dominated the national movement and were mainly members of the *Eesti Üliõpilaste Selts* (established in 1870), and social democrats on the other. Radicals blamed the former for their indifference regarding social matters (negating stratification in Estonian society), and the latter for being too cosmopolitan.[43] Through its search for an alternative, the radicals opened a path towards solidarism, a doctrine that sees members of a society as interdependent, subordinating their personal rights to the interests of the "social organism".[44] Several prominent members of *Raimla* (emerging in 1922 from the *Ühendus*) continued to support this worldview, also supporting the authorities during the years of authoritarianism. Autocratic Estonian president Konstantin Päts (1874-1956) was close to *Ühendus* in his student years.

Some radicals were complimented in Soviet historiography for their work in popularising materialism (Darwinism), but also for being "revolutionary" (supposedly leftist) and pro-Russian.[45] This was not completely true. It was even admitted that the movement was following German scientific tradition, as the Russian perception of natural sciences seemed too idealistic.[46] The radicals popularised the ideas of Charles Darwin (1809-1882), but their main "hero" – especially concerning their ideological goals – was Ernst Haeckel (1834-1919), whose philosophy, even if understood as flawed, was supposedly creating a system that must be considered as adequate, given the era.[47]

Vilms, a member of the group, was an Estonian political thinker who during one period was openly pro-fascist. In the middle of the 1930s, his writings began to include strong criticism of the socialist dogmatics of equality but also of liberalism, both of which he viewed as leading to degeneration.[48] Vilms demanded stronger control by the state and the medical profession over public health matters and related fields. In this context, he welcomed the authoritarian turn of 1934 – even comparing it to the dawn of Estonian independence in the War of Liberation (1918-1920).[49] (Later, Vilms became disappointed in the corporate system created by the political turn in 1934, and after emigrating he seemed to have returned to the leftist rhetoric once dear to him.[50])

Moderate liberals, mostly attacked by Soviet era historiography as "nationalists", also popularised evolutionary theory.[51] They were mainly linked to the previously mentioned *Eesti Üliõpilaste Selts* and to the Tartu newspaper *Postimees*. After 1905 they managed to organise the first Estonian political party – the Estonian Democratic Party of Development (*Eesti Rahvameelne Eduerakond*) – which worked in co-operation with the Russian Constitutional Democrats (*Конституцйонно-Демократическая партйя*). In independent Estonia, liberals constituted the People's Party (*Rahvaerakond,* later National Centrist Party / *Rahvuslik Keskerakond* with Tõnisson in the lead). After 1934, when political parties were banned, liberals formed the core of opposition.

Liberal circles did not form one solid entity on matters of eugenics; i.e. there was criticism by Professor Henrik Koppel (1863-1944, rector of the University of Tartu from 1920-1928) of Lüüs – also belonging to the group – for excessive propaganda in the field.[52] Among liberals, the biologised worldview, in fact, had emerged with difficulties. In 1902, at a time when radicals clearly supported materialistic approaches, the calendar of the *Eesti Üliõpilaste Selts* released an article against them.[53]

The role of Lutheran pastors in the society was evident, but these pastors were also active in the fight against alcohol, accepting in such a way the ideas of eugenics. As a result, radical journalist E. Laaman noticed ironically in 1915, and with a hint towards eugenics, that the ideology of moderate liberals was based on the ideas of cattle breeding. This comment related to liberals having declared Marxism gluttonous for its supposedly low ideals.[54]

Additionally, one could see leftist ideas among members of the *Eesti Üliõpilaste Selts*. Biologist Audova, much appreciated by Soviet-era historiography, in fact represented the ideas of *Naturheilärzte*. He could not find an academic job, was marginalised, became embittered and, as a protest, spent the final year of his life in the USSR. Another left-wing eugenicist and anti-alcohol activist, Villem Ernits (1891-1982), spent most of his energy on abstinence work. He must be recognised for his extensive work in organising support for anti-alcohol scientific research and propaganda.

Socialist parties that represented workers seemed to have been against eugenics, seeing it as contradictory to the goals of improving society.[55]

6. Was there Racism in Estonian Eugenics?

Given that eugenics became an important tool for Estonian nationalism, one must ask whether the issues of ethnicity were approached through race criteria, and whether racist concerns – stratification and discrimination on the basis of human morphology – were present in local eugenics ideology.

It seems that despite strong interest of Estonian society in race issues,[56] the strict category of race – understood here to be the morphological types of humans – did not become crucial in eugenics. The reason that qualities deriving from the scholarship of physical anthropology did not become a major concern in Estonian "race hygiene" could be linked to the possible racial inferiority that race concepts ascribed to Estonians at that time. For example, certain theories linked Finnic tribes to what was then considered an inferior "yellow race". (Admittedly, much of the work by inter-war Estonian physical anthropologists was to deny this theory, with the leading anthropologist Aul also being an active eugenicist.) [57]

There was also a concept that broadly classified Estonians into at least two different anthropological types.[58] The dominant population in the western and northern parts of the country – taller than average, dolichocephalic and blond people – supposedly represented the so-called Nordic race. Prevailing in the southern and eastern parts of the country, the so-called East Baltic race was considered by some scholars to be the "original" or "basic" type for Baltic Finns (i.e. – also Estonians).[59] In such a scenario – despite a temptation to believe that representatives of the Nordic race could be characterised by higher levels of intelligence[60] – it was difficult to identify a clear racial aim for eugenics.

Even if Estonian eugenicists did not aim for race improvement in the strict sense, the watchwords of their society nonetheless included a warning against intermarriage of Estonians and non-Estonians, calling it undesirable for both family and eugenic reasons. Vilms wrote a chapter on the hazards of mixed marriages in the 1927 eugenics textbook.

In fact there was a contradictory approach towards "race mixtures" in Estonia. Folklorist Villem Grünthal-Ridala (1885-1942) brought forward a Gobineau-style concern about "bastards" in his 1915 compilation dedicated to the Estonian "stock".[61] Additionally, the blending of distant racial types ("negroes" and "whites") was not viewed positively by Hellat in his *Tervise õpetus* (the second issue from 1913). On the other hand, Hellat believed that marriages between closer race types could produce healthy offspring (though warning, relying on Forel, that intermarriages involving Jews could become infertile within several generations).[62] In fact, Hellat, being ardently anti-German, accused the Baltic system of serfdom – "the 700-year slavery" – for creating conditions that favoured inbreeding, as peasants were denied the right of mobility and for centuries had to marry within their own parish.[63] The topic of inbreeding was also important to Luiga.[64]

The idea heralding "blood improvement" through intermarriages was first supported by journalist Ado Grenzstein (1849-1916). Grenzstein's views were rather eclectic and supposedly pro-Russian (and anti-Semitic). He was among the few from Estonian elite who were not afraid of Estonians losing their language, believing that the situation is in order when Estonian "blood" is still circulating.[65]

In independent Estonia, "mixed" populations were being discussed by Aul and Vilms. Aul had studied a group of children on the island of Saaremaa whose fathers had served in a local Russian garrison during World War I. Teachers there claimed that among these children, who were born outside of wedlock to non-Estonian fathers, seemed to be over-represented at both ends of the normal distribution – the more and less talented.[66] Aul states in his research that the mixed group was "constitutionally better developed"[67] and more heterogeneous somatically and mentally.[68] Aul did not take a stand on the issue from a eugenics viewpoint. In an article published in a popular journal, he supported the then-widespread theory according to which interbreeding between distant "stocks" was proved harmful. He also claimed to have witnessed positive outcomes of intermarriages in Estonia but he did not dare to draw conclusions, giving a hint that perhaps the "biological quality" of the involved nations is also important. Thus, future research was needed.[69]

Therefore it must be understood that the fear of intermarriage was based mainly on cultural concerns, first of all targeting Estonians' marriages with people from the so-called large nations – those "which have entered our land with the right of stronger", e.g. Germans and Russians.[70] Vilms stated that these marriages might not cause any biological harm but that they posed cultural threats, namely that the "stronger" cultures would denationalise children. Thus, mixed marriages were only to be allowed if it was granted that the children remain Estonian.[71] The chapter written by Vilms in the eugenics textbook accepted the idea that the "blood-improvement" marriages with close to Estonians ethnic groups – Finns, Livonians [72] and also Latvians (believed to possess 30% of Livonian "blood") – were welcome. Such a practice, which lacked cultural threats, could be biologically favourable.[73]

In 1934 the eugenics society approved a rather xenophobic document – a project for the Law of Eugenic Defence of the Future Generations (*Järeltuleva soo tõulise väärtuse kaitseseaduse eelnõu*) – demanding among other things that applicants for Estonian citizenship should be free of medical conditions such as mental deficiencies, alcoholism and venereal disease. The paper also demanded that children with one Estonian parent should be registered as Estonians, and that the children of national minorities should study either at schools using their native language or Estonian. A *numerus clausus,* based on the ratio of every nation in the country, was to be introduced in facilities that were important for the state (ethnic Estonians had to constitute 90% of their personnel).[74]

Concerning anti-Semitism in Estonia, Vilms, one of the leading eugenicists in the country, seems to have held these beliefs. His statements on the issue were mainly concerned with developments within the medical profession (competition with local Jewish colleagues), but also targeted foreign entrepreneurs who were supposedly indifferent to the national interests of Estonia.[75] On the other hand, Jews were looked at with particular sympathy by eugenicists, though with an accent of anti-Semitism still present. Both Estonians and Jews were viewed as "degenerate" nations due to the supposed negative selection over the millennia, Estonians suffering from serfdom, and Jewish people due to their fight for survival among gentiles.[76]

Placing Estonian eugenic thought in the wider context of Eastern European politics at the time, it seems that eugenicists were worried more about the possible German threat than the Russian (Soviet) one, especially after the Nazis came to power and demanded *Lebensraum*. For Madissoon, a strong Russia was welcomed as a counterbalance to Germany.[77] The Russian threat was not entirely forgotten. The relatively modest Russian population in the country (compared to modern Estonia) was believed to be threatening Estonian domination due to its slightly higher natural birth rate. This was to occur within 300 years.[78]

7. The Birth of the Sterilisation Law

On 1 April 1937 the Estonian Law of Sterilisation came into force, having been signed by the authoritarian head of state in 1936. It suggested obligatory sterilisation or abortion for a certain group of people ("mentally insane on hereditary grounds", or suffering "severe epilepsy", but also for persons "incurably physically handicapped on hereditary grounds"). It was stressed that the subjects must act sexually abnormally, be a threat on these grounds to society, and tend to breed hereditarily handicapped descendants. The right to propose sterilisation or abortion was limited to medical doctors, custodians, heads of residential care establishments and medical institutions. Also, individuals themselves were permitted to volunteer for sterilisation (if the medical requirements were fulfilled). The final decision for the operation was made by one of two commissions (established on a regional basis, in Tartu and Tallinn). The commissions were constituted by medical doctors along with officials nominated by local authorities. The decisions of the commission were to be approved by the head of the Welfare Department of the Ministry of Social Affairs. The police could be called on to take non-cooperating subjects to the commissions and to the operation.[79]

Following more than a decade of debate, the final version of the law targeted far fewer groups than was suggested by eugenicists and medical organisations. An initial proposal by the Estonian Doctors' Days of 1924 called for obligatory sterilisations for deaf-mutes, children with learning difficulties, so-called *vagabonds,* chronic alcoholics, syphilitics and others. Voluntarily sterilisation was suggested for people with tuberculosis.[80] Between 1924 and 1937 eugenicists also called for sterilisation for criminals and drug abusers. The Estonian Doctors' Days kept the issue on its agenda in

1925 and 1926. Particular attention was paid to the issue in 1927, when the first Congress of National Education was held and a new Criminal Law was being planned.

At the time, eugenicists accepted the reality that it was difficult to pass special eugenics legislation, so they tried to insert chapters into the Criminal Law. The idea was to decriminalise a sterilisation procedure performed by a doctor on eugenics grounds. "Pathologies" such as mental disability and criminal inclinations, as well as mental and physical deficiencies, were mentioned.[81] This plan was not fulfilled.

In 1933 the Estonian Doctors' Days proposed to lawmakers that people with genuine epilepsy should be sterilised for the sake of the health of future generations.[82] In 1933 doctors' organisations commented on eugenicists' proposed amendments to Family Law concerning sterilisations. It seems that the doctors were informed that the state had a ready-made proposal for a sterilisation law, but it was seen as too early to be sent to the Parliament.[83] At the Estonian Doctors' Days in 1934, the question of sterilisation emerged once more, with doctors voting to support a resolution demanding the introduction of the sterilisation law (with 20 votes in favour, two against and six neutral).[84]

That same year, sterilisation issues were discussed in connection with planned amendments to the Social Welfare Law.[85] Officials stressed that the amendments were based on examples from Scandinavian countries, and that there was no plan for massive sterilisations, but for single cases emerging from a need created by the social support system. The subjects were to be alcoholics, drug abusers and sexual criminals (in addition to those included in the final Sterilisation Law). The possibility of castrating sexual offenders was also mentioned. At the Welfare Department in the ministry where the law was discussed, it was acknowledged that the eugenic goals – reducing the number of insane people – could not be easily achieved by such measures.[86] Therefore it was stated that sterilisation could be seen as a solution if there was a danger of children being born who, due to heredity, could not survive without the help of social support. In this way, the arguments that sterilisations would have little effect on the quality of the population – i.e. they are medically senseless – were declared invalid, as the entire question was to be approached not scientifically but from a "practical position".[87] Even Estonian eugenicists themselves criticised the weakness of the Sterilisation Law in medical terms. It was legal to perform a eugenic abortion only when the mother was declared "unfit"; the law did not allow an abortion if the father was "unfit".[88]

During discussions on the planned Law of Mothers and Children Protection in 1937, it was declared that attention must be paid to the registration of defective children.[89] There should be special institutions for them, but in the future – when the Law of Sterilisation would be in effect –

the state should attempt to prevent the possibility of such children being born.[90]

From these examples, it becomes evident that both the eugenics society and professional medical organisations, but also state officials, were determined to introduce the sterilisation practice. Statements against it were modest but not as marginal as declared by Madissoon, according to whom only "Communists, Catholics and religious sects" did not accept sterilisations.[91]

8. Criticism of Sterilisations

The idea of negative eugenics was not accepted by everyone in Estonia, even among members of the eugenics society. The counter-arguments often stemmed from ideological reasoning. At the Congress of National Education in 1927, pro-sterilisation campaigners (radicals such as Vilms and Madissoon as well as Audova, who was leftist) seem to have been in conflict with politicians from the ranks of liberals.[92] These included professor of education Peeter Põld (1878-1930), who declared that all propositions concerning negative selection among the physically weak must be viewed as an "enormous misunderstanding," adding that only spiritual values, and not the practice of animal breeding, could improve the nation.[93]

The essence of the conflict was evident in an argument between the two parties over methods of the anti-alcohol campaign. Põld, representing so-called direct abstinence work (relying on the personal will of the subjects), accused his opponents, the supporters of so-called indirect abstinence (underestimating the will of the people, trying instead to create conditions under which alcohol consumption would decline): "you [...] want to create a monist culture, leaving people at the mercy of their drives".[94] This was a reaction to Vilms, who had stated previously that "the broader masses do not exist on the basis of mental understanding but form an organism living on emotions and egoistic hedonistic principles".[95] Such an organicist (solidarist) approach was also attacked during the years of authoritarianism by Tõnisson, denying the subordination of individuals to the state.[96]

On political grounds, negative eugenics practice was opposed by Dr. Viktor Hion (1902-1969). Hion was a rather eclectic thinker and private medical practitioner who finally had a career during the Soviet era as the Estonian People's Commissar of Health. In 1934 he rejected euthanasia – even if the patient and his/her relatives supported it – explaining that future medical advances could help such patients. Hion's writing is rather demagogic, mixing issues of euthanasia with sterilisation (he even accepted the idea that in some cases voluntary sterilisation could be tolerated). Hion ends his article in a popular health journal with a statement predicting his prominent position in the Lysenkoist realm of Stalinist Estonia. He writes that hereditary science is a false teaching that is robbing people of their courage to improve themselves. Thus, the natural sciences should be "freed" from genetic scholarship.[97]

The rejection of sterilisation also stemmed from professional expertise. Bacteriologist Karl Schlossmann (1885-1969) declared in 1927 that the problems that some want to treat by sterilisation were in fact social. The task of the medical profession should be to cure people of disease and not injure them, and to strive to solve problems in this way. Schlossmann declared that the human environment – society – has to be improved (to produce better and healthier people).[98] Schlossmann was rather opportunistic, however, and favoured eugenics after becoming a leading science administrator during the autocratic regime.

Statements that questioned the epidemiological outcomes of sterilisation also emerged in the 1930s from the medical profession. Professor of psychiatry Maximilian Bresowsky (1877-1945) declared that obligatory sterilisation would be a serious violation of civil rights and perhaps not effective in terms of the epidemiology of mental disease. Bresowsky thought that propagandists of sterilisations were huge "idealists and optimists".[99] When giving his opinion of the election of Madissoon – the leading eugenicist of the time – to a position of a docent of forensic medicine in 1934, Bresowsky did not support his radical approach towards heredity and medical ethics.[100] Another psychiatrist, Teodor Soans (1894-1961), takes a similar position as Bresowsky.[101] Psychiatrist K. Toomingas, who practiced illegal eugenic abortions before the law of sterilisations was approved (see below), stated in his thesis that when eugenics policy is under discussion, social and not medical methods must be favoured.[102] Also sceptical of hereditary matters within his field of medicine was the professor of ophthalmology Ernst Blessig (1859-1940).[103] All of these opponents endorsed eugenics positions and agreed in principle that eugenics should be recommended to prevent the birth of handicapped and diseased children, but that educational and other "soft" methods must be used to achieve this.

At the end of the 1930s, the most outstanding critic of sterilisation was Robert Silvester (1908-?), a businessman with a background in law and forestry who published an article on the issue in *Akadeemia*, a journal issued by circles of liberals in Tartu opposing the authoritarian regime. Silvester stressed that sterilisation propaganda was based on popular approaches and not on exact science, on which genetics is supposed to be grounded.[104] In a presentation given to medical doctors, he claimed that sterilisation could not be used as a means of "mass selection" within the human population, which is heterogeneous in its genetic composition. In a discussion after the presentation, several doctors supported Silvester's ideas. When a comment was made that sterilisations must be understood within the social context – that communities could be freed from supporting children of the feeble-minded – Silvester declared that then sociologists, not the medical profession, should deal with the issue (adding ironically: "then let's sterilise the poor!"). He also remarked that the Estonian law of sterilisation did not mention social indications, such as single mothers who could not support their children.[105] In general, Silvester's message was that sterilisation could be viewed as an act that strongly violates the rights of citizens of a democratic state. Estonia, however, was not a democratic state.

9. Estonian Sterilisation Practice

There is no evidence to indicate why the scaling back of doctors' initial plans concerning the subjects of sterilisation did take place. As a tendency, one could point out a sort of devaluation in the faith of biological predestination. At the end of 1930s, the leading sterilisation advocate Madissoon partially denied his earlier attitudes, acknowledging that new methods in the studies of heredity (the method of twins) could raise the importance of environmental factors in understanding individual human development. He accepted that positive eugenics in such circumstances must not to be forgotten.[106]

The number of sterilised people in Estonia was relatively small compared to other countries with similar legislation. According to official data, 32 people in 1937 were brought to the sterilisation commissions, which approved sterilisation for 21 of them.[107] Nineteen of them were women, among a proposed list of 26 women.[108] In 1938, 20 of the 22 people brought to the commissions were sterilised (and six abortions were performed) in all 18 women were affected.[109] During these two years, a total of 41 people were sterilised, four of whom (about 10%) were men.[110] There is no data for 1939 or 1940, due to the loss of statehood. (Hints suggest that the practice was carried out in these later years, and also in modest numbers.[111])

There is little information about the subjects of sterilisations, so the exact criteria by which the people were chosen cannot be determined. The few cases available suggest that they had been declared "feeble-minded".[112]

In Estonia, the law and its application were preceded by illegal practices.[113] Thus, the introduction of the sterilisation law in Estonia could be understood as legalising existing habits. According to data from one Estonian psychological ward (Seewald in Tallinn), three abortions made "on eugenic grounds" were performed in the seven years preceding 1935.[114] These abortions were carried out by the previously mentioned Toomingas. Another hint that the practice of sterilisation could have been agreed upon in some circles before the law was passed comes from a so-called personal book of a patient, compiled in 1927 by the head of the Udriku infirmary for feeble-minded children, Hugo Valma (1891-1977). The document includes a chapter for proposing possible sterilisation, with space to analyse the results.[115] (There were also reported cases of sterilisations made on medical grounds by doctors who were active in the eugenics movement, including Madissoon.[116])

In 1940, with the arrival of the Soviet regime, sterilisations ceased. In 1941, during the Nazi occupation, the pre-Soviet Estonian legislation was re-enacted in principle, but there is no data about the fate of the sterilisation law. Some information refers to several eugenic abortions taking place in Estonian hospitals during this period, and also about a campaign to register the country's mentally disabled people.[117]

It must be stressed that social considerations were clearly present when Estonia's sterilisation policy was enacted. Strong support for sterilisations came from the Union of Estonian Towns (*Eesti Linnade Liit*), which was worried about the economic costs associated with "idiots".[118] (Rather active in popularising eugenics was the journal *Linnad ja Alevid* (Towns and Hamlets), which represented local communities.) Developments in the Estonian welfare system supported the popularity of sterilisations. The final step involved reforms to the system in the latter half of 1937, when the duties of local communities to grant social support were increased.[119]

10. Pro-natalist Aspects of Estonian Eugenics

It was mainly demographic concerns that fuelled the activities of Estonian eugenicists. The consensus that viewed the demographic situation as catastrophic – threatening the extinction of the nation – had been employed for decades by various Estonian public figures representing different political parties. Participation in debates about issues of population growth made the group of eugenicists well known in society.

Estonia's population was 1,107,059 in 1922, and 1,122,440 in December 1939. Various public figures proposed target growth figures, including plans that foresaw 3 million or more Estonians inhabiting the country. The general idea was that population growth must be at least twofold. A viewpoint held that a certain quantity was needed for the sake of quality; just a single genius was believed to be born among a million. The geniuses, such as outstanding scientists, who could make the small nation heard among large nations, were seen as providing a stronger defence for the nation than the army.[120] The demographic warfare between the nations was declared to be fierce. Even if proving its right to exist through an actual war (winning the War of Liberation), Estonia seemed to be losing its position in a social-Darwinist competition taking place in peacetime.[121] Worries about limited economic capabilities of a small nation brought forward a xenophobic fear of immigrants.

The role of institutionalised eugenics in supporting pro-natalist sentiment in Estonia was evident. As mentioned, a major portion of the watchwords of the Estonian eugenics movement was devoted to demographic issues. A means for the eugenicist to influence public opinion were the so-called Congresses of National Education (*Rahvusliku Kasvatuse Kongress*, held in 1927 and 1935). At the congress of 1927, matters mainly related to cultural ethnicity were discussed, though the event also seems to have led to textbooks on genetics and the opening of eugenics advisory centres.

The congress of 1935 was quite "carnal",[122] in mainly stressing the physical body of the nation, and associated demographic concerns. During his opening speech at the congress, where at least a fourth of participants were school teachers, Madissoon said that Estonia was threatened by extinction.[123] The authoritarian state paid much attention to the congress, with the President of Estonia greeting the assembly by challenging the medical profession in particular to dedicate itself more to issues of population growth. The spectrum of issues addressed at the congress was broad, from population expansion throughout the country and sterilisations, to military strength and raising national pride.

The third congress of the series, planned for 1940, was to concentrate mainly on demography, but it was not held due to the Soviet annexation. The proposed programme stressed, among other issues, the role of women in demographic and eugenics work.[124]

Eugenicists had long discussed the position of woman in Estonian society. Estonian women were believed to possess good physical attributes for motherhood, yet they were blamed for not being eager to give birth, an indication that their "natural instincts" had been weakened.[125] Thus, the "threat" of the emancipation of women remained actual and was considered a major problem for a new, small nation aiming to become large in population and intellectual capability.

Luiga had greeted the emergence of the eugenics society by stating that the new institution should make women its main focus, as they must assume a fundamental role if the goal of the nation was rebirth.[126] Many Estonian eugenicists seem to have adopted the global perception that the highest achievement for a woman must be motherhood.[127] Citing Swedish activist Ellen Key (1849-1926), who, while praising family values, stated that the 20th century should become a "Child's Era" – there emerged an idea to label the coming decades or even centuries when Estonia would be working to strengthen its statehood as the "Woman's Era".[128] The "solving of the female question" witnessed new approaches during the autocratic years.

11. The "Silent Era"

The eugenicists' message was well known in Estonia. Their society claimed for itself a position as the main voice of demographic concerns and matters of public health. A significant breakthrough came in 1934, when a *coup d'état* turned Estonia into an autocratic regime. The state then became a strong partner in understanding the rhetoric of pro-natalist and health propaganda. This was due to the declaration by political powers for the means, criteria and goal of the national state – the so-called national entirety (*rahvusterviklikkus*).[129] This encompassed a supra-political union of

Estonians – a solidarist entity in which political and class struggle could no longer degrade the "national body". From the "Silent Era", as the autocratic period was popularly known, comes an example of how genealogy – linked institutionally to eugenics in Estonia – was seen as crucial for handling the social history of the nation within the ideological context of solidarism. It was also hoped that through genealogical data, people could be convinced that the class divisions within society were artificial and could be overcome by the "natural" ties of the population.[130]

During the parallel crises of democracy and the economy at the beginning of 1930s, the would-be autocratic president, Päts, had argued against the negative side of the democratic system, using the euphemism of a "sick society".[131] Similar arguments were made by Vilms.[132] As we can see from post-1934 events, the "curative" methods included ideologies derived from medicine, such as eugenics.

The "Silent Era" featured attempts to restrict political freedom. Political parties were banned, and the work of student organisations and the media was controlled. The authorities made efforts to establish a new system of political representatives, with the professional chambers substituting parliamentary parties. Among the 17 chambers that were created was one representing medical doctors, one representing women (the Chamber of Domestic Economy, *Kodumajanduskoda,* established, according to the official rhetoric, to value the domestic work of women). Both chambers had a strong potential in running eugenics discussions in an ideological environment in which political parties were missing.

Official propaganda praised the new autocratic era for raising the role of women in society. In fact, this role was stereotyped according to the ideology of solidarism. The task of women was to be fertile; the possibilities of controlling fertility became limited. During the first months of the autocratic regime, amendments to the planned Criminal Law banned abortions. [133] The result could have been opposite, but perhaps a memorandum from eugenicists (submitted a week after the *coup d'etat*) threatening a demographic catastrophe, stopped the process of legalising abortions in Estonia.[134]

Yet it was not only negative eugenics – the law of sterilisation – that should be stressed in regard to the Estonian state of the late 1930s. Much was done in terms of positive eugenics, mainly due to the high profile of the pro-natalist sentiment. The Congress of National Education in 1935 led the state to establish the so-called Board of Population Increase and Welfare (*Rahva Juurdekasvu ja Heaolu Komisjon*). [135] Thus, eugenics ideology was institutionalised on a state level. This basis supported new activities – for example, the founding of the Institute of Eugenics. The board was behind the declaration of the six-year plan, beginning in 1936, for promoting public health. The aim of the plan included developing and increasing the number of medical institutions in the Estonian countryside, expanding lower-ranking medical personnel, and fighting infectious disease. Much attention was paid to the mother-child welfare system.[136]

By the end of the 1930s, eugenics had become an official trend. This can be illustrated by the case of K. Schlossmann, one of the few medical scholars who rejected the idea of eugenic sterilisations in the 1920s, but who accepted the ideology after being named president of the Estonian Academy of Sciences (*Eesti Teaduste Akadeemia*), which was established in 1938.[137] (Still, the wish of the eugenics society to unite with the academy of sciences, as happened to many other scientific societies, was not fulfilled.[138])

The support of eugenic-minded individuals and organisations for the autocratic political system was evident. The new corporate system, under which the state organised professional chambers for restructuring power mechanisms in society, was a strong argument for those trying to include into politics the applications derived from professional knowledge. Vilms declared that the medical profession finally could spread its influence without incompetent politicians (who were dependent on voters) getting involved.[139]

12. Conclusions

The practice of eugenics did not provide rapid results, as the rate of population increase in Estonia remained rather stable. Still, the eugenics movement was well known. In fact, the ideology survived Stalinism far into the Soviet era. When it was again possible, Aul made hints that the genetics of the time should care more about its applications.[140] In this context, a notorious text by geneticist Toivo Orav (1932-1989) was published in 1967 in the journal *Eesti Komunist* (Estonian Communist), in principle calling for the rehabilitation of (positive) eugenics.[141] The reality developed otherwise. A three-volume history of Tartu University published at the beginning of the 1980s (during the late Soviet period) rather harshly denounced the eugenics teaching at the university.[142] Officially, the ideology, or at least the rhetoric linked to it, was to remain in the past.

Between the two world wars, one could witness in Estonia groups as disparate as nationalists, social democrats, theologians and agrarians participating in eugenically oriented discussions. A reason for the broad consensus among elites but also among the people could be that eugenics served as a lowest common denominator that united the realms of social reasoning, populism and scientific argumentation. There was no factor, such as the Catholic Church in predominantly Protestant Estonia, to resist sterilisation policy. Additionally, social democrats were losing their positions after the unsuccessful Communist putsch of 1924. Most importantly, the ideology of eugenics suited the rather defensive character of the small nation's self-perception.

When trying to place Estonia by its application of eugenics into the international context, it falls somewhere between Germany and Scandinavia. Despite planned improvements to the political system, Estonia was still an autocratic regime when the sterilisation law was introduced. The Estonian Law of Sterilisation mainly set eugenic goals and mandatory sterilisations were foreseen. What made Estonia different from Germany was that no racist reasoning was present in the ideological documents of Estonian eugenicists. The "loudest" Estonian eugenicists seem to have been even anti-German. Despite the goals stated in the law, the carrying out of sterilisations in Estonia included a clear social rationale, similar to Scandinavia. Also in Estonia, most of the people sterilised were women, and local communities – not so much the medical profession – seem to have been involved in selecting the subjects for sterilisations. The sterilisations could even be viewed simply as ballast attached to the general ideology of eugenics, fuelled in Estonia by demographic concerns. The strong pro-natalist sentiment present in Estonia, in fact, creates a temptation to make a connection with Italy or France, which were worried by similar concerns during this period.

Estonia shows that the positive, pro-natalist path of eugenics that was successful in promoting public health still could be rather controversial. The reason is authoritarianism. In pre-war Estonia, the demographic frenzy, added to the general eugenics-minded atmosphere and a still-developing social welfare system, led the society into a condition one could call biologised. Society was mobilised for fulfilling a biological task – to breed. Discussions about health issues led to undemocratic decisions, justified social stratification, minimised the role of individuals, etc. Social discussion in an autocratic state obtained a paternalist path and alienated different social layers, and the state and its citizens, from each other. Solidarism did not support solidarity. The Estonian elites who aimed for a bright future for the nation were faced with a controversy when it became evident that the goals of the national idea seemed to be hindered by the development of democracy, modernisation of the society and emancipation of its members.

Notes

[1] Weindling, 1999, 179-197; Weingart, 1999, 163-177; Roll-Hansen, 2001, 75-82.
[2] Laaman, 1915, 40-46.
[3] Karjahärm, 1993, 1347-1364.
[4] Kurvet-Käosaar, 2009, 729-745.
[5] Kalling, 1999, 31-42.
[6] Koppel, 1910, 3.

[7] Luiga, 1995a, 221.
[8] Karjahärm, 1993, 1347.
[9] The first thorough overview in Estonian on the matters of race was perhaps: Reiman, 1931, 9-13.
[10] Aul, 1933a, 15.
[11] Kalling, 2007a, 253-262.
[12] Hurt, 1989, 29-30.
[13] Luiga, 1995a, 213.
[14] Luyga, 1904, 14.
[15] Luiga, 1995a, 224-227.
[16] Laaman, 1938, 36.
[17] Ibid.
[18] Hellat, 1892, 175.
[19] Luiga, 1995a, 224-227.
[20] Kalling, 2007b, 59-78.
[21] Grenzstein, 1885.
[22] Villem, 1920, 28.
[23] Reiman, 1890.
[24] Tõnisson, 1907, 7-9.
[25] See: Ibius, 1939, 142.
[26] Üprus, 1938, 112.
[27] "Eesti Eugeenika Seltsi "Tõutervis" juhtlaused [The Watchwords of the Estonian Eugenics Society "Tõutervis"]" *Tulev Eesti*, no. 1, 1925, 2-6.
[28] Madissoon, 1930a, 203-204.
[29] "Kirjavahetus seltsi organisatsioonilise tegevuse kohta [Correspondence on the Organisational work of the Eugenics Society]": Eesti Riigiarhiiv (Estonian State Archive – further ERA), 4855/1/5, 18.
[30] Before independence was gained, at least two scholars of the University of Tartu – then officially: Yuryev – had been involved in the matters of eugenics. Nikolai Gamaleya (1859-1949), an outstanding Russian microbiologist, a lecturer at Tartu University in 1912-1913 gave a lecture on eugenics in Tartu which was also published in Estonian ("Tõutervenduse-õpetuse, eugeenika põhjusmõtteist ja ülesannetest [On the Principles and Tasks of Stock Hygiene / Eugenics]" *Postimees*, nos. 257-260 (November 5th-8th), 1912). Evgenii Shepilevskii (1857-1920), the professor of hygiene at the University of Tartu in 1904-1918 issued a book on eugenics in Russian: Shepilevkii, 1914.
[31] "Tartu Ülikooli Arstiteaduskonnast [From the Medical Faculty of the University of Tartu]" *Eesti Arst*, no. 5, 1940, 318-319.
[32] Klein (Aul), 1926; *Pärivus ja valik* [Heredity and selection], A. Audova, A. Lüüs, H. Madissoon and J. Vilms (eds.), Tartu: Loodus, 1927.

[33] "Seltsile saabunud kirjad koostöö kohta teiste eugeenikaorganisatsioonidega välismaal [Correspondence of the Eugenics Society with Foreign Eugenics Organisations]": ERA, 4855/1/6, 20.
[34] Ibid., 114-126.
[35] "Seoses Eesti Eugeenika ja Genealoogia Seltsi töö lõpetamisega Eesti Loodusuurijate Seltsile vara üleandeaktid [Documentation concerning the Handing Over the Materials of the Closed Eugenics Society to the Estonian Naturalists Society]": ERA, 4855/1/10, 6-7.
[36] "ENSV Rahvakomissaride Nõukogu määrus Steriliseerimise seaduse kehtetuks tunnistamise kohta [A Decree by the Committee of the Peoples' Commissariate of the ESSR on the Abolishing of the Sterilisation Law]" *ENSV Teataja*, no. 32, 1940, art. 368, 346-347.
[37] "Kroonika [Chronicle]" *Eesti Arst*, no. 3, 1941, 191.
[38] "Seoses Eesti Eugeenika ja Genealoogia Seltsi töö lõpetamisega Eesti Loodusuurijate Seltsile vara üleandeaktid [Documentation concerning the Handing Over the Materials of the Closed Eugenics Society to the Estonian Naturalists Society]": ERA, 4855/1/10, 16.
[39] Auväärt, Maripuu, 1994, 10-20.
[40] Aul, 1949.
[41] Kalling, 1997, 174-180.
[42] *Tööerakonna üleriiklik kongress Tallinnas, 2. ja 3. märtsil 1929* [The Congress of the Labour Party, Tallinn March 2-3 1929], 9-10.
[43] Vilms, 1989, 57.
[44] Jans, 1914, 10.
[45] See: Laul, 1956, 164-193.
[46] Jans, 1931, 28-30.
[47] Jans, 1914, 20.
[48] Vilms, 1933a; Vilms, 1936.
[49] Vilms, 1933b, 8-9.
[50] Vilms, 1945?, 127.
[51] See: Kalling, Tammiksaar, 2008, 217-229.
[52] "[Correspondence of Oskar Kallas / Henrik Koppel to Oskar Kallas]": Estonian Museum of Literature, Department of Manuscripts 186/M65/27, 21/39.
[53] Tischler, 1902, 56-79.
[54] Laaman, 1915, 41.
[55] "Sotsialistide leht sülitab tuld [The Newspaper of Socialists Spits Fire]" *Postimees*, no. 8 (January 9th), 1927.
[56] See also the article by Leiu Heapost and Ken Kalling in this book.
[57] Kemiläinen, 1998.
[58] Aul, 1936.
[59] Kemiläinen, 1993, 234-252.

[60] Tork, 1940, 302.
[61] Ridala, 1913, 353.
[62] Hellat, 1913, 352-353.
[63] Hellat, 1913, 352-353.
[64] Luiga, 1995b, 145.
[65] Karjahärm, 1993, 1347-1348.
[66] Aul, 1932, 291-292.
[67] Aul could not tell whether the phenomenon was to be explained by laws of molecular genetics or was it just the women who had chosen bigger and taller men, i.e. the sailors, belonging to the then Russian elite troops. See: Aul, 1933b, 17-18.
[68] Aul, 1938, 83.
[69] Aul, 1932, 291-292.
[70] *Eesti rahva tulevik. II Rahvusliku kasvatuse kongressi kõned ja läbirääkimised* [The Future of Estonian Nation. The Speeches and Discussions Held at the II Congress of National Education]. Tartu: "Tuleviku" kirjastus, 135-138.
[71] Vilms, 1927a, 281.
[72] A Finnic tribe, inhabiting once parts of Western and Northern Latvia.
[73] Vilms, 1927a, 280-282.
[74] Madissoon, 1934a, 344-354.
[75] Vilms, 1927b, 33; Vilms, 1933a, 103.
[76] Klein (Aul), 1926, 99.
[77] *Eesti rahva tulevik*, 10.
[78] Üprus, 1937, 93.
[79] *Riigi Teataja* (1936) no. 98, 1936, article 776, 2078-2081. (Translation into English: "The Estonian Law of Sterilisation" *Annual Report 1998*, Tartu University History Musem, vol. 3,1999, 15-19).
[80] "Öigeenikakomisjoni tegevusest [On the Activities of the Eugenics-Commission]" *Eesti Arst*, no. 2, 1925, 61-62.
[81] Madissoon, 1927, 340-341.
[82] "Ettekannete kogumik. XI Eesti Arstide Päevad [Presentations at the 11th Estonian Doctors' Days]" *Eesti Arst*, Appendix, 1933, 63.
[83] "Kirjavahetus seltsi organisatsioonilise tegevuse koht [Correspendence on the Organisational work of the Eugenics Society]": ERA, 4855/1/5, P, 58.
[84] "Vastuvõetud resolutsioonid ja sooviavaldused. XII Eesti Arstidepäev [Accepted Resolutions and Demands at the 12th Estonian Doctors' Days]" *Eesti Arst*, Appendix, 1934, 54.
[85] "Hoolekande seaduse muutmise ja täiendamise seadus (steriliseerimise küsimuse osa) [The Amendment Bill for the Public Health Law (Concerning Sterilisations)]" *Eesti Arst*, no. 1, 1935, 69-74; See also: "Seltsile tutvumiseks saadetud hoolekande seaduse muutmiste kava, Koduse

Kasvatuse Instituudi kodukord, 1936/37. aasta töökava, Kehakultuuri sihtkapitali inspektori aruanne, III Kehalise Kasvatuse kongressi päevakord [Different Legal Projects, Reports, Drafts etc. sent to the Eugenics Society]": ERA, 4855/1/4, 24-29.

[86] See i.e.: "Steriliseerimise kava. Milline võib olla steriliseerimise tulemus? [Plans for Sterilisations. What could be the Outcome of Sterilisation?]" *Rahva Sõna,* 8 (26. jaanuar) 1935, 7.

[87] "Hoolekande seaduse muutmise ja täiendamise seadus (steriliseerimise küsimuse osa)", 70.

[88] Madissoon, 1938a, 131-137.

[89] "Arstide Koja tegevusest"[On the Activities of the Chamber of Doctors]" *Eesti Arst* (1937), no. 6, 1937, 506.

[90] "Saabunud kirjad ja muud materjalid sihtasutuse "Eesti Lastekaitse" tegevuse korraldamise kohta [Correspondence of the Eugenics Society on the Estonian Child Protection Organisation]": ERA, 4855/1/8, 68.

[91] Madissoon, 1938b, 664.

[92] Vilms, 1938, 386-387.

[93] *Referaatide juhtmõtted ja resolutsioonid. Rahvusliku Kasvatuse Kongress Tartus, 2. ja 3. jaanuaril 1927. a.* [The Watchwords and Resolutions of the Congress of National Education in Tartu January, 2nd and 3rd, 1927], 12.

[94] Põld, 1923, 62.

[95] Vilms, 1923, 37.

[96] Kruus, 1938, 578.

[97] Hion, 1934, 87-90.

[98] "Steriliseerimine oleks vigastamine – intervjuu Schlossmanniga [Sterilisation would be Injuring – an Interview with Schlossmann]" *Postimees*, no. 101 (April 12th), 1927, 4.

[99] Bresowsky, 1935, 536-539.

[100] "Gerhard Rooks [Personal File of Gerhard Rooks]": Eesti Ajolooarhiiv (Estonian History Archives – further: EAA), 2100/2/1031a, 48.

[101] Soans, 1935, 535.

[102] Toomingas, 1925, 173.

[103] Blessig, 1930, 409-413.

[104] Silvester, 1938, 383.

[105] Silvester, 1940, 392-393.

[106] Madissoon, 1938b, 665.

[107] *Valitsusasutiste tegevus 1937/1938* [The Activites of State Institutions in the Year 1937/1938]. Tallinn: Riigikantselei, 1938, 103.

[108] Raid, 1939, 273.

[109] *Valitsusasutiste tegevus 1938/1939* [The Activites of State Institutions in the Year 1937/1938]. Tallinn: Riigikantselei, 1939, 108.

[110] For general background – the number of mentally insane, epileptics and feeble minded in Estonia – the potential subjects of sterilisations – in 1938/1939 was 3358, of whom 1220 were taken care outside the psychological wards (*Valitsusasutiste tegevus 1938/39* [The Activites of State institutions in the Year 1937/1938]. Tallinn: Riigikantselei, 1939, 108).
[111] "Operatsioonide žurnaalid [The Journal of Operations of the Womens' Clinic of the University of Tartu]": EAA, 2451/2/186, passim.
[112] "Kirjavahetus..." [Correspondence of the Social Welfare Dperatment on Numerous Matters...]: Tallinn Linnaarhiiv (Tallinn Town Archive – further: TLA), 137/1/142, 1288.
[113] Mattila, 1999, 43-75.
[114] Toomingas, 1935, 513.
[115] Valma, 1927, 14.
[116] Madissoon, 1930b, 210-215.
[117] Kalling, 2007c, 95-104.
[118] Eesti Linnade Liidu ja Eesti Maaomavalitsuste Liidu esindajate ühise nõupidamise protokoll, märgukirjad ning kirjavahetus Haridus- ja Sotsiaalministeeriumi ja linnavalitsustega hoolekandelise abi (sealhulgas arstiabi) korraldamise ja hoolekandeseaduse muutmise kohta [Minutes of the Estonian Union of Towns and Local Communities Meeting, Correspondence on Different Matters, incl. on Social Welfare and Health Topics]: ERA, 923/1/100, pages not numbered.
[119] *Valitsusasutiste tegevus 1937/38. a*, 105.
[120] See: Laid, 1932, 12-16.
[121] See: Kalling, 2003, 175-192.
[122] With some reservations, the Estonian language understands the word *kasvatus* in two ways – it means „education" but it derives from a stem meaning „growth". There was a tendency in the topics of the congresses towards the latter, from educating of the population to the increasing of it.
[123] *Eesti rahva tulevik,* 10.
[124] "Miks rahvaarv ei kasva [Why There is no Population Increase?]" *Päevaleht*, no. 4 (January 5th), 1940, 7.
[125] Põld, 1922, 9.
[126] Luiga, 1995c, 137.
[127] Vilms, 1920, 39.
[128] Vilms, 1920, 5.
[129] Kruus, 1940, 143-152.
[130] *Eesti rahva tulevik*, 103-104.
[131] See: *Riigikogu V koosseis. III ja IV istungjärk. Täielikud protokollid ja stenograafilised aruanded 1933–1934* [The 5th *Riigikogu* /Parliament. 3rd and 4th Sessions. Full Protocols and Stenogrammes 1933-1934]. Tallinn, 1934, 1438.

[132] Vilms, 1945?, 4.
[133] Kalling, 2005, 359-365.
[134] Madissoon, 1934b, 449-450.
[135] "Ülevaade Vabariigi Valitsuse 1935. a. 6. veebruari otsusega moodustatud rahva juurdekasvu ja ühenduses seega kodanikkude üldise heaolu komisjoni tegevusest [A Report on the Activities of the Board of Population Increase and Welfare, Established by the Government on Feb. 6th 1935]" *Eesti Arst*, no. 11, 1935, 875-880.
[136] "Seletuskiri tervishoiu ala arendamise kava juurde. [Letter of Explanation to the State Plan of Promoting Public Health]" *Eesti Arst*, no. 11, 1935, 886-894.
[137] Schlossmann, 1938, 708-716.
[138] "Kirjavahetus seltsi organisatsioonilise tegevuse kohta" [Correspendence on the Organisational work of the Eugenics Society]: ERA, 4855/1/5, 103.
[139] Vilms, 1933a, 92-93.
[140] Aul, 1968, 2.
[141] Orav, 1967, 22-30.
[142] K. Siilivask and H. Palamets (eds.), *Tartu ülikooli ajalugu* (1982). Tallinn: Kirjastus "Eesti Raamat", vol. 3, 152.

References

Aul, J. (1932), "Andekus ja ristamine [Talentedness and Crossbreeding]" *Olion*, nos. 10-12, 291-292.

Aul, J. (1933), "Eestlaste tõulisest kuuluvusest [On Estonians' Racial Classification]" *Eesti Hõim*, nos. 1-2, 15-22.

Aul, J. (1933), "Maailmasõja antropoloogilisest mõjust saarlastele [On the Anthropological Influence of the World War on the Inhabitants of Saaremaa]" *Tartu Ülikooli Zooloogia-Instituudi ja -Muuseumi tööd*, vol. 13

Aul, J. (1936), Anthropologische Forschungen in Eesti. *Tartu Ülikooli Zooloogia-Instituudi ja -Muuseumi Tööd*, vol. 19.

Aul, J. (1938), "Eesti antropoloogilise uurimise senine viljelemine ja tulevikuülesanded [The History and Future Tasks of Estonian Anthropological Research]" in: *Omariikluse süvendamisel*. ÜS Raimla koguteos. Tartu: ÜS Raimla Kirjastus, 74-101.

Aul, J. (1949), "Eksimustest minu senises teaduslikus ja pedagoogilises tegevuses [Mistakes in My Previous Scientific and Pedagogical Work]" *Rahva Hääl*, no. 21 (January 27th), 2.

Aul, J. (1968), "Meisterlikkus, töö ja andekus: Geneetika mõisteist [Mastery, Work and Talent: On the Meaning of Genetics]" *Sirp ja Vasar*, no. 9 (March 1st), 2.

Auväärt, L. & Maripuu, I. (1994), "Hans Madissoon. Kurikalduvustega alaealiste esmauurija Eestis [H. Madissoon – the Pioneer in the Studies of Juvenile Deliquency in Estonia]" *Eesti Riigikaitse Akadeemia Toimetised*, no. 2, 10-20.

Blessig, E. (1930), "Pärilikkusest oftalmoloogia alal [On Heredity in the Field of Ophtalmology]" *Eesti Arst*, no. 11, 409-413.

Bresowsky, M. (1935), "Vaimuhaigete sterilisatsioonist [On the Sterilisation of Mental Patients]" *Eesti Arst*, no. 7, 536-539.

Grenzstein, A. (1885), "Joomise vastu [Against Drinking]" *Olevik*, no 21 (May 20th), 1.

Hellat, P. (1892), *Tervise õpetus* [Textbook of Health] (1st edition). Tartu: Mattiesen.

Hellat, P. (1913), *Tervise õpetus* [Textbook of Health] (2nd edition.) St. Petersburg: "Ühiselu" kirjastus.

Hion, V. (1934), "Parandamatute haigete surmamisest ja steriliseerimisest [On Euthanasia and Sterilisation of Incurably Sick]" *Tervis*, no. 6, 87-90.

Hurt, J. (1989), "Meie koolitatud ja haritud meestest [On Our Educated Men]" (Speech given in 1870) *Kõned ja kirjad. Loomingu Raamatukogu*, no. 1/2.

Ibius, O. (1939), "Eesti karskusliikumise ajalugu [The history of the Estonian Anti-alcohol Movement]" in: *Manuscript*, Tartu: University Library, Department of Rare Books and Manuscripts, vol. 2.

Jans, J. (1914), "Ernst Haeckel ja tema elutöö [Ernst Haeckel and His Work]" in: *Loodus ja inimene: Viis päätükki Ernst Haeckeli töödest ühes [J. Jansi] sissejuhatusega tema 80. sünnipäeva mälestuseks* [Nature and Man. Five Chapters from the Works of E. Haeckel to Celebrate His 80th Anniversary, Added with an Introduction by J. Jans]. Tartu: Noor-Eesti, 1-20.

Jans, J. (1931), "25 aastat [25 years]" in: *Eesti Üliõpilaste Selts "Ühendus" 1906-1931*. Tartu: Postimees.

Kalling, K. (1997), "Prof. Juhan Aul and Eugenics" *Papers on Anthropology*, vol. 7, 174-180.

Kalling, K. (1999), "Introduction to the History of Estonian Eugenics" in: *Annual Report 1998*, Tartu University History Musem, vol. 3, 31-42.

Kalling, K. (2003), ""Sigimisvõitlus" toimub ilma relvadeta... [The Demographic Warfare is Held Without Weapons]" *Toimetised*, vol. 2, 175-192.

Kalling, K. (2005), "Abort sõjaeelses Eestis – legaliseerimisest keelustamiseni [Abortions in Pre-war Estonia – from Legalisation to Prohibition]" *Eesti Arst*, no. 5, 359-365.

Kalling, K. (2007), "The Self-Perception of Small Nation: The Reception of Eugenics" in: P. Weindling and M. Turda (eds.), *Interwar Estonia in Blood and Homeland. Eugenics and Racial Nationalism in Central and Southeast Europe, 1900–1940*. Budapest: Central European Univ. Press, 253-262.

Kalling, K. (2007), "Karskustöö sünnitab eugeenikaliikumise [The Abstinence Movement Creates Eugenics Movement]" *Mäetagused*, vol. 36, 59-78.

Kalling, K. (2007), "Estonian Psychiatric Hospitals During the German Occupation (1941-1944)" *International Journal of Mental Health*, vol. 36, no. 1, 95-104.

Kalling, K. and Tammiksaar, E. (2008), "Descent versus Extinction: The Reception of Darwinism in Estonia" in: E. M. Engels and T. F. Glick (eds.), *The Reception of Charles Darwin in Europe*, vol. 1. London-New York: Continuum Publishers, 217-229.

Karjahärm, T. (1993), "Tõuküsimus Eestis iseseisvuse eel: Historiograafiline referaat [Racial Issue in Estonia Before the Independence Gained. A Historiographic Overview]" *Akadeemia*, no. 7, 1347-1364.

Kemiläinen, A. (1993), *Suomalaiset, outo Pohjolan kansa. Rotuteoriat ja kansallinen identiteetti*. Helsinki: SKS.

Kemiläinen, Aira (1998). *Finns in the shadow of Aryans. Race theories and racism.* Helsinki: SKS

Klein, J. (Aul, J.) (1926), "Pärivus ja rahvas [Heredity and Nation]". Tartu: Loodus.

Koppel, H. (1910), "Eessõna [Foreword]" *Eesti Üliõpilaste Seltsi Album*, no. 8, 1-4.

Kruus, H. (1938), "Jaan Tõnisson kaugvaates [Jaan Tõnisson from the Distance]" *Jaan Tõnisson töös ja võitluses*, 573-589.

Kruus, H. (1940), "Rahvusterviklusest [On National Entirety]" *Akadeemia*, no. 3, 143-152.

Kurvet-Käosaar, L. (2009), ""Algupärase talupojavere jõud." Eesti tõu küsimus Aino Kallase loomingus ["The Vitality of the Native Peasant Blood." The Role of Race in the Work of Aino Kallas]" *Keel ja Kirjandus*, no. 10, 729-745.

Laaman, E. (1915), "Tõutervendus Eestis [Running Racial Health in Estonia]" *Vaba Sõna*, no. 2, 40-46.

Laaman, E. (1938), *Juhan Luiga: elu ja mõtted* [The Life and Ideas of Juhan Luiga]. Tartu: Noor-Eesti Kirjastus.

Laid, E. (1932), "Tartu Ülikooli osa meie välissuhtluses [On the Role of the University of Tartu in Estonian Foreign Policy]" in: *Kultuuri teaduse teilt. Mõtteid ja uurimusi Tartu Ülikooli 320-nda mälestusaasta puhuks*. Tartu: EÜS „Veljesto" Toimetised, vol. 2, 12-16.

Laul, E. (1956), "Loodusuurijate-materialistide võitlusest religiooni ja idealismi vastu Eestis (XIX sajandi lõpust 1917. aastani) [On the Fight of Naturalists-Materialists against Religion and Idealism in Estonia from the End of the 19th Century until 1917]" in: *Religiooni ja ateismi ajaloost*, vol. 1. Tallinn: Eesti Riiklik Kirjastus, 164-193.

Luiga, J. (1995), "Rahvaste tõus ja langemine [The Rise and Fall of Nations]" (Speech given in 1909) in: *Mäss ja meelehaigus*. Tartu: Ilmamaa, 210-227.

Luiga, J. (1995), "Pranglisaare kuriteod. (Vaimuhaigused.) [The Crimes of the Island Prangli. (Mental Diseases)]" in: *Hingejõu ilmed*. Tartu: Ilmamaa, 141-145. (First published 1925)

Luiga, J. (1995), "Küsimusi arstidepäevalt. [Issues Raised on the Doctors' Days]" in: *Hingejõu ilmed*. Tartu: Ilmamaa, 137-140. (First published 1925)

Luyga, Y. (1904), "Prizrenie dushevnobol'nych' v pribalstiyskom kraye. Istoriko-statisticheskiy ocherk' [Guardianship of Mentally Insane in the Baltic provinces. An Historico-statistical Overview]" Yur'yev': Tipografiya gaz. "Postimees".

Madissoon, H. (1927), "Alaväärtuslikkude ja kurikalduvusega isikute sigivõimetuks tegemisest [On the Sterilisation of Inferior and Criminal Individuals]" *Pärivus ja valik,* 301-341.

Madissoon, H. (1930), "Eesti Eugeenika seltsi "Tõutervise" abielu-nõuandla tegevusest. [On the Activities of the Marriage-Consulting Office of the Estonian Eugenics Society "Tõutervis"]" *Eesti Arst*, no. 5, 203-204.

Madissoon, H. (1930), "Über einen Fall von Sterilisierung des Mannes durch Röntgenstrahlen" *Deutsche Zeitschrift für die Gesamte Gerichtliche Medizin*, vol. 16, No 3., 210-215.

Madissoon, H. (1934), "Järeltuleva soo tõulise väärtuse kaitseseaduse eelnõu [Project for the Law of Eugenic Defence of the Future Generations]" in: *Tööd ja artiklid 1921-1934*. Manuscript at the Tartu University Library, 344-354

Madissoon, H. (1934), "Märgukiri h. Riigivanemale abordi vabaduse vastu [Note to the State Head against the Freedom of Abortions]" in: *Tööd ja artiklid 1921-1934*. Manuscript at the Tartu University Library, 449-450.

Madissoon, H. (1938), "Kas tervel naisel on õigust eugeenilise abordi peale, kui mees on pärilikult vaimuhaige? [Does a Healthy Woman Have a Right for an Eugenic Abortion if Her Husband is Genetically Mentally Ill?]" *Õigus*, no. 3, 131-137.

Madissoon, H. (1938), "Eugeenilise mõtte levingust, eriti seadusandluses [On the Spreading of Eugenical Thought, Especially Concerning Legislation]" *Eesti Arst*, no. 9, 664.

Mattila, M. (1999), "The alegal eugenic sterilisations in Finland – an international perspective." *Annual Report 1998*, vol. 3, 43-75.

Orav, T. (1967), "Pärilikkus ja ühiskond [Heredity and the Society]" *Eesti Kommunist*, no. 11, 22-30.

Pärivus ja valik [Heredity and selection], A. Audova, A. Lüüs, H. Madissoon and J. Vilms (eds.), Tartu: Loodus, 1927.

Põld, P. (1922), *Eesti riigi tulevik ja karskus* [The Future of Estonian State and the Abstinence Work]. Tartu: Eesti Karskusseltside Kesktoimkond.

Põld, P. (1923), "Sõnavõtt [A speech]" in: *Eesti XIII karskuskongress Tartus 24. ja 25. veebruaril 1923. a.* Tartu: Eesti Karskusliit, 58-63.

Raid, K. (1939), "Kastratsioon ja sterilisatsioon [Castration and Sterilisation]" *Politseileht*, nos. 11/12, 273.

Reiman, H. (1931), *Eestlaste rassiline koostis* [The Racial Composition of Estonians]. Tartu: "Loodus".

Reiman, V. (1890), "Kas elutilk või surnumeri [A Drop of Life or Sea of Death]" *Postimees*, nos. 22-23 (February 24th and 27th).

Ridala, V. (1913), "Tõu küsimus [The Stock Issue]" *Eesti Kultura*, vol. 3, 199-363.

Roll-Hansen, N. (2001), "Eugenic Practice and Genetic Science in Scandinavia and Germany" *Scandinavian Journal of History*, vol. 26, no. 1, 75-82.

Schlossmann, K. (1938), "Haiguste osa ajaloo ja kultuuri arengus [The Role of Diseases in the Development of History and Culture]" *Eesti Arst*, no. 9, 708-716.

Shepilevkii, E. (1914), *Ocnovy I Sredstva Rasovoy Gigieny. (Gigiena Rasmozheniya)* [Basic Principles and Means of Racial Hygiene. Hygiene of Proceation.]. Yur'yev': K. Mattisen.

Silvester, R. (1938), "Sterilisatsioon pärivusõpetuse seisukohalt [Sterilisation from the Viewpoint of Hereditary Studies]" *Akadeemia*, no. 6, 383-389.

Silvester, R. (1940), “Sigivõimetustamine geneetilisest seisukohast [Sterilisation from a Genetic Viewpoint]” *Eesti Arst*, no. 5, 392-393.

Soans, T. (1935), “Vaimuhaiguste pärilikkusest [On the Heredity of Mental Disease]” *Eesti Arst*, no. 7, 522-535.

Tischler, J. (1902), “Materialismus – aineline ilmavaade [*Materialismus* – a Materialist Ieology]” in: *Sirvilauad 1903*. Jurjev (Tartu): Eesti Üliõpilaste Seltsi kirjastus, 56-79.

Tõnisson, J. (1907), “Sõnavõtt [A Speech]” in: *Kuues Eesti karskusseltside kongress*. Tartu: Eesti Karskusseltside Kesktoimkond, 7-9.

Toomingas, K. (1925), *Shitsofreenne konstitutsioon ja selle suhe shitsofrenia'ga. Väitekiri arstiteaduse doktori saamiseks* [Scizophrenic Constitution and its Relations to Schizophrenia, a M.D. Thesis]. Tartu: Tartu Ülikool.

Toomingas, K. (1935), “Vaimuhügieeni ülesandeist Eestis [The Tasks of Mental Hygiene in Estonia]” *Eesti Arst*, no. 7, 505-515.

Tork, J. (1940), “Eesti laste intelligents. Pedagoogiline, psühholoogiline ja sotsioloogiline uurimus [The Intelligence of Estonian Children. A Pedagogical, Psychological and Sociological Study]” *Eesti Vabariigi Tartu Ülikooli Toimetised, B (Humaniora)*, vol. XLVIII.

Üprus, V. (1937), “Väljavaateid rahvusriigi püsimiseks Eestis rahvusbioloogilisil kaalutlusil [Perspectives for a National State in Estonia Deriving from Ethnic-biological Assumptions]” *Akadeemia*, no. 2, 80-93.

Üprus, V. (1938), “Perekondade uurimise tähtsus sotsiaal-bioloogiliste küsimuste selgitamisel [The Importance of Family-studies in Solving Socio-biological Issues]” in: *Omariikluse süvendamisel.* Tartu: Ü.S. Raimla Kirjastus, 102-112.

Valma, H. (1927), *Isikuraamat. Paranduskasvatuslik isiklikkude andmete ja vaatluste kogu. Teine kurikalduvusliku ja kriminaalse küljega täiendatud väljaanne*. [A Personal Book. A Correctional Collection of Private Data and Examinations. Second Issue with Amendments Concerning Criminality]. Tallinn: Eesti Kirjanduse-Ühisus.

Villem, E. (1920), *Karskus ja Haridus* [Abstinence and Education]. Tallinn: Ühiselu.

Vilms, J. (1920), *Poligaamiline abielu Eestis. Mõtted mitme naise pidamise lubamise üle* [Polygamic Marriage in Estonia. Thoughts on the Legalizing of Polygamy]. Tallinn: Varrak.

Vilms, J. (1923), "Sõnavõtt [A speech]" in: *Eesti XIII karskuskongress Tartus 24. ja 25. veebruaril 1923. a.* Tartu: Eesti Karskusliit, 35-45.

Vilms, J. (1927), "Sega-abielude ärahoidmine muulastega Eestis. Abielu hõimurahvastega [Avoidance of Intermarriages in Estonia with Foreigners. Intermarriages with Kin-nations]" in: *Pärivus ja valik.* Tartu: K/Ü. "Loodus", 280-282.

Vilms, J. (1927), *Suguhaigused – Tegeliku arsti seletused suguhaigustest hoidumisest ja nende arstimisest* [Veneral Diseases – Explanations by a Practicing Doctor on Veneral Diseases their Avoidance and Cure]. Tartu: Loodus.

Vilms, J. (1933), *Erakondadeta ühiskonna poole. Eesti kui korporatiivne riik* [Towards a Society Without Political Parties. Estonia as a Corporative State]. Tallinn: Vaba Maa.

Vilms, J. (1933), "Kolmas rahvuslik tõus [The Third National Rise]" in: *Töötute Haritlaste Ühingu Aasta-album.* Tallinn: Estotrükk, 8-9.

Vilms, J. (1936), *Korporatiivne ühiskond. Fašism ja demokraatia. Kutseesinduslik korra uuestisünd* [The Corporative Society. Fascism and Democracy. The Rebirth of the Corporative System]. Tartu: Eesti Kirjanduse Selts.

Vilms, J. (1938), "Sirgjooneline võitleja [A Straightforward Fighter]" in: *Jaan Tõnisson töös ja võitluses.* Tartu: Koguteose "J. Tõnisson" komitee, 385-387.

Vilms, J. (1945?), "Rahvaste edasielamise alustest. Rahvastikupoliitiline uurimus [On the Survival of Nations. A Populational-political Study]" in: Manuscript, ERA, 4006/2/11.

Vilms, J. (1989), “Eesti naiivne kosmopolitism ja rahvuslus [The Estonian Naive Cosmopolitism and Nationalism]” *Akadeemia*, no. 1, 47-60. (First published in 1915)

Weindling, P. (1999), “International Eugenics: Swedish Sterilization in Context” *Scandinavian Journal of History*, vol. 24, no. 2, 179-197.

Weingart, P. (1999), “Science and Political Culture: Eugenics in Comparative Perspective” *Scandinavian Journal of History*, vol. 24, no. 2, 163-177.

Racial Identity and Physical Anthropology in Estonia 1800-1945

Ken Kalling and Leiu Heapost

Abstract
At the turn of the 19th and 20th centuries, emergent Estonian national thought was strongly influenced by the natural sciences, and especially by racial theories. A challenge was the assumptions of Western scholars to classify Estonians (as other Finnic people) as racially "inferior". The local reception of the race concept thus obtained a strong taint of defensiveness, the race-linked prejudices becoming part of self-stereotypes.
In such a way physical anthropology received an important place among the so-called national sciences practised in the post-the World War I Estonian state. Physical anthropology attempted to answer questions about Estonians' descent and their kindrednations in Europe and further afield. The race concept became involved in the local eugenics discussion.

1. Introduction

At the end of 1930s, Juhan Aul (1897-1994), after rising to be the highest authority in the field of physical anthropology in Estonia, looked back on the history of his discipline. His judgment on the work performed before him and his Estonian colleagues – i.e. before Estonia became independent in 1920 – was critical. Aul pointed to four main faults that were characteristic of non-Estonian scholarship: 1) incorrect descriptions of Estonians' physical habitus; 2) incorrect statements concerning different anatomical particularities allegedly characteristic to Estonians (and marginalising the nation in this way); 3) incorrect information concerning race types in Estonia; and finally (4) incorrect positioning Estonians as part of the "Yellow race". During a period of nearly 20 years of independence, much work was done to erase these misunderstandings. Finally, Aul could state that Estonians do not possess most of the traits important for the racist prejudices of the era.[1]

In the same article, Aul also gave directions for future work. First, there was a need to create the so-called *Anthropologia estonica*, a database (including a vast photo collection) of the Estonian population. He planned deeper regional studies and studies of women. A separate challenge was palaeoanthropology, further development was needed in the fields of so-called social and applied anthropology (both understood by Aul in a biologised fashion). To achieve these goals, teaching this topic was to be organised, and the Institute of Anthropology was established at the University of Tartu.[2]

This article aims to shed light on the historiography of Estonian physical anthropology, a tool in the creation of the self-perception of the modern Estonian nation. Mainly, the period from 1920 to 1944 is outlined, though preceding periods are touched upon to present the background of certain cases.

Historical overviews of this topic can be extracted from early compilations that sought to describe the Estonian "race", for example, by Hugo Reiman (1887-1957), an autodidact.[3] The first dedicated historiography of physical anthropology in Estonia was the previously cited article by Aul. Recently, an important work on the adoption of race stereotypes among Estonian educational and political elites was written by Toomas Karjahärm.[4] Shorter contributions to the history of physical anthropology have been provided by Linda Kongo, Erki Tammiksaar, Jaan Kasmel and the present author.[5]

The wider context concerning the classification of Estonians according to race theories has been touched on by Finnish scholars, most significantly Aira Kemiläinen[6] and Pekka Isaksson.[7] Their works serve as a general background for this article, supported by the fact that Estonian belongs to the so-called Baltic-Finnic language group of Finno-Ugric languages. Such a linguistic basis has played an important role in the race theories of the era under discussion.

2. Preconditions for the Emergence of Racial Self-Perception among Estonians

Research on race discussions conducted in Estonia must be introduced with a brief overview of ethnic and social relations in the region. In his study on mental disease in his homeland from 1904, Estonian psychiatrist Juhan Luiga (1873-1927) documented the inhumane conditions for the feeble-minded and blamed this on the local *Herrenmoral*, the noblemen who spoke a different language and who were culturally alien to most of the population. For Luiga, the class difference in the society also included race differentiation.[8] Luiga shared the stereotypes of his era, namely the biologising of ethnic and social stratification.

The Baltic political discourse was strongly influenced by linguistic divisions based on social hierarchy. The upper layers of the three so-called Russian Baltic provinces (Estonia, Livonia and Couronia, now Estonia and Latvia) were German speaking. The lower social layers, the so-called *Undeutsche* (non-Germans), comprised of indigenous peasants, began to evolve into the modern Estonian and Latvian nations only during the 19th century. By the end of the century, a third player entered the Baltic political stage: the Russian central authorities, who escalated the policy of Russification.

The Estonian national consciousness was based to a large degree on opposing other players in the local political arena, mainly Germans. In 1868 Carl Robert Jakobson (1841-1882), one of the early leaders of Estonian national movement, coined what would later become the paradigmatic concept of the "700-year night of serfdom" (the suppression of Estonian peasants by German landlords), which transformed existing feelings into a nationalist ideology.

For Estonians, nationality was defined mainly by language. The particularity of the linguistic division – Germans being Indo-European, and Estonians being Finnic – may have supported the acceptance of race theories. Historian Hans Kruus (1891-1976), referring to similar phenomenon in Germany, described the process of shifting to race theories in Estonia, where the metaphysical "national spirit" was increasingly challenged by a new entity – the allegedly science-based "national body". Instead of Hegel and Kant, Estonia welcomed Darwin and Mendel, Kruus wrote.[9]

The concept of race was introduced to Estonian language discussions at the turn of the 20th century. The words "*tõug*" [(animal)stock] and "*ra(a)ss*" were used, both of which, according to the spirit of the era, contained not only morphological and taxonomic meanings, but also a much broader realm of psychological, mental and other aspects.[10] Biologised ideologies such as eugenics and pro-natalism spread rather broadly within the realm of the Estonian language. This was due to worries about the size and quality of the emerging nation, an outcome of the so-called self-perception of a small nation.[11] Knowledge about the concept of racial hierarchy was also introduced to the Estonian public.[12]

Given this, it is understandable that when Johann Friedrich Blumenbach (1752-1840) hinted in 1795 that Estonians could be racially "Mongols",[13] it would lead to a scientific discussion among Estonian scholars during the 20th century. This more than 100-year lag time can be explained by the fact that Estonians as a political nation emerged only in the second half of the 19th century. It was in the context of national emancipation that topics associated with race also found a place in the process of self-definition. Race was viewed as being among the key factors in explaining the supposed position of Estonians among fellow Europeans, and evaluating the potential of them as nation-builders. Folklorist Villem Grünthal-Ridala (1885-1942), influenced by the Social Darwinist tendencies of the era, saw race as a crucial factor in explaining the vital characteristics of a nation and its capacities for survival.[14]

The possibility of "racial inferiority" was viewed as a potential obstacle to national emancipation. Therefore it had to be studied and, depending on the position of those involved, either questioned or justified. Before the birth of their own state, Estonians lagged behind more developed nations in discussions concerning their biological characteristics, being capable of only making comments and compilations. The theme became important for the national scientific establishment. Race studies, believed to be based on the methodology of exact sciences, were seen to possess advantages over methods such as ethnography, when different aspects regarding the life and development of the nation were at stake. Additionally, the popular perception of race concepts remaining strong, it was acknowledged that race-related topics should be understandable to people without specialist education.[15]

3. Estonians as Seen by Craniologists

Even though Blumenbach mentioned only "Northern" Finns as being "Asians" (Estonians possibly "saved" from such a fate[16]), this did not prevent other theories from seeing Estonians as being racially "alien" to Europe. Nineteenth century anthropologists Jean Louis Armand de Quatrefages (1810-1892), Paul Broca (1824-1880), Paul Topinard (1830-1911) and others believed that Estonians differed from other Europeans. A morphological argument for this could serve the supposedly stronger prognathism (the positional relationship of the jaw) of Estonians,[17] and also the so-called nasal index.[18] As the cranial samples to support such arguments were few from the viewpoint of modern methodology, one could suggest that there was a wish to confirm the existing hypothesis, based on scientific ignorance (inertia) or created by social prejudice.

In the middle of the 19th century, a new (and long-standing) stratifying criterion for humans emerged – the cephalic (cranial) index[19] – introduced by Swedish anthropologist Anders Retzius (1796-1860). From this point on, people and populations were divided by scholars into long-headed (dolichocephalic), medium-headed (mesocephalic) and short-headed (brachycephalic). A theory emerged according to which dolichocephalic individuals were more capable of progress. Both A. Retzius and his son Gustaf (1842-1919) declared Finnish tribes to be dominantly short-headed.[20] The idea of human stratification was supported by the Darwinian theory of evolution. Another important Darwinian aspect was the general atmosphere that his teaching established, which supported the biologisation of social theories. Academic scholarship was amplified by textbooks and encyclopaedias describing Finnic people as racially alien to Europe.[21] Even at home in Estonia, in teaching materials on physical anthropology issued in 1912 by Eber Landau (1878-1959, a lecturer in anatomy at the University of Tartu), Estonians were placed into the Ural group of the Uralo-Altaic branch of "Mongoloid race".[22]

At the end of the 19th century, the Mongoloid theory began to fade. New data was spreading by which Finnic people, even if short-headed, were blonde. In such a way, they could be ascribed with an important trait that was characteristic to the "White" race.

An important step in the development of race theories determining the Estonians' position in the race hierarchy was the definition of a new "East Baltic race" type by a Swedish commentator of Finnish origin Rolf Nordenstreng (1878-1964) in his treatise on European races, which was published in several gradually amended editions from 1917-1926.[23] This new form of classification occupied a central place in discussions by Finnish and Estonian anthropologists in explaining the ethnic history of their nations. As a result, the "Mongoloid" question was marginalised. The issue of the possible inferiority of the Finnic people still remained, as the East Baltic race (especially when compared to the so-called Nordic race) was viewed to possess fewer moral and cultural characteristics of positive value.[24]

The beginning of Estonian physical anthropology as a distinct discipline is traditionally pinned to the 1814 dissertation by Karl Ernst von Baer (1792-1876), which discussed endemic diseases among Estonians. Baer declared Estonians to be of "lymphatic constitution" and receptive to chronic disease: harsh living conditions and physical work during childhood made Estonians age prematurely.[25] Baer did not address the race issue; he simply stressed the relatedness of Estonians to Finns.

Later, Baer became one of the founders of a unified methodology for anthropometric measuring, and the initiator and organiser of anthropological research in Russia and Germany.[26] In 1850, despite supporting Blumenbach's viewpoints on the race division in Europe, he comments that Finns, though short-skulled, differ from a typical "Mongol".[27]

The University of Tartu's first study on the physical anthropology of Estonians based on anthropometric methodology was an 1838 thesis by Alexander von Hueck (1802-1842). Studying 10 skulls, Hueck declared that Estonians did not belong to the so-called European variety (a concept of race as established by Blumenbach) or to the Asian, but constituted a separate one with "less beautiful body build and fewer strength in their muscles".[28]

The next thorough study on Estonians was published in 1878 by Oscar Grube (1853-??), who studied 100 people living around Tartu using the methodology of Paul Broca. In 1879 a study by Hugo Witt (1852-?) on 86 skulls from the University of Tartu's collection was published.[29] Witt paid attention to the cephalic index and compared Estonians with their neighbours and other Finnic people. He declared Estonians to be mesocephalic. The work by Witt, like Grube's study, was done under the supervision of anatomy Professor Ludwig Stieda (1837-1918), who worked in Tartu (or Dorpat as it was then known) until 1885 and founded the anthropological collection and the academic school of physical anthropology in Tartu.[30]

At the end of the 19th century, new methods such as cerebral anatomy were put into use to characterise Estonians as a race. Active in this field in Tartu were Richard Weinberg (1867-1927), E. Landau and Nikolai Goryainov (1879-?). Landau[31] and Goryainov[32] did not establish any significant anatomical particularities to distinguish Estonians from other races.

The most important brain studies on Estonians – in the sense that they were being discussed in Estonian newspapers – were by Weinberg.[33] An anatomist, Weinberg in 1894 defended a dissertation in Tartu on the brain gyri of Estonians, a work supervised by anatomist August Rauber (1841-1917).[34] Studying eight brains, Weinberg concluded that by the shape and number of gyri, Estonian brains could not be distinguished from European samples.[35]

Using data from more than 1,000 conscripts, Weinberg placed Estonians into the mesocepahlic type with inclination towards brachycephaly. He mentioned the so-called Finnish facial type and describes it as "low and wide" with strong cheekbones (zygmata). The latter, along with other details of facial anatomy, gave Estonians and other Finns a look similar to the "Mongoloid" race, but also Weinberg denied that Estonians and other Finns are "Mongoloids".

In independent Estonia, Hans Madissoon (1884-1956), a forensic scientist and one of the most prominent eugenicists in the country, compared brain sizes from individuals of different social layers and different periods of Estonian history. He was convinced that he had found a correlation between education and brain size. He also declared that Estonians living in the 20th century had larger brains than their medieval ancestors. Despite admitting that brain size should not correlate directly to intelligence, Madissoon was still tempted to suggest some biological emancipation – an increasing brain size – of Estonians during the previous centuries.[36]

4 .The Body of Estonians in the Light of Biologised Ideologies

There are other morphological traits besides brains and skulls useful for the purpose of racial distinction and social stratification. An important work in the historiography of Estonian physical anthropology was published in 1867 by Johannes V. Holst (1823-1906), professor of obstetrics and gynaecology at the University of Tartu. He studied Estonian women and declared them to have a strong body-build and project a healthy impression.

Holst pointed to an anthropometric trait that remained a topic of discussion for decades. According to Holst, the so-called pelvic angle of Estonian women was smaller than that of other European women.[37] Later, Aul was critical of the theory, claiming it incorrect. The particular angle could be used as a means to position Estonians along the racial hierarchy. Aul knew that such a trait was believed to be characteristic of "primitive" people.[38]

Also Weinberg measured Estonian women. Contrary to previous information, by which he denied any stratifying traits concerning Estonians' brains (including those of females), his findings on the morphology of women can be seen as supporting existing racial stereotypes. Weinberg stressed the strong physical build of Estonian women and supported the idea of the infamous pelvic angle. He also believed it was possible to conclude that Estonian women tended to not notice their menstruation and that they gave birth with ease, able to return to their work the next day.[39] Such characteristics could be used to form a link to "primitive nations"; still, Weinberg's statements on Estonian women were met with agreement by the local press. A newspaper article stressed that according to scholars, Estonian women were suitably equipped for fulfilling their reproductive task by having a broad pelvis.[40]

The idea that Estonian women give birth with an ease persisted in the independent state. One of the national politicians from the beginning of the 20th century, the educationalist Peeter Põld (1878-1930), wrote in regard to demographic matters that Estonian women were armed by nature (as established by anthropologists) with a suitable physical build for giving birth easily.[41] In 1924 the Estonian Health Board rejected the international labour convention that called for six weeks of time off for pregnant women before giving birth. In Estonia, women received only four weeks, as "Estonian women possess strong health".[42]

In 1884 Bernhard Körber (1837-1915) concluded that Estonian women gave birth to smaller children than Russians.[43] In independent Estonia, forensic scientist Gerhard Rooks (1901-1975) declared the opposite,[44] confirming the 19th-century data of August V. Schrenck (1852-?).[45] H. Reiman wrote in 1928 Estonian children were born strong, which was confirmed by their size and relatively lower death rates in the first days of life.[46] In 1936 the long-time chair of the Estonian eugenics society, paediatrics Professor Aadu Lüüs (1878-1967), published a study on the size and weight of children. Lüüs concluded that Estonian newborns were among the largest in "civilized nations", comparable to babies in Scandinavia.[47]

Despite these studies supporting the idea that there was nothing wrong with Estonians' inborn characteristics concerning their vitality, there was still a problem, as the natural rise in the population was low.

An issue that bothered the national elites was the possible extinction of Estonians, a topic that emerged within the so-called small nation's self-perception[48] and was supported by the degeneration theory. Signs of the degeneration could be seen in various epidemiological and anthropological characteristics of the population. Estonian scholars remembered Baer's statements about Estonian soldiers' weak resistance to stress, injury and disease during the Napoleonic campaign.[49] Luiga thought that he witnessed this same phenomenon during World War I when his countrymen in the Russian army did not demonstrate any instinct to fight their arch-enemy – the Germans – and instead hid in hospitals. Regarding matters of degeneration among Estonians, Luiga said that traits of "sicklyness" could be explained by the fact that Estonians are "a nation of indigents".[50]

The flaws in vitality seemed to be confirmed by different physical markers. The weak physical constitution of Estonians was suggested by Leonid Malinovski (1854-?), the head physician of the Estland province (now northern Estonia) at the end of the 19th century.[51] A thorough study on Estonians regarding the thoracic region – the body part supposedly most suitable for evaluating one's fitness – was conducted by Alexey Kharuzin (1864-1933), a Russian anthropologist, archaeologist and ethnographer. In 1892, Kharuzin used a representative number of military draftees (including 2,253 ethnic Estonians) from the province of Estland. He declared that the so-called vital capacity[52] placed Estonians in a rather negative position, with the supposedly weakly developed thorax serving as a sign of an asthenic constitution. This, in turn, was seen as a strong hint for tuberculosis, which was more widespread in the Baltic provinces than in the other parts of the Russian Empire. (About a fourth of Estonian draftees were rejected by the army at the end of the 19th century; for 10 percent, the reason was a rachitic thorax.)[53] On the other hand, some data proved Estonians to be strong (even though of medium stature and possessing a slim physical build).[54]

The first Estonian-born scholar, Heinrich Niggol (1890-1918) to conduct anthropological research, concentrated on the so-called thoracic index[55] of Estonians, comparing it to groups supposedly belonging to other races (Armenians, Georgians, Jews and Germans from the Caucasus). As a result, he could declare that Estonians had the flattest chest (the smallest anterioposterior diameter of thorax).[56] Niggol's work confirmed the data from 1845 of Georg Schultz (1808-1875) on the broad chests of Estonians. (In the study, Estonians were declared to have the largest dimensions of breadth among the ethnic groups in the Russian Empire.[57])

An important trait for judging nations was the stature of its people. Some authors declared during the 19th century that Estonians were among the tallest people in the Russian Empire.[58] In the independent state, studies on stature were conducted by Nikolai Köstner (1889-1959),[59] Aul and others. Touching upon the topic in several studies, Aul concluded that as a criterion of racial identity, stature is not beyond criticism; in Estonia, clearly not all Nordic people were tall. Stature could not serve as the main criterion in judging the strength of people and nations, the assessment of which required more complicated methods. Also weight – which Aul was the first to study in Estonians – could be influenced by various non-biological factors (ecological and social). Aul was still tempted to declare, according to the Rohrer index,[60] that Estonians must be considered a "bodily excellent" nation. In such a way, he could declare incorrect or outdated the opinions that Estonians are "thin" and "slack".[61]

5. The Race of Estonians and Juhan Aul

Discussions by Estonians themselves on their own racial background were initiated by Weinberg. Henrik Koppel (1863-1944), who introduced the public to his work, commented that in "older times, elderly people must have stopped their work startled when the young ones told them about the weird creatures – the Estonians – written about in geography and history textbooks in foreign countries, where people must thank God that they are not similar to Estonians".[62] After such an introduction, Koppel welcomes Weinberg's studies for changing the stereotypes. Weinberg's studies were cited by several Estonian newspapers, which were pleased by his conclusion that the Estonian brain should be viewed as an average European one.

The advent of the Estonian state sparked new concerns about the racial constitution of the nation. The new member of the international community felt itself keenly observed from outside, as international institutions requested information on the Estonian population. In 1926 anatomy Professor Alfred Sommer (1858-1939) and hygiene Professor Alexander Rammul (1875-1949) of the University of Tartu, with their colleague the physician Richard Villems (1887-1940), applied to the Cultural Endowment of Estonia (*Eesti Kultuurkapital*) for support for the anthropological study of Estonians. The applicants relied on a letter from the League of Nations to the Estonian Department of Statistics, asking for information concerning the anthropology of Estonians. The scholars stressed that in Latvia similar research under the supervision of Gaston Backman (1883-1964) had already found support.[63] Until then, such research in Estonia had been supported by the Anatomical Institute of Tartu University, with Villems having already measured 2,000 people.[64]

Villems' 1925 dissertation was the first thorough study of the anthropology of Estonians done during the era of independence.[65] Aul praised this work, mainly for declaring invalid the theories that categorised Estonians as members of "Mongoloid" race. On the other hand, Aul was critical of the dissertation for only containing simple statistical data with no broader generalisations and comments, making it not possible for anthropologists to use. Additionally, it was only a manuscript, with some of its data published in 1931 in a booklet by H. Reiman.

Reiman's work was a compilation containing studies ranging from morphology to serology seeking to discuss the racial background of Estonians. Reiman's conclusion was that Estonians are a racially mixed population with the presence of different components, and with Nordic and East-Baltic types dominating. He ridiculed the popular "ethnic romanticism" – the idea of a common Finno-Ugric ancestry and the "kinship movement" – saying that one should not search for a "blood connection" to justify international cooperation, as this first of all should be primarily based on mutual (economic) profit.[66]

Aul declared Reiman's booklet to be unscientific and was unsatisfied with Reiman's knowledge of hereditary science; Aul himself stressed that "anthropology is a science of genetics". He used the word "anthropogenetics" to characterise his own studies.[67] Aul also accused Reiman of raising a redundant question in his work – to prove that linguistic relationships within human populations do not always parallel racial (biological) relationships. The issue here was the previously mentioned Finno-Ugric language group, including people from distant geographic regions and racial types. For Aul, it was self-evident that language and race are not linked, but he accused Reiman of disputing any solidarity among Finno-Ugric nations.[68]

Here, a few words should be said about Aul (known as Johan Klein until 1931), the patriarch of Estonian physical anthropology.[69] Aul graduated from the University of Tartu as a zoologist in 1928. He was taught techniques of anthropological measurement by Madissoon, who had learned these skills in Finland from Yrjö Kajava (1844-1929).[70] Both Estonians came to anthropology through their interest in genetics. Madissoon worked in criminal anthropology; as for Aul, there is information regarding studies of twins he planned to conduct according to suggestions by the educationalist Juhan Tork (1889-1980). Like Madissoon, Aul was keen on eugenics. According to Aul, eugenics dealt with pathologies within a human population, while the role of physical anthropology was to study "normal" humans.[71] Aul's work can be divided into three categories. First, he studied the racial constitution of Estonians (including ethnic history). Second, his works were linked to eugenics, for example, attempts to extract morphological knowledge that was useful for classifying people for pedagogical, psychological, economic or other purposes. Third, he attempted to contribute to the theory of human morphology (i.e. age-related changes in human morphology).

Aul's involvement in anthropological research began in 1927, when he made an expedition to the Sõrve peninsula on the south-western part of the island of Saaremaa. Aul viewed Sõrve as important because it was a meeting point where Western anthropological types must have influenced Eastern types.[72]

In the following years, Aul published several papers on the Estonian race. In 1938, for example, he criticised a compilation by Swedish anthropologist Sten de Geer (1886-1933) on the spread of the so-called Nordic race.[73] Aul's argument was that de Geer had declared almost the whole of Estonia to stand outside the range of the Nordic race, and accepted the Nordic race to be present only in some regions partly inhabited by ethnic Swedes. Aul presented his own map of race types in his homeland. It was based on an enormous body of research: the measurements of more than 15,000 conscripts. Aul used a methodology by which each person was classified into a particular race type. Aul admits that he slightly changed the criteria for race determination used by Scandinavian scholars, who said that a man belonged to the Nordic type if he was more than 168 cm tall, had a cephalic index below 78, and had bright eyes and blonde hair. Aul added face shape to these criteria. For him, Nordic people had to have a facial index[74] either at least 84 and a cephalic index no higher than 78, or a cephalic index no higher than 81 if their facial index was no higher than 89. Aul's idea was that the Nordic race must be more varied than what previously had been believed. As a result, Aul said that 25% of the Estonians he studied could be classified as members of the Nordic race. According to Aul, the proportion in Sweden was 31%.[75]

More difficult was the case of the so-called East Baltic race, the second most prominent type among Estonians. This type seemed to be especially polymorphic and thus difficult to strictly define. More studies were needed, and Aul's dream – *Antropologia estonica* – was his dedication to studying this race.

The racial dualism of Estonians created possibilities for racial stratification within the nation. During the inter-war period, when Aul began his studies, various racial prejudices filtered into the field, clearly exceeding the limits that future generations would tolerate. Reiman, for example, supported the era's racial stereotypes by writing that in south-east Estonia, where the "East Baltic race" dominated, the mental capabilities of schoolchildren were lower than in "Nordic"-dominated areas.[76] A similar belief came from J. Tork, who studied the intelligence of Estonian schoolchildren.[77] Aul himself seems to have avoided making statements about the possible ranking of the two types.

Still it cannot be denied that Aul believed that biology shaped the fates of humans as societies. He participated in spreading ideologies, such as eugenics, that today would be denounced. On the other hand, despite the rather high popularity of physical anthropology in Estonia, this field managed to maintain its distance from populist and racist ideologies, especially those characteristic of National Socialism. In Estonia, there seems to have been no particular racial goal or ideal as an agenda of popular discussions. Besides the unclear racial status of Estonians, the absence of such a goal also must have been due to the anti-German position among Estonian elites, which developed into an anti-Nazi sentiment. Nazis were seen to embody the ultimate in German aggressiveness.

Aul, who attacked Germans in several of his works, published an article in 1933 blaming Nazi Germany for practicing so-called "political anthropology". Aul ridiculed the concepts of "Aryan stock" and "German blood". According to him, it was equally as possible to speak about a "blonde dictionary" or "long-headed grammar". Besides being unscientific, Aul said the Nazi-German approach intruded into the private spheres of other nations, searching for German footprints everywhere where Teutonic traits could be found in history, explaining cultural development through this.[78]

It should be mentioned that the article discussed here was published in a journal of the so-called Club of Estonian Nationalists (*Eesti Rahvuslaste Klubi).*[79] This organisation, mainly comprised of intellectuals, could be viewed as rather reactionary and even pro-fascist, as it supported corporatist ideology. For example, its founding charter declared that the success of Estonia as a nation depended on the fusing of different social layers into one entity, and subsuming them to the interests of the state.[80]

Aul made his second important anti-German statement in 1935 in a journal of student organisations, in which he blamed several German scholars for their humiliating imperialist and paternalist approach toward Estonians.[81] The only study on physical anthropology that he mentioned was by the Baltic-German Sophie Ehrhardt (1902-1990), who was a graduate of Tartu University and who was working at the time at the Anthropological Institute of the University of Munich. She conducted research on one particular Estonian parish population (Kõpu).[82] Aul blamed Ehrhardt for scientific inadequacy, measuring too few people from too wide of an age spectrum, which was of no use for making broader generalisations. Later, Ehrhardt became infamous for her race studies on German gypsies in Nazi Germany.[83] Aul's main worry was not the unreliable numbers but the introductory text of the article, in which the author diminished the role of Estonians in history and in the present day. Aul believed that such an approach – besides Germans being unwilling to accept Estonian achievements – could be explained through Oswald Spengler's (1880-1936) work, *Jahre der Entscheidung* (Years of Decision). Here, Aul saw praise for war and reflections on the German "barbarian mentality", with the *Third Reich* simply being a manifestation of Prussian aggressiveness.

6. Prehistory in the Light of Race Studies

Stereotypes about modern populations were included in discussions of the past. The so-called "Gothic theory" emerged in the middle of the 19th century, which added a racial aspect to discussions on Estonian early history. It posed the idea that during various eras of pre-history, the Estonian territory must have been settled by Germanic tribes. The question was whether "Goths" were the first settlers (before the Finns and Balts) or did they arrive later and constitute just part of the Estonian population? Despite such questions, the main idea remained the same: archaeological sites with more elaborate finds were explained through the heredity of "Goths". It was presupposed that the direct ancestors of ethnic Estonians must have been incapable of creating the more complex sites found by archaeologists. This approach remained prominent until the beginning of the 20th century.[84] (Aul accepted that some Goths (Eastern-Germanic people) arrived to the coast of northern Estonia in the first centuries after Christ.[85]) Following the context established by the "Gothic theory," scholars concentrated on the cephalic index, which was an important scientific category in distinguishing race types. Initially, attention was limited to Neolithic skeletal remains.

Weinberg said that a Neolithic skull from Võisiku, which he said belonged to the so-called Nordic (dolichocephalic) type, seemed to have been different from the Estonian population of his time.[86] Swedish anatomist Carl M. Fürst (1854-1935), in an article on skeletal finds from the island of Saaremaa, said that a female skull clearly belonged to the brachycephalic type (typical for modern Estonia). In such a way, both types of humans, short- and long-headed one – which were also present in modern Estonia – must have inhabited the territory in Neolithic times. Fürst said that short-skulled people, with roots in the South (moving into Estonia from the East), entered the scene earlier than Nordic types who came from the West.[87] The Nordic type was also said to be present in skeletal remains from the Bronze Age and later.[88]

Russian archaeological scholarship became interested in the Baltic region during the period of Russification (from the last decade of the 19th century). Kharuzin, who studied Iron Age burials from north-eastern Estonia, denied the theory of the "Goths". Disappointed that most of the skeletons he collected were of women, whose skulls he said were less valuable than those of men, Kharuzin compared the remains with modern Finnic people. On the basis of cephalic indexes, he declared that the skeletons he studied could not belong to Finnic tribes because the skulls were predominantly dolichocephalic. Although there is a temptation to believe that Kharuzin was suggesting that the people were Slavic – the site was near a newly built Orthodox monastery – he does not hint to which nationality the former inhabitants may have belonged.[89]

Aul became involved with archaeological remains in the 1930s, focussing first on Neolithic finds. In 1942 he commented on six Stone Age skeletons, those described by Weinberg and Fürst plus three from northern Estonia and one from the south.[90] Aul's conclusions about the first inhabitants of Estonia seem to have depended on the political environment present.

During the years of independence, Aul believed that the Estonians' substrata must be viewed as racially East Baltic, with later Indo-European and Germanic migrations adding Nordic elements.[91] The Indo-European type (supposedly ancestors of modern Baltic-language speakers) were believed to have arrived from the South, and the Finnic tribes from the East. The Finns either possessed some "Mongoloid" influences or carried a separate race character similar to the "yellow race". The Stone Age brachycephalic East Baltic type was related to the Finnic tribes. Aul suggested that this population could be under-represented in archaeological remains because Finnic people supposedly cremated their dead.[92]

During the Nazi occupation, Aul changed slightly his mind, writing that even if some Stone Age skulls belonged to mixed types, most of them had Nordic traits, suggesting that the Nordic type was prevalent in the region. He believed the transition from the Nordic to the East-Baltic type to be gradual. This process could have been hastened by invasions of neighbouring Russians, leaving the most difficult to access parishes untouched, inhabited by the original Nordic people.[93]

During the Stalinist era, Karin Mark (1922-1999) – a former student of Aul who went on to become the main palaeoanthropologist for decades in Soviet Estonia, and whose research areas included the so-called Mongoloidness index[94] – blamed Aul for attempting to prove that Stone and Bronze Age Estonian populations were more Nordic than they were, attempting in such a way to link the country's inhabitants to the "West".[95] These cases show how the interpretation of anthropological findings could easily change for political reasons.

7. The Social Applications of Physical Anthropology

Aul's wish was to bring anthropology from the theoretical realm into practical use. There were several possible means to achieve this. The emergence of criminal anthropology at the turn of the 20th century opened possibilities for research that compared Estonian and Latvian criminality. This approach was based on the fact that the nations had a very similar social history, but according to the ideas of the era, they belonged to different races (because of belonging to different language groups). Any differences between the two thus could be explained by race. Criminal anthropologists attempted to obtain information on supposedly racial indicators of crime patterns between the two nations. The pioneer of this methodology was a psychiatry professor at the University of Tartu, Vladimir Chizh (1855-1922). This topic was further developed by an amateur, Russian meteorologist Alexander Klossovsky (1846-1917) in Odessa, and later by Estonian scholars.

Both Chizh and Klossovsky claimed that they observed distinguishable patterns in the criminal behaviour of Estonians and Latvians. Analysing court files from the Province of Livonia (*Livland*) from 1894-1897, Chizh found that Estonians were more inclined to commit emotional and violent crimes, and that their community seemed to be less empathetic. Estonian single mothers were said to be more inclined to commit infanticide than Latvians.[96] Klossovsky's work concentrated on crimes against property, finding broadly that crime rates increased moving north toward Estonia (except for Saaremaa county, inhabited by Estonians). Klossovsky believed that in addition to biological factors, economic factors also could have played a role, as Latvian areas were more prosperous.[97]

Mihkel Pung (1876-1941), a lawyer who commented on the Chizh study in Estonian, questioned the approach. Pung could not challenge the Chizh's factual data, so he questioned the racial approach, especially the theory of Cesare Lombroso (1836-1909). Pung questioned whether people could be judged based on the "length of their hands and breadth of chest". Pung declared it impossible that physical traits could determine the moral character of an individual; thus, the entire concept of criminal anthropology must be invalid because it denied people the free will to live as they desired.[98]

The ideas of Chizh were also discussed by J. Luiga, who had been concerned about Chizh's 1908 work that compared the revolutionary activities of Estonians and Latvians during the 1905 Russian Revolution. Chizh said that he observed Latvians' greater inclination towards rebellious activities. He could not understand the reason for this, as Latvians generally seemed to be more wealthy and less criminally inclined than Estonians.[99] According to Luiga, Chizh's approach was too simplistic. Luiga suspected that wealth might translate into higher awareness of social issues (i.e. – social factors must not be forgotten while discussing such issues).[100]

In 1927, Estonian psychiatrist Konstantin Lellep (1888-1958) made a final attempt to characterise the "race"-related criminal inclinations of Estonians. Providing an overview of previous studies in the field, he cited the research by H. Niggol on the thorax of Estonians. According to Lellep, the results of this study could explain how C. Lombroso could attribute the relatively high level of criminality among Estonians to a trait of "primitiveness".[101] Relying on Ernst Kretschmer (1888-1964), Ernst Rüdin (1874-1952), Ignaz Kaup (1870-1944), Eugen Fischer (1874-1967) and others, Lellep declared that according to the race and anthropological theories of the time, Estonians in general should fit the "Nordic (leptosomic) type, be high-grown and asthenic, with cold and contemplative character".[102]

Lellep studied different crimes (e.g. murder, infanticide, physical assault) as well as suicide and alcohol consumption. He pointed to the relatively high ratio of affective crimes (infanticide, murder, physical assault) among Estonians, which he said could suggest a different "racial" background from their neighbours.[103] According to Lellep, Estonia had the characteristics of a so-called agricultural nation, and the population's criminal inclinations could even qualify Estonia as a "noble nation". That was because criminal behaviour associated with supposedly submissive types – crimes of possession – was rare. Lellep accepted also socio-economic factors as explanations for his data – i.e. high suicide rates could be explained by the rapid emancipation of Estonians as an independent nation.

Racially oriented statements in criminal anthropology were also important for Madissoon, in his studies of juvenile delinquency in Estonia. Convinced that there were innate causes of criminal behaviour, Madissoon used the genealogical method to declare, that, for example, brown eyes could suggest Roma ancestry – whose criminal inclinations were not in doubt for him.[104]

Aul's planned *Anthropologia estonica*, a broad database of anthropological information on Estonians, was important for different applications. In the beginning of the 1940s, when Aul began collecting genealogical (ancestral) material concerning anthropological traits, he saw his work as being important for purposes such as determining paternity.[105]

Aul believed that physical anthropology could be useful also in pedagogics. As a eugenicist, he had an argumentation with one of the authorities of Estonian pedagogy, Peeter Põld (1878-1930). Aul blamed Põld for devaluing the importance of inherited traits in pedagogic work.[106]

Aul believed that physical anthropology could help predict the mental capabilities of people. Already a clear authority in the field in Estonia, Aul said that his scholarship should be viewed as crucial toward understanding the individual development of school children. The know-how from other countries on this topic was believed to be invaluable, as Estonian children had a different pattern of development.[107]

For Aul, pedagogic anthropology would include studying the physical constitution of children in collaboration with the medical profession, the relationship between their biological and chronological age, and divergence from the average. Aul wanted to follow parallels in the somatic and mental development of children. He also hoped to increase children's interest in their own body.[108]

Anthropologists also played a role in the "inner colonisation" of the country. This topic was viewed as important for eugenicists, as the vitality of the rural population was believed to support an increase in the population. Additionally, Aul believed that urban life diminished several vital anthropological characteristics of people. In 1940 he published an overview of his research into the problems of slenderness and stuntedness (characterised in different ways, i.e. by the relative sitting height).[109] Aul discussed the influence of urbanisation on the human body, stating that it increases longitudinal measurements and decreases thickness measurements.[110] Aul believed that he had observed the acceleration of sexual development and the growth in stature of urban children. He concluded that cities "disharmonise" the growth of the human organism, weaken physical proficiency, and lower the physical capabilities and endurance of the entire "national body". According to Aul, so-called "anthropological urbanisation" mainly affects the upper layers of society, meaning that the process must be viewed socially (caused, for example, by long periods in school). This trend was not seen positively, as it was not caused by developments in hygiene nor improvements in social conditions, but mainly by alienation from physical work. "Anthropological urbanisation" was to be viewed as a general problem, centred in the cities.

Aul presented his ideas at the Assemblies of Estonian Naturalists, where he had supporters as well as opponents. Voices from the audience remarked that the Nazi-German practice of obligatory service in physical work for office workers – it had seemed to the listeners that Aul was hinting at something of a kind – had proved to be not good for the fertility of (urbanised) women.[111]

In 1938 Aul showed initiative in founding an Institute of Anthropology at the Estonian Academy of Sciences, though the plan did not come to fruition. The academy's natural sciences section discussed a programme of "ethnobiological research", but it was mainly dedicated to demographic issues, such as eugenics.[112] Aul still found support for his other projects. In 1938 the Estonian Society of Naturalists, which was associated with the Academy of Sciences, opened an anthropology section chaired by Aul.[113]

8. The Nazi Occupation

In pre-war Estonia, two debatable scientific fields – eugenics and race studies – were institutionally separate, although several leading eugenicists including Aul and Madissoon had good relationships with physical anthropology. During the brief Soviet period before the German occupation, the eugenics movement and its institutions were banned. Physical anthropology as a discipline, however, did not suffer. Aul was even nominated as head of the zoology section at the Institute of Scientific Research at the University of Tartu, which the Soviets established to replace the dissolved Academy of Sciences.[114]

In summer 1941 the Bolsheviks were replaced by the Nazis due to the outbreak of the war. The University of Tartu continued its work according to the demands of war and Estonia's new masters. The Germans had a well-known interest in physical anthropology. At the end of 1942, Aul declared the arrival of a historic moment in the history of the university: race studies and physical anthropology were declared obligatory disciplines in the Faculty of Natural Sciences. Aul wrote that financial support for his field of scholarship was never better.[115] The chair of anthropology at the University of Tartu was established in March 1943, and Aul received a professorship in July. The Institute of Anthropology and Racial Studies (*Antropoloogia ja Rassiteaduste Instituut*) was established that same year, with Aul being named its head in October.[116]

Aul lectured on national studies and race theory at the School of Security Police under the Directorate for Internal Affairs of the Estonian Self

Administration (*Eesti Omavalitsuse Sisedirektooriumile alluv Julgeolekupolitsei kool*). He had contacts with the Office of Racial Policy (*Rassenpolitischen Amt)* in Germany.[117] In 1943 Aul participated in an expedition of Estonian scholars to the occupied Russian territories east of Estonia, for the purpose of studying the Estonian and other Finnic populations in the area. Although organised by the Directorate of Public Education of the Estonian Self Administration (*Eesti Omavalitsuse Haridusdirektoorium*), one might surmise a link with regard to the ideology of the *Generalplan Ost*, as there is a possibility that the expedition was connected to the plans to re-settle the people of Eastern Europe. (In fact, Aul did not conceal his participation in the event at a time when one would expect that he would, and he used the results in his Soviet-era publications.[118]) At the end of 1943, Aul was nominated to be the "general co-ordinator" of the application of sciences to military use (*üldkorraldaja teaduste sõjalise rakendamise alal*) at the University of Tartu.[119]

In April 1943 an exhibition and a conference dedicated to contributions by Estonian science to the German war effort were held. A newspaper that covered the events included an article by Aul in which he once more drew conclusions about the anthropological work conducted in Estonia. Aul stressed the high ratio of Nordic-type people in the country and pointed out that this was not due to later (medieval) Swedish influences but that it could be traced to pre-historic times. Aul stressed that both the Nordic and East Baltic types in Estonia differed from the similar types in surrounding areas. This stressed the borders with Russians, who also carried East Baltic traits. According to Aul, the fate of nations is shaped not only by economic and natural factors but also "human" (i.e. – biological) factors. Recalling his previous studies, Aul was proud to say that based on their physical build index[120] and the Pignet index,[121] Estonians comprise a strong nation.[122]

Judging by his CV, Aul engaged in collaboration. On the other hand, his writings from the period discussed here reveal limits in his willingness to accept German political pressure. In a newspaper article discussing the concept of human capacity, he declared it to be an important field of anthropology (in addition to race matters).[123] Aul told the readers that the term of capacity well known in cattle-breeding must be introduced also to the valuation of people. At this point, Aul's criticism of Nazi ideology may be apparent, as he writes that at the time, some nations with "good" racial traits seem to possess weak characters of capacity (and that even the Nordic race could include "idiots").

According to Aul, human capacity must be measured by mental and physical potential (i.e. body weight, thoracic circumference, lung volume, several other bodily traits and, finally, intelligence). Aul was convinced that capacity traits could be improved, and that studies in the field would be valuable toward identifying cases and causes of imperfection.

On the other hand, he shared the basic eugenics view, writing that special attention had to be paid to regions populated by people with better bodily traits. Aul ended his article with a statement characteristic of the tragic period in Estonian history – that Estonians must not be ashamed of their biological quality, which exceeded that of most European nations. This would guarantee that Estonians would survive current and future ordeals.

A 1944 manuscript by Aul can be found in the archives of the library of the University of Lund amongst the papers of Edgar Kant (1902-1978), the rector of the University of Tartu during the Nazi occupation. These papers must have been created with the hope that after the war, Estonia's independence could be re-established. They had to convince would-be post-war decision-makers that Estonia should be considered distinct from Russia (USSR).

Aul's paper, "Estonians and Russians Anthropologically", is a summary of his studies on the race and capacity characteristics of Estonians. The document, which is rather inadequate because it is too keen on human morphology for the new era, ends with a statement acknowledging the Russian influence on Estonians' (biological) type, but declaring that it had not changed the physical and mental structure of the Estonian "stock" to a point that it could be considered inseparable from the Russian. Thus, Estonians were to remain amongst the "family of civilised nations" and resist any larger racial influxes in the future.[124]

9. Conclusions

After the war, during the Stalinist repressions, Aul felt that his involvement with Nazi-era physical anthropology could bring him trouble. Although Russian (Soviet) colleagues wrote for him a letter of support in 1948 that praised his scholarly work and supported its continuation,[125] Aul had to write letter of apology during the witch-hunt within academic circles in 1948-1950. It was published in a newspaper, and mentions physical anthropology as well as eugenics.[126] It is difficult to say which field cost him his position at the university in 1950. Until then, he had been an assistant zoology professor; the chair of anthropology was never refilled. The repressions concerning Aul in 1950 were relatively modest; he could stay in Tartu but he remained unemployed. He regained his status in 1954 and in 1957 was named professor of zoology. Aul remained the leading physical anthropologist in Estonia in the following years, as the discipline flourished.

Aul, who sought not only to study human morphology but also the applications of physical anthropology on a much broader level, succeeded in publicising his field of science. Still, he and his works may only be viewed as the outcome of the factors that supported the biologisation of national thought. In part, the interest in race issues was provoked by European scholarship in physical anthropology, which attempted to classify Estonians as racially inferior. The main roots were still in the national emancipation of Estonians. The nationalist rhetoric in the Baltic region created ethnic prejudices and stereotypes. One may notice, especially during the inter-war years of independence, that the openly ideologised attitudes of Estonian physical anthropology were linked to Estonian defensiveness towards the Germans. Despite this, it must be acknowledged that the traditions of Estonia's scientific schools were too weak to prevent opportunism in adapting scientifically based ideologies to the atmosphere created by the German occupation of World War II.

Most race-linked prejudices in Estonia during the era were linked to self-stereotypes. Physical anthropology was important toward answering questions about Estonians' ancestors and its kin-nations in Europe and further afield. The awareness of the nation's tragic history, and the threats that it faced, created a phenomenon generally known as the small nation's self-perception. Being biologised, it contained concerns related both to the quantity and quality of Estonian people. In such a context, race served as a broadly used term.

Notes

[1] Aul, 1938, 79.
[2] Ibid., 94-100.
[3] Reiman, 1931.
[4] Karjahärm, 1993, 1347-1364.
[5] See: Kongo, 2002, 73-76; Kasmel, 2008, 152-157; Tammiksaar, 1999, 191-202.
[6] Kemiläinen, 1998 and 1985.
[7] Isaksson, 2001.
[8] Luyga, 1904, 14.
[9] Kruus, 1921, 89.
[10] Karjahärm, 1993, passim.
[11] See the article on the history of Estonian eugenics by Ken Kalling in this book.
[12] Hiiemaa, 2009.
[13] See: Blumenbach, 1865, 265.
[14] Ridala, 1913, 199-200.
[15] Annist, 1923, 28.
[16] Reiman, 1931, 127.
[17] Topinard, 1872, 661.
[18] Broca, 1872, 35.

[19] The ratio of the maximum width of the head multiplied by 100 divided by its maximum length.
[20] Kemiläinen, 1998, 166-167.
[21] Aro, 1985, 195-208; Halmesvirta, 1995, 209-247; Vihavainen, 1995, 269-295.
[22] Landau, 1912.
[23] Kemiläinen, 1993, 237-242.
[24] Kemiläinen, 1993, 241-242.
[25] Baer, 1976a, 29-34.
[26] See: Tammiksaar, 1999; Ottow, 1966, 43-68.
[27] Baer (1850), 585.
[28] Hueck, 1838, 10-11.
[29] Witt, 1879, 51-53.
[30] See: Kongo, 2002; Kasmel, 2008.
[31] See i.e.: Landau, 1909, 350.
[32] Goryainov, 1912, 173.
[33] See also: Kongo, 2001, 93-97.
[34] Weinberg, 1894.
[35] Veynberg, 1901, 19, 20, 22, 24. "Gyri" is a medical term for the superficial structure of ther brain.
[36] Madissoon, 1925, 233-238.
[37] Holst, 1867.
[38] Aul, 1938a, 75.
[39] Veynberg, 1901, 30-31.
[40] [Anonyme] (1904), "Eestlaste inimeseteadusline seisukoht [Estonians from the Anthropological Viewpoint]" *Linda*, XVI vol, no. 12/13: 237.
[41] Põld, 1922, 9.
[42] "Haiguse vastu kinnitamine meil ja mujal. [Insurance Against Illness in Estonia and in Other Countries]" *Postimees*, no. 53 (23 February), 1924.
[43] Körber, 1884, 240-241.
[44] Rooks, 1925, 133-136.
[45] Schrenck, 1880.
[46] Reiman, 1928, 525.
[47] Lüüs, 1936, 497-510.
[48] See: Kalling, 2007, 253-262.
[49] Baer, 1976b, 35.
[50] Laaman, 1938, 90.
[51] Malinovsky, 1891.
[52] Thorax circumference divided by the stature.
[53] Kharuzin, 1894a, 292, 310-311.
[54] Aarma, 1987, 7.

[55] The anteroposterior diameter of the thorax times 100 divided by the transverse diameter of the thorax.
[56] Niggol', 1913.
[57] Schultz, 1845, 224.
[58] Aarma, 1987, 8.
[59] Köstner, 1927, 431-445.
[60] Body weight in grams times 100 divided by the cube of height in centimeters.
[61] Aul, 1936, 49-53.
[62] Koppel, 1902.
[63] On Backman and his research in Lativa see the chapter of Björn Felder on Eugenics in Lativa in this volume.
[64] "Kirjavahetus isikute ja kultuurorganisatsioonidega Eesti Kultuurkapitali Vabariigi Valitsuse osast toetuse määramise asjus [Correspondence of the Estonian Cultural Endowment with Persons and Organisations on Support from the State Fund of the Endowment]": Eesti Rigiiarhiiv (Estonian State Archives – further: ERA) 1108/5/404, 121-126.
[65] Villems, 1926.
[66] Ibid., 166.
[67] "Auli antropoloogiliste uuringute läbiviimist kajastavad paberid ... 1930-1979 [Documents on the Scientific Work by Aul ... 1930-1979]": The Fund of Juhan Aul. Tartu University Library. The Department of Rare Books and Manuscripts, 147/54, 3.
[68] Aul, 1931, 259-262.
[69] See: Põldvere, 1997, 7-14.
[70] Ibid., 9.
[71] Aul, 1938a, 100.
[72] Aul, 1937, 157. On the basis of particular studies there emerged rather interesting discussions on race mixtures. The issue is discussed in the chapter by Ken Kalling on eugenics in Estonia in this volume.
[73] De Geer, 1926, 162-171.
[74] The ratio of facial length to facial width multiplied by 100.
[75] Aul, 1938b, 81-82.
[76] Reiman, 1930, 155-164.
[77] Tork, 1940, 302.
[78] Aul, 1933, 190-192.
[79] In the academic year 1934/1935 Aul also gave a presentation at the club with a title "The Politization of German Science", see: Eesti Rahvuslaste Klubi tegevuse aruanne [Report on the Activities of the ERK]: ERA, 4394/1/14, 1.
[80] "Eesti Rahvuslaste Klubi" [Club of Estonian Nationalists]: ERA, 14/11/59, 3.

[81] Aul, 1935a, 34-37.
[82] Ehrhardt, 1933, 143-148 and 190-198.
[83] http://www.uni-tuebingen.de/frauenstudium/daten/biograhpien/Biogramm_ _SophieEhrhard.pdf, (accessed 1 December 2010).
[84] Jaanits, Laul, Lõugas and Tõnisson, 1982, 10-12.
[85] Aul, 1935b, 42.
[86] Weinberg, 1905, 110-113.
[87] Fürst, 1914, 43-44.
[88] Friedenthal, 1931, 1-39.
[89] Kharuzin, 1894b, 249-250.
[90] "Eesti on antropoloogiliselt parimini uuritud maa maailmas. Jutuajamine dr. phil. nat. Juhan Auliga [Estonia – Anthropologically the Best Researched Country in the World. A discussion with Aul]" *Eesti Sõna*, no. 278 (2 December), 1942, 4.
[91] Aul, 1935b, 42.
[92] Aul, 1935c, 158-162.
[93] Aul, 1944?, 6.
[94] Mark, 1975.
[95] Mark, 1953, 507.
[96] Chizh, 1901, 41-59.
[97] Klossovskiy, 1905, 75-78.
[98] Pung, 1902, 1.
[99] Chizh, 1908, 149-162.
[100] Luiga, 1909.
[101] Lellep, 1927a.
[102] Lellep, 1927b, 145.
[103] Lellep, 1927b, 149.
[104] Madissonn, 1929, 1.
[105] „Juhan Aul": ERA, 2100/2/47, 180.
[106] Klein [Aul], 1927, 58.
[107] "Eesti Arstideseltside Liidu peakoosoleku protokoll (14. mail 1933) [The Minutes of the General Assembly of the Unions of Estonian Doctors' Societies (May 14th 1933)]" *Eesti Arst*, vol. 12, no. 6, 1933, 321. One encounters a self-stereotype by which Estonian children stayed behind when compared to the local ethnic minorities, and that for such reason there should be avoided the entrance examinations to the University.
[108] Aul, 1938b, 98.
[109] Aul, 1940a, 26-31.
[110] Aul, 1940b, 15.
[111] "Neljanda Eesti Loodusteadlaste-päeva aruanne [A Report of the 4th Estonian Days of Naturalists]" *Eesti Loodus*, vol. 8, no. 3, 1940, 164.

[112] "Loodusteaduste osakonna koosolekute protokollid [Minutes of the Section of Natural Sciences]": The Archives of the Estonian Academy of Sciences, 40/1/24, 6-9.
[113] Kasmel and Kasmel, 2009, 14-38.
[114] "Mitmesugused määrused ja juhendid ülikooli juures asuva Teadusliku Uurimisinstituudi kohta [Documents Concerning the Institute of Scientific Research]": Eesti Ajolooarhiiv (Estonian History Archives – further: EAA), 2100/20/46, 5.
[115] "Eesti on antropoloogiliselt parimini uuritud maa maailmas. Jutuajamine dr. phil. nat. Juhan Auliga. [Estonia – anthropologically the best studied Country. A Discussion with dr. phil. nat. Juhan Aul]" *Eesti Sõna*, no. 278 (2 December), 1942, 4.
[116] "Juhan Aul": EAA 2100/2/47, 190 and 211.
[117] "Juhan Aul": EAA 2100/2/47, 146 and 168-171.
[118] Aul, 1964, 118-164.
[119] "Juhan Aul": EAA, 2100/2/47, 208.
[120] Weight divided by the stature squared.
[121] This is calculated by a formula: stature in cm - (weight in kg + chest circumference in cm).
[122] Aul, 1943, 2 and 4.
[123] Aul, 1944, 2.
[124] Aul, 1944?, 6-8.
[125] "Juhan Aul": EAA, 2100/2/47, 227-228.
[126] Aul, 1949, 4.

References

Aarma, L. (1987), *Põhja-Eesti meeste pikkus* [The Stature of North Estonian Men]. Tallinn: "Eesti Raamat".

Annist, A. (1923), "Meie ilmast ja meie ilmavaatest [About Our World and Worldview]" in:, *Mõtteid valmivast intelligentsist*. Tartu: Odamees / Carl Sarap, 7-53.

Aro, T. (1985), "Suomalaisten rotu saksalaisissa ja pohjoismaisissa tietosanakirjoissa [The Finnish Race in German and Scandinavian Encyclopaedias]" in: Kemiläinen, A. (ed.), *Mongoleja vai germaaneja?*. Helsinki: SKS, 195-208.

Aul, J. (1931), "Mõningaid märkmeid H. Reimani "Eestlaste rassilise koostise" kohta [Some Comments on V. Reiman's "The Racial Composition of Estonians"]" *Olion*, vol 1, no. 7/8, 259-262.

Aul, J. (1933), "Poliitilisest antropoloogiast [On Political Anthropology]" *ERK (Please give the full journal – for some reason I can not write a comment... so – it's the name of it (it was issued by the **E**esti **R**ahvuslaste **K**lubid / The Clubs of Estonian Nationalists)*, vol. 1, no. 7, 190-192.

Aul, J. (1935a), "Natuke tendentsist ja Spengleri filosoofiast. Mis saksa teadlased meist "arvavad"? [Some Words About Tendencies and Spengler's Philosophy. What German Scholars "Think" About Us?]" *Üliõpilasleht*, no. 2, 34-37.

Aul, J. (1935b), "Eestlaste tõuline kuuluvus [The Racial Affiliation of Estonians]" *Tähiseid. Eesti Rahvuslaste Klubi Toimetised*, no. 4, 42-45.

Aul, J. (1935c), "Noorema kiviaja inimluude leidudest Eestis [On the Finds of Human Bones from younger Stone Age]" *Eesti Loodus*, vol. 3, no. 5, 158-162.

Aul, J. (1936), "Eesti mehe pikkusest [On the Stature of Estonian Man]" *Eesti Loodus*, vol. 4, no. 2, 49-53.

Aul, J. (1937), "Eesti antropoloogilisest uurimisest [On Anthropolical Research of Estonia]" *ERK*, vol.5, no. 7/8, 157.

Aul, J. (1938a), "Eesti antropoloogilise uurimise senine viljelemine ja tulevikuülesanded [The History and Tasks of Estonian Anthropological Research]" in: (ed. Parijõgi, J.), *Omariikluse süvendamisel*. Tartu: Ü.S. Raimla Kirjastus, 74-101.

Aul, J. (1938b), "Põhja tõu osatähtsusest ja levikust Eestis [On the Spreading of Nordic Race in Estonia]" in: *Liber saecularis Litterarum societas esthonica 1838-1938*. (Commentationes Litterarum societatis esthonicae, XXX). Tartu: Õpetatud Eesti Selts, 81-82.

Aul, J. (1940a), "Sihvakuse ja jässakuse probleemist [On the problematics of slenderness and stuntedness]" *Eesti Loodus*, vol. 8, no. 1, 26-31.

Aul, J. (1940b), "Antropoloogilisest linnastumisest [On Anthropologic urbanisation]" in: *Neljas Eesti loodusteadlastepäev 18. ja 19. märtsil 1940 Tartus: ettekannete kokkuvõtteid*. Tartu: ELUS, 15.

Aul, J. (1943), "Eesti antropoloogilise uurimise tulemused ja väljavaated [The Results and Perspectives of Estonian Anthropological Research]" *Eesti Sõna*, no. 88 (16 April), 2, 4.

Aul, J. (1944?), "Eestlased ja venelased antropoloogiliselt [Estonians and Russians Anthropologically]": Library of the University of Lund. Department of Manuscripts. 2/077 (Edgar Kant collection), 6.

Aul, J. (1944), "Jõudluse mõistest antropoloogias [On the Concept of Capacity in Anthropology]" *Postimees*, no. 50 (1 March), 2.

Aul, J. (1949), "Eksimustest minu senises teaduslikus ja pedagoogilises tegevuses [Mistakes in My Previous Scientific and Pedagogical Work]" *Rahva Hääl*, no. 21 (27 January), 4.

Aul, J. (1964), "Antropologicheskie issledovaniya vodi i izhortsev na zapade Leningradskoy oblasti. [Anthropologic Data on the Vodyans and Isurians from the Western Part of Leningrad District]" Tartu Riikliku Ülikooli Toimetised, vol.. 155, 118-164.

Baer, K. E. (1976b), "Eestlaste endeemilistest haigustest [On the Endemic Diseases of Estonians]" *Loomingu Raamatukogu*, no. 33.

Baer, K. E. (1850). "Chelovek v estestenno-istoricheskom otnoshenii" [A Man in the Natural-Historical Context]. Russkaya fauna, I. Sankt-Peterburg: K. Vingeber, 389-623.

Blumenbach, Friedrich (1865) *De generis humani varietate nativa.* Editio tertia. Cited from: *The Anthropological Treatises of Blumenbach and Hunter.* London: Longman, Green, Longman, Roberts & Green, 65-344.

Broca, P. (1872), "Recherches sur l'indice nasal" *Revue d'Anthropologie*, vol. 1, 1-35.

Chizh, V. (1901), "Vlijanie nacional'nosti na prestupnost' [The Influence of Nationality on the Criminality]" *Vestnik Prava*, vol. XXXI, no. 9, 41-59.

Chizh, V. (1908), Znachenie politicheskoj zhizni v' etologii dushevnykh' boleznei [The influence of political life on the etiology of mental diseases]. *Obozrenie psichiatriu, nevrologii i eksperimantl'noy psichologiy* no. 1: 1-12; no. 2: 149-162.

De Geer, S. (1926), "The kernel Area of the Nordic race within Northern Europe. An attempt at synthetic mapping, on the basis of the distributional areas of three racial characters" in: Lundborg, H. and Linders, F. J., *The Racial Characters of the Swedish Nation*. Uppsala: Almqvist & Wiksell, 162-171.

Ehrhardt, S. (1933), "Die Rassenzusammensetzung des estnischen Volkes" *Volk und Rasse*, vol. 6, no. 4 & no. 6, 143-148 & 190-198.

Friedenthal, A. (1931), "Ein Beitrag zur vorgeschichtlichen Anthropologie Estlands" *Zeitschrift für Ethnologie*, vol. 63, no. 1/4, 1-39.

Fürst, C. (1914), "Neolithische Schädel von der Insel Oesel [Saaremaa]" in: *Baltische Studien zur Archäologie und Geschichte*. Riga: Gesellschaft für Geschichte und Altertumskunde der Ostseeprovinzen Russlands, 43-44.

Goryainov, N. (1912), *Insula Reili u estov* [Insula Reili among Estonians]. Dissertation. S.-Peterburg: Tip. "Zhizn'".

Halmesvirta, A. (1985), "Anglo-amerikalaisen antropologian, etnologian ja kielitieteen näkemyksiä suomalaisesta rodusta ja sen kultuuritasosta evolutsionistisen kultuurikäsityksen valossa n. 1820-1930 [Anglo-American Anthropology, Ethnology and Linguistics on the Finnish Race and its Cultural Level in the Light of Evolutionary Theory during 1820s-1930s]" in: Kemiläinen, A. (ed.), *Mongoleja vai germaaneja?*. Helsinki SKS, 209-247.

Hiiemaa, K. (2009), *Aafrika imago eestikeelses trükisõnas* [The Image of Africa in Estonian Press]. Tartu: Tartu Ülikooli Kirjastus.

Holst, J. (1867), "Die Estin in gynäkologischer Beziehung" in: Holst, J., *Beiträge zur Gynäkologie und Geburtskunde*. Tübingen: Verlag der H. Laupp'schen Buchhandlung, no. 1.

Hueck, A. (1838), *De craniis Estonum commentatio anthropologica*. Dorpat: F. Schlater.

Isaksson, P. (2001), *Kumma kuvajainen. Rasisimi rotututkimuksessa, rotuteorioiden saamelaiset ja suomalainen fyysinen antropologia* [An Odd Reflection. Racism in Racial Studies, the Sami in Racial Theories and Finnish Physical Anthropology]. Pohjoiset Historiat 1. Jyväskylä: Kustannus Puntsi.

Jaanits, L.; Laul, S.; Lõugas, V. and Tõnisson, E. (1982), *Eesti esiajalugu* [Estonian Prehistory]. Tallinn: "Eesti Raamat".

Kalling, K. (2007), "The Self-Perception of a Small Nation: The Reception of Eugenics in Interwar Estonia" in: Weindling, P. and Turda, M. (eds.), *Blood and Homeland. Eugenics and Racial Nationalism in Central and Southeast Europe, 1900–1940*. Budapest: Central European Univ. Press, 253-262.

Karjahärm, T. (1993), "Tõuküsimus Eestis iseseisvuse eel: Historiograafiline referaat [Race Question Before Estonia's Independence: An Historiographic Report]" *Akadeemia*, vol. 5, no. 7, 1347-1364.

Kasmel, J. (2008), "On Prof. Ch. H. L. Stieda and the Beginning of Systematic Anthropological Research at the University of Tartu." *Papers on Anthropology*, vol. 17, 152-157.

Kasmel, J. and Kasmel, T. (2009), "Seventy Years of the Anthropology Section of the Estonian Naturalists' Society." *Papers on Anthropology*, vol. 18, 14-38.

Kemiläinen, A. (ed.) (1985), *Mongoleja vai germaaneja? Rotuteorioiden suomalaiset* [Mongols or Germanics. The Finns of the Race Theories] Helsinki: Suomen Historiallinen Seura.

Kemiläinen, A. (1993), *Suomalaiset, outo Pohjolan kansa* [Finns, a Weird Northern Nation]. Helsinki: Suomen Historioallinen Seura.

Kemiläinen, A. (1998), *Finns in the shadow of Aryans. Race theories and racism.* Helsinki: SKS;

Kharuzin, A. (1894a), "K antropologii naseleniya Estlyandsoy gubernii [On the Anthropology of the Estland Province]. *Vremenik Estlyandskoy gubernii*" Kn. I, 1893 god. Trudy Estlyandskago gubernskago statistichekago komiteta, T. 9. Revel' [Tallinn]: Nasledniki Lndforsa, 287-346.

Kharuzin, A. (1894b), "Zametka o cherpach' Pyuchtitskago mogil'nika [Notes on the Skulls from Pühtitsa Cemetery]" *Vremenik Estlyandskoy gubernii*" Kn. I, 1893 god. Trudy Estlyandskoy gubernskago statisticheskago komiteta, T. 9. .Revel' [Tallinn]: Nasledniki Lindforsa, 245-260.

Klein [Aul], J. (1927), "Kasvatus või pärivus? Vastuväiteid prof. P. Põllule rahvusliku kasvatuse kongressil [Education or Inheritance? Counter-arguments Against Prof. P. Põld at the Congress of National Education]" *Külvaja*, vol. 2, no. 6, 58.

Klossovskiy, A. (1905), "Sravnitel'naya prestupnost' sredi estov i latyshey Liflyandsoy gubernii (etnograficheskaya zametka) [Comparison of Criminality between Estonians and Latvians in the Livonian Province. Etnographic Note]" in: *Zapiski Novorossiyskago obshchestva estestvoispytateley*, no. 27, 75-78.

Kongo, L. (2001), "Richard Jacob Weinberg – Karl Ernst von Baeri preemia laureaat [R. J. Weinberg – a Laureate of K. E. v. Baer Prize]" *Eesti Antropoloogiaregistri aastraamat 2001*, 93-97.

Kongo, L. (2002), "Chr. H. L. Stieda tegevusest antropoloogias [On the Activities of Chr. H. L. Stieda in the Realm of Anthropology]" *Eesti antropomeetriaregistri aastaraamat 2002*, 73-76.

Koppel, H. (1902), "Eestlased [Estonians]" *Postimees*, no. 22 (26 January), 1.

Körber, B. (1884), "Die Durchschnittmasse ausgetragener Neugeborener und ihre Lebensfähigkeit, berechnet aus den Jahresberichten der Findelhäuser in St. Petersburg und Moskau" *Vierteljahrsschrift für gerichtl. Medizin und öffentl. Sanitätswesen*, vol. 40, 240-241.

Köstner, ? (1927), "Eesti sõdurite pikkus 1924 [The Stature of Estonian Soldiers in 1924]" *Eesti Statistika*, vol. 69, no. 8, 431-445.

Kruus, H. (1921), *Jaan Tõnisson Eesti kodanluse juhina* [Jaan Tõnisson as the Leader of Estonian Bourgeois]. Tartu: Odamees.

Laaman, E. (1938), *Juhan Luiga: elu ja mõtted* [The Life and Thought of Juhan Luiga]. Tartu: "Noor-Eesti".

Landau, E. (1909), "Über die Orbitalfurchen bei den Esten" *Zeitschrift für Morphologie und Anthropologie*, vol. 12, no. 2, 350.

Landau, E. (1912), Kratkoye rukovodstvo k izucheniyu Antropologiyu [A Short Directory into the Studies of Anthropology]. Yur'yev' [Tartu]: K. Mattizen.

Lellep, K. (1927a), "Kuritegevus Eestis [Criminal activities in Estonia]" *Eesti Arst*, vol. 6, no. 1/2, 1-16.

Lellep, K. (1927b), "Der Einfluss der Rasse auf die Kriminalität der Esten" *Monatsschrift für Kriminalpsychologie und Strafrechtsreform*. vol. 18, no. 3., 145.

Luiga, J. (1909), "Mäss ja meelehaigus [Mutiny and Mental Disease]" *Päevaleht*, vol. IV, no. 226 and 232 (3 and 10 October), 1-2 and 5-6.

Lüüs, A. (1936), "Antropoloogilisi andmeid eestlaste vastsündinute kohta [Anthropological Data on Estonian Newborns]" *Eesti Arst*, vol. 15, no. 7, 497-510.

Luyga, J. (1904), *Prizrenie dushebnobol'nykh v pribaltyskom kraye. Istoriko-statisticheskiy ocherk.* [The Guardianship of Insane in the Baltic Provinces. A Historico-statistical Overview.]. Yur'yev' [Tartu]: Tipografiya gaz. "Postimees".

Madissoon, H. (1925), "Sugemeid eestlaste ajukaalust [On the Brain Weight of Estonians]" *Eesti Arst*, no. 4, 233-238.

Madissoon, H. (1929), "Pärivusest kurikalduvusega perekonnas [On Inheritance in a criminal Family]" *Tartu Ülikooli juures oleva Eesti Looduseuurijate Seltsi Aruanded*, vol. 30, no. 3/4, 1.

Malinovsky, L. (1891), *K izuchenyu v mediko-topografichskom I statisticheskom otnoshenyi g. Revelya. Serija dissertatsii, dopushchennych k zashchite v Imper. Voenno-Meditsinkoy Akademii v 1890-91 akademicheskom godu* [Medico-Topographic and Statistic studies of Revel. The Series of Dissertations at the Imperial Military-Medical Academy in the 1890-91 Academic year]. Revel': Nasledniki Lindforsa,

Mark, K. (1953), "Paleoantropoloogilistest uurimistest Eesti NSV-s [On Palaeoanthropological Research in the E.S.S.R]" *Eesti NSV Teaduste Akadeemia Toimetised*, vol. 2, no. 4, 505-523.

Mark, K. (1975), *Antropologiya pribaltiysko-finskich narodov* [Anthropology of Baltic-Finnish Nations]. Tallinn: Valgus.

Niggol', G. (1913), *Ob odnoy osobennosti grudnoy kletki u estontsev* [About a Particularity of Thorax among Estonians]. Yur'yev': K. Mattizen.

Ottow, B. (1966), "K. E. von Baer als Kraniologe und die Anthropologen-Versammlung 1861 in Göttingen." *Sudhoffs Archiv*, no. 1, 43-68.

Põld, P. (1922), Eesti riigi tulevik ja karskus [The Future of Estonia and Abstinence]. Tartu: Eesti Karskusseltside Kesktoimkonna väljaanne. no. 95.

Põldvere, K. (1997), "Juhan Auli elust ja tegevusest [On Life and Activities of Juhan Aul]" in: Mikelsaar, S. (ed.), *Professor Juhan Aul: kirjanduse nimestik 1919-1997*. Tartu: Tartu Ülikooli Raamatukogu, 7-14.

Pung, M. (1902), "Kuritööd eestlaste ja lätlaste juures [Crimes among Estonians and Latvians]" *Teataja*, vol. I, no. 40-43 (18 and 22 January), 1.

Reiman, H. (1928), "Vastsündinute pikkus, kaal ja peamõõdud Tallinnas 1926-27 [The Stature, Weight and Head Measures of Newborns in Tallinn in 1926-1927]" *Eesti Statistika*, vol. 7, no. 9, 525-531.

Reiman, H. (1930), "Õpilaste edasijõudmine koolides [The success of children at schools]" *Kasvatus*, vol. 12, no. 4, 155-164.

Reiman, H. (1931), *Eestlaste rassiline koostis* [The Racial Constitution of Estonians]. Tartu: "Loodus".

Ridala, V. (1913), "Tõu küsimus [The Stock Issue]" *Eesti Kultura*, vol. III, 199-200.

Rooks, G. (1925), "Mõnda vastsündinud laste ja nende päramiste mõõtudest [Data on the Measures of Newborns and Their Placenta]" *Eesti Arst*, vol. 4, no. 5, 133-136.

Schrenck, A. (1880), *Studien über Schwangerschaft, Geburt und Wochenbett bei der Estin*. Dissertation. Dorpat: C. Mattiesen.

Schultz, G. (1845), "Bericht über Messungen an Individuen von verschiedenen Nationen, zur Ermittelung der menschlichen Körperverhältnisse" *Bulletin de la classe physico-mathematique de l'Academie Imperiale des Sciences de Saint-Petersbourg*, vol. I, no. 15-16, 224.

Tammiksaar, E. (1999), "Jooni Karl Ernst von Baeri tegevusest antropoloogina [Karl Ernst von Baer as an Anthropologist]" *Eesti Antropomeetriaregistri Aastaraamat 1999*, 191-202.

Topinard, P. (1872), “Du prognathisme alveolo-sous-nasal” *Revue d’Anthropologie*, vol. 1, 628-668

Tork, J. (1940), “Eesti laste intelligents. Pedagoogiline, psühholoogiline ja sotsioloogiline uurimus [The Intelligence of Estonian Children. A Pedagogical, Psychological and Sociological Study]” *Eesti Vabariigi Tartu Ülikooli Toimetised*, vol. 48.

Veynberg, R. (1901), *Esty* [Estonians]. Moskva: A. V. Vasil’yeva i Ko.

Vihavainen, T. (1985), “Venäläisiä ja neuvostoliittolaisia käsityksiä suomalaisista ja heidän rotuominaisuuksistaan [Russian and Soviet Approaches to the Finns and their Racial Particularities]” in: Kemiläinen, A. (ed.), *Mongoleja vai germaaneja?* Helsinki: SKS, 269-295

Villems, R. (1926), *Eestlaste antropoloogiast = Zur Anthropologie der Esten*. Dissertation. Tartu: University of Tartu.

Weinberg, R. (1894), *Die Gehirnwindungen bei den Esten. Eine anatomisch-anthropologische Studie.* Inaugural-Abhandlung. Jurjew (Dorpat): C. Mattiesen.

Weinberg, R. (1905), “Der Schädel von Woiseck [Võisiku]” *Sitzungsberichte der Naturforscher-Gesellschaft bei der Universität Jurjew*, Band 14, Heft 2, 110-113.

Witt, H. (1879), *Die Schädelform der Esten.* Dorpat: Schnackenburg.

"God forgives – but Nature never will" Racial Identity, Racial Anthropology, and Eugenics in Latvia 1918-1940

Björn M. Felder

Abstract
Physical anthropology in Latvia is strongly connected to anatomist Jēkabs Prīmanis who enlarged the discipline to a national science and embedded it in the national eugenic programmes of the authoritarian state after 1934. Funded by state money Prīmanis developed broad anthropometric measurements seeking to find the "original" Latvian racial type emphasising the "Nordic" character of the Latvians. Race and racial identity became a means of political legitimation and national self-perception. Further racial science in Latvia became also an important trajectory for the process of biologisation of the nation and for forming the utopia of the racial state. Prīmanis became in the 1930ies an outstanding figure in implementing applied eugenics in Latvia. With his research agenda including social, genetic and pathological besides anthropological data, he laid the foundation of broad eugenic surveillance of the nation and the basis for further biological engineering that was ended by Soviet occupation. Without labelling racial science in Latvia a pseudo-science the chapter outlines the heavily politicisation and political instrumentalisation of this discipline.

1. Introduction

Employing the slogan "God forgives – but nature never will", anatomy and histology Professor Jēkabs Prīmanis reminded Latvian athletes of the eugenics goals of the Latvian nation at a sports festival in Riga in 1938.[1] The newspaper headline, taken from the final sentence of his speech, was a reference to the Darwinist "struggle for existence" as a natural law, at least according to the conviction of Prīmanis. He claimed that in "difficult times", it was the case that only "the healthy and strong" survived, creating a "valuable selection" for the nation. Prīmanis emphasised to the athletes the importance of having a mentally and physically healthy nation, a nation that must be "biologically" of "high value". Therefore, the birth of "inferior" people and the circumstances that hinder "us" from "inheriting genes of high value" must be "liquidated". At the largest sports event in Latvia, he stressed in particular that without "proper genes the best sport will not help". Besides citing an increase in "inferior" people among Latvians, Prīmanis also pointed out the danger of decreasing birth rates, which would jeopardise the process of "natural selection". This could lead to the extinction of a nation due to biological decay. But at the end of his speech, he was optimistic about the

Latvian nation's struggle for survival. He underlined that Latvians have a high value in terms of their "biological perspective", being a "Nordic nation" like Scandinavians and others. Following Prīmanis, the Latvians' "Nordic" character was a good starting position in the struggle for existence.

In his speech, Prīmanis emphasised the connection between racial anthropology and eugenics: he described the situation of the Latvian nation as a social Darwinist struggle among nations for survival. With the biological quality of Latvians decreasing, as well as their total number, the nation's future was in danger. The only response that could help would be a national eugenics project to increase both biological quality and the birth rate. The initial situation was not particularly disadvantageous, since Latvians belonged to the anthropological category of the "Nordic race" and therefore were of "high biologic value".

The connection between race and eugenics, not an unusual phenomenon within the global eugenics movement, was of significance, as Prīmanis himself was no less than one of founding fathers of the Latvian eugenics project. His speech reflected the official eugenics propaganda. In 1937 a national project of eugenics was established that included broad, public eugenics propaganda; implementing a programme of abortions and voluntary sterilisations following eugenics-based diagnoses; and founding a national research institute for eugenics (*Tautas dzīvā spēka pētīšanas instituts* – Institute for Research of the National Vitality, TDSPI). The driving force was the authoritarian regime established after a *coup d'état* by Latvian Prime Minister Kārlis Ulmanis in May 1934. The eugenics programme reflected the biological self-image of nation fostered by the regime and articulated by the concept of a racial state, a nation-state that sought biological rejuvenation and ethnic homogenisation through a strong eugenics agenda.[2]

In 1938 Prīmanis became the director of the national eugenics institute. The leading physical anthropologist at this time in Latvia, Prīmanis created a Latvian identity as a "Nordic nation" that consisted of a "Nordic race". Since Latvia became independent in 1918 and the University of Latvia opened in Riga in 1920, Prīmanis had a very active role in the task of creating a Latvian racial identity. With other Latvian racial anthropologists, Prīmanis was eager to measure the "national body", searching for the "pure" Latvians in ancient tombs and identifying the nation's biological parameters to categorise Latvians in the biological and racial context of the current discourse. Creating a racial identity meant not only placing Latvians into the framework of European anthropological categories of race, but also into a hierarchy of values. Labelled as "Nordic", Latvians not only became biologically and culturally part of "civilised" Europe, but they were also described biologically and genetically as being "worthy". This genetic or "racial" value gave Latvians a solid position in the process of "natural selection", at least for Prīmanis, the struggle for survival among nations – and it legitimised the young nation and its nation state.

This chapter provides the example of creating a racial identity for one of the "young" European nations – the Latvians – who established their independent state after World War I. The racial identity that formed a biological definition of nation was not simply a concession to the biologised understanding of nations at that time, but it also resulted from the biological and exclusive self-definition of the Latvian national movement since 1905. The article will portray the close relationship between physical anthropology and eugenics in Latvia, especially the role that racial identity played in implementing eugenics in Latvia, and how physical anthropology provided the scientific framework for bio-politics in Latvia that had the form of a "racial state".

2. Latvians in the Eyes of Anthropologists

The anthropometric measurement of the Baltic population began in the middle of the 19th century. J. C. Baehr conducted craniological studies of skulls from local Baltic regions. Following the anthropological approach of his time, he also categorised skulls. At this time, "doliocephalic" (longheaded) characteristics – as opposed to "brachicephalic" (broadheaded) – were seen as signs of the "Indo-Germanic" race. Baehr was the first to classify the Baltic population as "Germanic". For Baehr, Baltic peoples stand between the Slavic, Finnish and Germanic types. "Though they [Latvians] might join the line of the latter [Germans], as one can see very often a formation of the face with long aristocratic forms".[3] An important impulse for anthropology in the region was provided by the Baltic-German scientist Karl Ernst von Baer, who graduated from the University of Dorpat (Tartu, Yur'yev') and became a professor in Königsberg and later in St. Petersburg, where he conducted craniological research. Under the anatomy Professor Ludwig Stieda and later August Rauber, physical anthropology flourished in Dorpat in the late 19th century.[4] Students of Stieda and Rauber began examining living people from the Baltics, and made body and skull measurements. Otto Gustav Friedrich Waeber and Richard Weinberg focused on Latvians as well as Estonians and Livonians, though they did not use racial classifications.[5] Rudolf Virchow published palaeoanthropological research on skulls from tombs in Livonia in the 1870s. Having studied living Latvians, he described them as follows: "As far as I could see, most Latvians have blonde and brunette hair colour, and blue and grey eye colour, and are strong folk with a long skull formation and a wide, squarrose, pointy nose".[6]

This research steered the racial studies that followed with a subtle direction toward the "Nordic" character of the Latvians. At the turn of 20th century, there was a general tendency to form and elaborate the definition of race, and to create a clear system according to Carl von Linné. Some influential Western scholars such as William Z. Ripley, Madison Grant and Joseph Deniker were eager to create a racial system. All three mentioned the Baltic population. In his 1899 book, Ripley referred to the work of Waeber and to the research of Virchow concerning the "Letto-Lithuanian" population and its racially "Nordic" characteristics.[7] Concerning the Baltic provinces, it seems that Madison Grant simply followed Ripley's arguments.[8] Both saw the "Nordic" influence in the Baltics, and both distinguished Latvians and Lithuanians from the Finnish and Slavic population, classifying them as "Nordic" but with a strong influence from the surrounding non-Nordic population. Deniker obviously placed the Latvians into an even stronger "Nordic" framework, but he did not give any literature sources.[9] All three authors popularised racial science. Though Deniker favoured a six-race theory, unlike Ripley and Grant, who stuck to the common three-race system (*Nordic, Alpine, Mediterranean*) for the population of Europe, the picture of the tall, blonde, blue-eyed Nordic race – Grant's "great race" – was popularised and manifested in public debates – as was the racial image of the Baltic population as being near the "Nordic" race. Hans F. K. Günther, an important German racial anthropologist who, like Madison Grant, favoured the "Nordic thought" glorifying the Nordic race, and who later became the leading racial anthropologist in Nazi Germany after 1933, classified the Latvians as "Nordic with Eastern Baltic admixture". Günther followed the five-race system – *Nordic, Western, Dinaric, Eastern* and *Eastern Baltic* – which was internationally accepted by anthropologists during the 1920s.[10]

Recently, Rita Grāvere has given a general overview of the anthropological research concerning Latvia in the 19th century.[11] Though historians had not reflected on the interaction of race and eugenics in Latvia, or even recognise the phenomenon, as late as 2004 Andrew Ezergailis insisted upon the absence of applied eugenics and eugenics propaganda in Latvia before World War II.[12] The first scholarly studies on Latvian eugenics during the post-war era by Björn Felder questioned this assumption. They also revealed the interaction between racial anthropology and eugenics.[13] Latvian scholars , such as Vita Zelče, did not see this connection. Zelče also underestimated the role of racial anthropology, focusing on the alleged "pro-natalist" characteristics of the eugenics project. Further, Zelče did not consider the importance of the Latvian eugenics institute, *Tautas dzīvā spēka pētīšanas institūts*.[14] Recently, Latvian historians including Aivars Stranga continued to deny the importance and popularity of race and biological nationalism among Latvians under the Ulmanis dictatorship.[15]

3. Creating a Racial Identity for Latvians

After declaring independence in November 1918 and turning back the Soviet army in 1920, the newly emerged Republic of Latvia attained the ultimate goal of Latvian nationalism: independent Latvian statehood. Like the inhabitants of other Eastern European states that were established after World War I, Latvians had to legitimise their state and their nation. They had to prove that they were also a "civilised" nation – a state that Western European should value as being on the same social, political as well as biological and racial levels. Because the Western perspective was often shaped by racist stereotypes and racial patterns of thinking, Eastern European nations were seen as unequal and inferior to Western nations.

An important instrument for self-legitimising the new nations was the merging of national science with the new nation-states. The performing of national science, and especially the foundation of national universities, had several goals. First, the universities sought to educate the new academic elite and become a fundamental resource for further social and political development. Second, national universities were seen an obligatory element of a European nation. Academic education was prestigious. In the 1930s Latvians were proud that they had the highest rate of university students of all European countries.[16] Third, national sciences had to create the new national identity. The most important role in this process was played by historical science, as historians had the important task of proving the historical roots of the nation, and creating a historical identity and national remembrances. Latvian historians, especially scholars in archaeology and mediaeval studies, had to search for evidence of the Latvian nation from the past.[17] This was particularly important because Baltic-German historians had been arguing since the 19th century that it was German missionaries and crusaders who brought "culture" to the region and prevented the local population from being influenced by Russian rule.[18] German aristocrats had ruled the area since the 13th century. German language and culture had dominated aristocratic and bourgeois society. Latvians only had the position of slave-like farmers in the feudal system. As a post-colonial strategy, Latvian historians were eager to illustrate the Latvian past before the appearance of the Germans, while they also emphasised the high standard of Latvian cultural life under German domination. This began with the start of the Latvian national movement in second half of 19th century. Latvian intellectuals, for example, collected Latvian "dainas", traditional poems and folk songs that even today are seen as an important part of Latvian cultural heritage.[19]

National identity was developed not only by historians who followed the European tendency of a biologically defined nation, but also by scholars of bio-medical sciences. Those physical anthropologists, whose interests focused on racial anthropology at the time, constructed a biological-racial identity in the context of the anthropological system of races. In addition to a political history, a biological history also had to be "reconstructed". Palaeoanthropologists searched for the biological origins of the nation, looking for "pure" ancestors and the nation's biological genesis.

The history of a nation was no longer a subject only for historians. The newly emerged school of bio-historians expanded the realm of historical sources. In addition to texts and artefacts, bones from ancient tombs became archives unto themselves. The national body itself became an archive for national history. Physical anthropology became the leading role in bio-history. By measuring skulls and bones, they created a racial history and identity. Comparing the data from craniological studies with the anthropometric measurements of students', pupils' and recruits' bodies, anthropologists were able to describe the racial genesis of their nation.[20]

The discourse of bio-history and racial identities were brought to Latvia via scientific transfer from northern Europe. The first scientist who began to construct a racial identity for Latvians was a non-Latvian, the Swedish anatomist and anthropologist Gaston Backman. Several scientists had conducted craniological studies on ancient Latvian skulls, including the renowned physician Rudolf Virchow from Berlin. Living Latvians had also been measured. Backman was not only the first Latvian professor of anatomy, he also systematically measured thousands of Latvians under the auspices of physical anthropology. Backman brought racial anthropology from Sweden to Latvia. Born in Rejmyre in southern Sweden in 1883, Backman began studying at the University of Uppsala under Vilhlem Hultkrantz and Ivar Broman in 1901, specialising in anatomy, anthropology and psychiatry. From 1905-07 he was a research assistant in anatomy, and later also in psychiatry. In 1913 he completed his Ph.D. in medicine and went on to work as a physician in cities including Uppsala, Stäter and Västervik. He also conducted ethnographical studies for the Riksmuseet in Stockholm. In 1920 he became a professor of anatomy and histology at the University of Latvia in Riga, but he was eager secure a position back in Sweden. In 1925 he became a professor at the Karolinska Institut, the royal medical academy in Stockholm. When he left Latvia, he received a Ph.D. *honoris causa* and the Latvian Three-Star-Order, the country's highest civil order. In 1933 he became a professor of anatomy at the University of Lund.[21]

He had been interested in racial anthropology and eugenics since his time as a student in Uppsala. His university teacher, Professor V. Hultkrantz, was a leading member of the Swedish eugenics movement. Backman was also an avowed eugenicist who supported efforts to establish a Swedish eugenic institute, which opened in 1921 as the Institute for Racial Biology in Uppsala.[22] In the 1930s and 1940s, Backman was an active member of the Swedish pro-Nazi movement *Riksföreningen Sverige-Tyskland* (National Society Sweden-Germany). One of its leading activists was Backman's mentor, Ivar Broman. Backman may have been a friend of Sven Hedin, with whom he stayed in close contact during the 1940s.[23] Prior to 1914, Backman published works in the field of racial anthropology as well as psychiatry. During World War I, following the example of Sven Hedin, Backman published an article that described the war as a "racial war" between Germanic and Slavic "races", and which supported the German war effort.[24] In his *Ymer* article of 1915, Backman described the Baltic region as being the frontline of racial warfare. In his position in Riga, he took the opportunity to investigate Latvians in order to prove that they were of the "Nordic" race, verifying not also assumptions of other racial scientists, but also finding allies in the racial war that he constructed.

Backman built the professorial position in Riga from the ground up. He brought anatomical specimens and current anthropological measurement tools with him from Sweden. His twin brother Louis, a professor in Uppsala, donated specimens for the new chair of anatomy at the University of Latvia. Backman also established a museum of anatomy that was connected to the chair. In taking the opportunity to introduce racial anthropology to the academic landscape in Latvia, Backman's aim was to begin a programme of systematically measuring Latvians in order to categorise the nation in terms of race. At the time, Backman must have had in mind the idea that most Latvians have to be "Nordic" or of Nordic "admixture", following the position of Deniker and Ripley, who had discussed the Baltic population in their work. Only after arriving in Riga did Backman learn about the early research of anthropologists, particularly from Dorpat, who conducted anthropological measurements of Latvians in the late 19th century. Though these scholars did not use the racial categories of the 20th century, their results could be used as comparative material.[25] Backman encouraged his students and assistants – including Jēkabs Prīmanis, Jānis Vilde and Lucija Krastiņa (later Jeruma-Krastiņa) – to become interested and even specialise in racial anthropology. His first assistant, Prīmanis, followed him as professor and pursued Backman's approach until World War II.

Backman began his anthropological project in 1922 by measuring about 11,000 Latvian army recruits with support from the Latvian Minister of Defence, General Jānis Balodis. The measurements were actually taken by 13 army physicians, not by Backman personally. In 1924, Backman sent N. Jērums and T. Vitols to the east-central region of Cesvaine in the Cēsis district for additional measurements. After analysing the results of the first large-scale investigation, Backman stated that the individuals were the highest percentage of "pure" Latvians, with only small scale of "foreign" influence. To contrast to Backman's measurements of the last remaining Livonians in Latvia, an ethnicity linked to Finno-Ugric descendants, Prīmanis in 1922 asked his student J. Vilde to conduct measurements of the Livonians. Prīmanis and Jerums also examined the supposed descendants of Curonians in western Latvia. Backman was convinced that the ancient Curonian tribes were not of Baltic but of Finno-Ugric origin, which resulted in strong arguments with Latvian historians. In additional research on Latvia's bio-history, skulls of ancient and medieval tombs were examined. Prīmanis and

Georg von Knorre conducted craniological studies on skulls from the early Middle Ages.[26] The Latvian scholars found – following their assumptions - tall, blonde, blue-eyed individuals with brachiacephalic skulls among the living, but more mesocephalic and doliocephalic skulls from mediaeval times. The studies of the living Latvians, with their short and broad skull shapes, did not fit perfectly into the image of the "Nordic race", Backman concluded that the "original" (*ursprünglichen*) Latvians, at the time they entered the Baltic region, must have been doliocephalic, referring to the research on supposedly Latvian medieval skulls. This would mean that Latvians were originally of the Nordic race. Only by mixing with Finno-Ugrians, namely Livonians and Curonians, were they blended, thus losing their "Nordic" features such as long skulls. The brachicephalic characteristics of the living Latvians was the biological heritage from the Livonians, according to Backman. He did not directly refer to racial paradigms. But using parameters such as growth, eye and hair colour, and cephalic index, Backman concluded in 1925 that "the Latvians originally were moderately tall, clearly doliocephalic, very blonde and blue-eyed people". His intention was obvious: this was the first clear contribution by modern physical anthropology to the racial identity featuring "Nordic" Latvians.[27]

Backman stuck to this theory, as did his students. In the mid-1930, referring to the latest racial anthropological research from Latvia, he repeated and enlarged his thesis. At the time, current anthropological measurements were strengthening his argument, Backman was arguing due to the latest research results in archaeology, palaeoanthropology and prehistory. Following the narrative of Backman, various Baltic tribes that were forefathers of the Latvians, who he categorised as "Nordic" (*nordide*), had been assimilating the Finno-Ugric tribes of Livonians (who were settling by the river Daugava and in the coastal area of Curonia and Livonia), and also the Curonians living in Curonia. With the help of archaeology, palaeoanthropology and contemporary anthropology, Backman further stated that there was a strong „Nordic" influence on the Livonians, through contact with Sweden in mediaeval and in early modern times. According to archaeological research, Swedish settlements have continuously been in Curonia from 1500 B.C. through the early Middle Ages. There were also settlements of Swedish Vikings around 900. As a result, the Livonians, as members of the alleged "Asiatic" Eastern Baltic race, were biologically influenced by the Nordic races, namely the Swedish settlers. The genetic material was passed to the Latvians, who were assimilating the Livonians. The result was re-bastardisation (*Rebastardisierung*) and the strengthening of "Nordic" characteristics among the Latvians.[28] It is remarkable the Backman did not mention the possible genetic influence of the Baltic Germans, who ruled the region for about 700 years. Backman also marginalised the Russian and Slavic influence, even though the Baltics had been under Russian rule since the early 18th century. With the Livonian theory, Backman found the perfect explanation for the "weak" "Nordic" head shape of contemporary Latvians and the "Nordic" origins of Latvian bio-history. It is not surprising that Latvian anthropology after Backman did not focus on ethnic minorities other than the Finno-Ugric population in Latvia, though there was a considerable Russian minority of 300,000 in Latvia after 1918. Latvia also has significant German and Jewish minorities, though none of these populations became the subject of racial anthropological research until 1945.[29]

Backman's anthropological work was responded to very positively in Latvian society at the time. Latvians expressed a strong demand for a racial identity. Not only did the Latvian government through the Ministry of Defence support anthropometric measuring of its soldiers, there were also private initiatives for large-scale measurements of the Latvian public. After writer and politician Janis Rainis died, the director of the University of Latvia asked a Latvian anthropologist to measure his body. Further, national funds such as the "Kultura Fonds" regularly supported anthropological projects until World War II.[30]

4. Racial Anthropology as a National Science under Jēkabs Prīmanis

Backman left Riga in 1924 for a position in Stockholm. His position in Riga may have been problematic and non-prestigious, not only because of the difficulties in founding the chair, but also because of its place on the periphery of Europe. Amid the nationalistic mood at the founding of the Latvian Republic, Backman may well have had problems with his Latvian colleagues. He did not speak Latvian, even though he understood the language at the time he left Riga, and his lectures were conducted in German and translated by his assistants. This alone must have been seen as an affront to the increasingly nationalistic Latvian academic elite. Some of his theses were also opposed, namely the Finno-Ugric origin of the Curonians. Professor Jānis Enzelīns, a famous philologist, insisted that the Curonians were an ancient Latvian tribe, referring to language monuments.[31]

In 1925, Russian anatomist Arsenij Starkov followed Backman in his position. Because Starkov was not an anthropology specialist, further research stagnated. After his death in 1927, Backman's former assistant Prīmanis was elected as the new professor of anatomy and histology at the University of Latvia. Prīmanis began his academic career in 1911, studying medicine at the University of Dorpat for two years, where he became interested in anthropology when listening to the final lectures of Professor August Rauber. He then transferred to the Military Medical Academy in St. Petersburg, took part in World War I and the Latvian War of Independence, and became an assistant of Backman in 1920.[32] He conducted several anthropological studies in Riga and managed to obtain a fellowship from the

Rockefeller Foundation with the help of Backman. He then spent a year at the University of Michigan working on bacteriology and embryology with Professor Charles Huber. After returning from the US in 1926, he spent a number of months at universities in Munich, Vienna and Prague. Prīmanis was appointed to the institute in January 1928, but was not a professor until 1932 due to internal politics at the university.[33] In the field of anthropology, Prīmanis was clearly a follower of Backman. He also used Rudolf Martin's methodology and followed the Nordic-*völkisch* direction of racial anthropology. Consequently, his students were always citing Hans Günther.

Racial biology appeared to be emphasised more under Prīmanis than Backman. At the beginning of the 1930s, for example, Prīmanis was lecturing about the racial paradigms of anthropology to first-term anatomy students. [34] Prīmanis was obsessed with the idea of wide-ranging measurements and recording, and the racial categorisation of the Latvian nation. In 1929 his assistant Lucija Krastiņa began working on a project to measure 1,000 Latvian women. Following the direction of Backman and Prīmanis in her 1935 thesis, Krastiņa discussed her results extensively, including examining eye and hair colour, height and cephalic index referring to racial categories, and classifying Latvians as "Nordic" and "Eastern Baltic" with dominant "Nordic" characteristics.[35] Krastiņa's noteworthy project, one of many conducted under Prīmanis, was the first mass-scale measurement of women, not only in the Baltics but in all of Europe. Further, Krastiņa was among the first women to earn a Ph.D in Latvia. With the 1934 *coup d'état* of Kārlis Ulmanis and the installation of his nationalist authoritarian dictatorship, the new government strongly pushed and enlarged racial biological science. Prīmanis was able to expand his programme and his chair together with the Museum of Anatomy to become the Institute of Anatomy.[36]

In 1935 Prīmanis set out to complete his racial anthropological study of Latvians after spending 10 years covering the entire nation, following the anthropological examples of other small nations such as Finland.[37] Most anthropological projects were funded by the "Kulturas fonds", the largest government fund for scientific research in Latvia. Prīmanis expanded his efforts, reportedly organising 25 anthropological expeditions before 1940. In 1940, on the 20th anniversary of the anatomy institute, a report stated that his institute had collected anthropological data on about 160,000 people.[38]

Another major step was the establishment of the TDSPI, the national eugenic research institution, which mainly was a product of Prīmanis' political endeavours. Prīmanis himself headed the TDSPI, and one of its three departments was dedicated to racial anthropology. Again, this meant more funding and resources for Prīmanis' anthropological research. His students collected anthropological data from every corner of the country, including from the Finno-Ugric minority. Ancient human remains were also researched in order to create a bio-history.[39] Beginning in the 1930s, Prīmanis' researchers obtained blood samples during their expeditions in order to investigate the "racial character" of Latvians and create a racial blood index. A publication on this issue was not published until 1943.[40] Prīmanis also tried to prove his theories on Latvian racial heredity through literary documents. In 1929 he compared the description of bodies from the ancient Latvian "dainas" to his racial research, which resulted in clear evidence of the original "Nordic" character of Latvians. Clearly he was looking for the last remaining "pure" Latvians who had not yet lost their "body characteristics" by mixing with people from "foreign" nations.[41]

In 1937 Prīmanis resumed anthropological research at his institute by authoring a book, *The Latvian Anthropological Character*, which was more of a political treatise than an anthropological study. His book describes Latvians as being "dominated" by the "Nordic race", though with a strong percentage of the "Eastern Baltic race". Following Backman's approach, Prīmanis categorised ancient Baltic tribes as "Nordic" who mixed with local Livonian and Estonian populations responsible for the Eastern Baltic racial influence and the brachicephalic characteristics of contemporary Latvians. Prīmanis acknowledged the biological influence of Russians on Slavic tribes, calling it "negative". But he marginalised this influence even though Latvia had been under Russian rule for more than 200 years. Prīmanis also found the influence of Baltic Germans as not important, which he mentioned, unlike Backman. According to Prīmanis, Germans and Latvians lived separately and generally were of similar racial origin. So he did not observe any relevant influence of importance. This may be understood to be contribution to the general Germanophobia, Latvian nationalism and the open conflict with Baltic German minority. Prīmanis also followed Backman's Livonian thesis, emphasising the strong Swedish ("Nordic") influence to Livonians from the coastal areas. Empirical data supported this thesis, as there was a significant decrease in height and a shift from mesocephalic to brachiocephalic skull forms shifting from the Western coastal areas to the Eastern region. These arguments seemed to be strongly motivated by the political discourse at the time: Russians and Slaves were often seen as "inferior" by Western anthropologists, and Germans – being the former ruling nation and the colonisers – were somehow the main enemy of Latvian nationalism. In contrast, the Swedish rule of the 17th century was perceived as a "good era". Taken together, Prīmanis again "confirmed" the "racial" value of the Latvian nation.[42]

As Prīmanis was eager to establish anthropology as a science that defined and created a national identity, his goals were far beyond pure anthropometric recording and the analysing of results in the context of racial science. Backman and Prīmanis in fact created a racial identity that was important for legitimising the Latvian state in order to become accepted by

Western European nations. Further, Prīmanis had the idea of an ethnic or a genetic anthropology. This was evident from his anthropological research during the late 1930s. Prīmanis changed Backman's questionnaire, adding questions concerning genealogy, hereditary pathology, hereditary diseases, and family and social issues such as profession, the number of children, and the age when people became parents.[43] These changes were due to the fact that Prīmanis became active within the national eugenics project at the end of the 1930s. This genetic approach searched for the "anthropological characters" of Latvians, as Prīmanis repeatedly stated: the biological-genetic definition of the nation, a form of biological identity and nationalism.

Prīmanis' project was beyond racial identity and the racial paradigms of anthropology as the latter tend to be supranational, only described and framing large groups of people – races – among several nations. His genetic anthropology as a national science sought to describe the biological framework and the genetic foundation of Latvians. The shift toward genetics was described by Prīmanis himself, citing a meaning of "race" that differed from the anthropological system and instead focused on the genetic foundation of a nation.[44] This evoked a general biological definition of a nation. In the context of genetic anthropology and the clear lack of interest in ethnic minorities, this form of biological nationalism seemed to exclude non-Latvians on the basis of genetics or "race". In the 1930s Prīmanis began to collect national genetic data in terms of anthropology. Suddenly, this information could have become crucial in Prīmanis' effort to biologically form the Latvian nation by implementing a national project of eugenics that intended to achieve genetic "purity".

5. Jēkabs Prīmanis as Main Promoter of Latvian Eugenics

On 15 May 1934, Prime Minister Kārlis Ulmanis usurped power and became dictator. His regime combined authoritarian and totalitarian elements in terms of its organisation, ideology and rituals (i.e. a state corporate system, his designation as *vadonis* [leader] and the fascist salute), but it avoided the use of extreme forms of violence. Indeed, Latvia under Ulmanis did not become a one party-state to the same extent as Nazi Germany, nor did it have a one party-state militia equivalent to the SA. In terms of ideology, the Ulmanis regime was wedded to a form of *völkisch* biologised nationalism. Following the racial approach of a national self-image, and propaganda and efforts to "improve" and homogenise the national body, this regime can be described as a racial state.

After Ulmanis's putsch, support for racial research increased and articles about race and eugenics multiplied in the censored press. Although Ulmanis himself had not contributed any fundamental writings or programmes, several intellectuals close to him tried to establish a *völkisch* ideology for the new state under the label of "new nationalism." One such intellectual was Ernests Brastiņš, director of the Latvian War Museum. Brastiņš was searching for a true "Latvian" religion to serve as the foundation for a national identity. He discovered this in a reconstructed Latvian paganism, which he used as the basis for the creation of a new church, *Dievturība*.[45] He also wrote a position paper on the *völkisch* state – a mélange of classic *völkisch* literature on mythology, culture and national history. The *raison d'être* of the nation – the future Latvian state – would be situated between authoritarian and totalitarian forms of rule, he believed. Brastiņš saw the nation as a biological entity or a "race," and when describing the Latvians, he used the term *Aryan*.[46] Consequently, he promised that the future Latvian nation ultimately would become a distinct biological group, the product of systematic "breeding" – clearly implying a future eugenics agenda.[47] Another intellectual supporter of Ulmanis was Jānis Lapiņš, editor of the journal *Sējējs*. In the foreword to the programmatic volume *New Nationalism*, which included contributions on racial psychology, racial anthropology, religion and other subjects, Lapiņš called for "racial purity" of the nation.[48] The Latvian elite shared a consensus that race was an essential characteristic of a nation, and engaged in a debate on the biological traits of the national body. In the late 1930s, articles on "race" appeared not only in intellectual journals but also in media outlets more generally, from daily newspapers to the journal of the Latvian railway. These popular articles advanced not so much racial anthropological theories as science blended into a common understanding of race, for example, as a blood-related community embedded in folk myths.[49]

The media campaign on race can be seen as the instrument of Ulmanis' propaganda to promote ideas of racial identity and a biological nationalism among broad layers of society. Further, as many articles were referring to or even asking for eugenics, racial hygiene or "racial improvement", its propaganda was a tool for promoting eugenics ideas within the population. Finally, in 1937 a national eugenics project was enacted that included applied "negative" eugenics in the form of voluntary abortion and sterilisation by eugenics-based diagnoses, marriage bans for the mentally ill and probably also obligatory genetic certification for marriage. "Positive" eugenics was disseminated in the form of marriage counselling, eugenics propaganda and eugenics research.[50] In total, some 648 abortions and 63 sterilisations were performed in 1938 and 1939 in order to "reduce the inferior" in Latvia.[51] The programme was abandoned during the Soviet occupation but it continued under Nazi occupation from 1942 to 1944.[52] The main victims of the sterilisations were so-called "feeble-minded".

Prīmanis was one of the main promoters of the Latvian eugenics project under Ulmanis. He not only propagated eugenics in public but he was also responsible for eugenics courses held at the Latvian University of Riga. As the head of the Society for Promoting Health *(Veselības veicināšana biedrība*, VVB), together with his colleague Pauls Stradiņš, he also had great influence in the field of politics. Karlis Arājs, an anthropologist, former student and later an assistant of Prīmanis, wrote in his memoirs that Prīmanis became interested in genetics while studying in the US and Western Europe in the mid-1920s. After 1935, Prīmanis regularly attended eugenics lectures by Professor Reinhards at the University of Latvia.[53] But it seems that Prīmanis' interest in eugenics developed earlier. Evidently, Prīmanis was a true believer in eugenics since his time as a medical student in 1911 at the University of Dorpat, where new eugenics ideas were being discussed by students and lecturers. Since eugenics was often related to the anti-alcohol movement, it was the Estonians and Latvians who became interested in this issue. Because of their beer-drinking tradition at fraternities, Baltic German students were hesitant in adopting eugenic theories.[54] Prīmanis must have discussed the issue with fellow student Hermanis Buduls, as they were at the same time members of the same fraternity in Dorpat, *Atauga*.[55] Buduls was a eugenics expert and was keen on promoting eugenics ideas among Latvians. As an undergraduate in 1909, he published *Marriage and the Goal of Human Life: A Biological and Ethical Essay*, which was the first monograph on eugenics in Latvian and the first monograph ever published on eugenics in the Baltics or Russia. As a psychiatry professor in the 1930s, Buduls became an elder statesman in the drafting of the Latvian eugenics law.[56] At the University of Dorpat, the Professor Evgenij Shepilevskii of hygiene and bacteriology and lecturer Nicolaj Gamaleja held public lectures on eugenics. Shepilevskii wrote what apparently was the first monograph on racial hygiene in Russian, published in Dorpat in 1914.[57] In 1913 Prīmanis was drafted into the military and had to leave Dorpat for St. Petersburg. His time in Dorpat was just as crucial for his interest in racial anthropology as for eugenics.

There is no information about any discussion of eugenics during lectures at the Military Medical Academy in St. Petersburg, where Prīmanis was stationed during the First World War in addition to serving in the field. The second important influence came from his mentor Gaston Backman, who was deeply involved in the Swedish eugenics movement.[58] Ultimately, Backman was important in the formation of racial anthropology as well as eugenics in Latvia. Prīmanis did not publish any material on the subject until the late 1930s. After the eugenics law was enacted and the eugenics research institute (TDSPI) was founded, he frequently issued propaganda on the issue in articles and during public speeches. As an assistant to the secretary of the medical faculty in 1935, Prīmanis organised the first lecture on eugenics at the University of Latvia. These non-obligatory lessons were held by Gustavs Reinhards.[59] After Reinhards' death in 1939 the young Privatdozent Verners Kraulis continued the eugenics lectures, while Prīmanis, as dean of the faculty of medicine, was again involved. Prīmanis' most influential position was with Professor Pauls Stradiņš as the director of the Society for Promoting Health (Veselības veicināšanas biedrība), which consisted of several charities and physicians' organisations. This society was extremely important in lobbying Ulmanis and his ruling circles to support eugenics. In 1937, for example, it is said that Prīmanis and Stradiņš proposed an obligatory genetic certificate proving the genetic value of all couples wanting to be married. This would have been in addition to the marriage ban on the mentally ill and people with acute syphilis in the new civil law of 1937.[60]

The most important action of the society was the founding of a eugenics institute – the Institute for Research in National Vitality (Tautas dzīvā spēka pētīšanas instituts - TDSPI) – by the Society for Promoting Health in 1938. Prīmanis became the director of the institute and the head of its anthropology department. The department for population research was headed by V. Salnītis, and the eugenics department by Verners Kraulis. Like its European counterparts – the Institute for Racial Biology in Uppsala and the Kaiser-Wilhelm-Institute for Anthropology, Human Genetics and Eugenics in Berlin – the institute aimed to conduct research in the fields of racial anthropology, genealogy, population policy, eugenics and genetics. The research results were intended to contribute to the national eugenics project. In relation to "positive eugenics", the TDSPI opened a bureau for young couples planning to marry to counsel them in genetics and eugenics issues. Headed by Dr. Neiburgs, the bureau had about 200 visitors in 1939. Another objective was to popularise eugenics ideas among the population. Public lectures on eugenics and related fields articles were regularly published in all major newspapers from 1938 until summer 1940.[61] Latvian radio broadcasts also regularly featured eugenics beginning in 1938. The Latvian government also took eugenics very seriously. Civil servants such as Oskars Alks, the head of the health department, regularly publicised the work of the TDSPI and the eugenics programme. Alks introduced the new eugenics law, which he personally wrote, and its abortions and sterilisation agenda to physicians, nurses and health officials throughout Latvia. In September 1937, before the eugenics law was enacted, he informed doctors in Jēkabpils about the law's contents and their future duties. He did the same with the association of Latvian physicians in Riga. Beginning in 1938, physicians and nurses throughout Latvia were specially trained in eugenics practices.[62] The Society for Promoting Health and the eugenic institute promoted their eugenics agenda by contributing to several public exhibitions on health and hygiene from 1937 to 1940. For example, in 1939 the TDSPI contributed several displays for the health exhibition "Work and recreation" (*darbs un atputa)* at the Congress Hall in Riga, which included eugenics and racial anthropological sections promoting issues as "population crisis", racial biology and eugenics concepts. Further, in cooperation with the government health administration and the University of Latvia, the TDSPI organised regular lectures on issues including eugenics and population policy. Films on eugenics issues were shown in public; in 1939 the TDSPI imported movies from Nazi Germany such as "To Prevent is Better than to Cure" (*Vorsorgen ist besser als heilen*). The TDSPI also made its own eugenics propaganda movie: in 1938 the institute's Verners Kraulis produced a film at the VEF factory studios in Riga with director R. Kalniņš that premiered in spring 1939.[63]

6. From Racial Identity to Racial Improvement of the Latvians: the Racial State

Prīmanis' influence was felt both in science and in politics. He was the second director of the Society for Promoting Health, the director of the eugenics institute, the director of the Institute of Anatomy, and beginning in 1939 the dean of the medical faculty at the University of Latvia in Riga. He created the racial identity for Latvians, and he was also notable for designing the eugenics agenda and creating a form of ethnic eugenics to achieve an ethnically and biologically "pure" nation, amounting to a type of racial state.

Government propaganda efforts in praise of eugenics began in 1936. Two years after the putsch, the Cabinet ruled that the Ministry of Public Welfare had to "make propaganda on a law concerning national hygiene or eugenics". Consequently, a campaign was launched to popularise eugenics and "racial hygiene".[64] The official propaganda campaign that backed Latvia's national eugenics project was launched in late 1937 when the Ministry of Public Welfare issued a special journal – *Tautas Labklājibas Ministrijas Mēnešraksts* – to inform the public about its achievements in national health and eugenics. The campaign was eager to popularise both the new medical law and its eugenics agenda, as well as the TDSPI's eugenics research. The goals of eugenics were defined as increasing the "quality" as "quantity" of the Latvian nation.[65] Experts found evidence for the "degeneration", an increase in biologically "inferior" individuals, and a "biological crisis" in the form of lower birth rates and a decrease in the Latvian population.[66] With the number of Latvians standing at fewer than 2 million before World War II, the urgency attached to the demographic situation was not surprising. Similar to Estonia, increasing the birth rate and other instruments of pro-natalism had been an important role on Latvia's eugenics agenda. One of the TDSPI's three research departments was dedicated to demographic research, headed by A. Salnītis. In 1939 the institute published a proposal to increase the birth rate by supporting families with many children.[67] A number of popular and scientific articles addressed this issue. This was not unique to the Baltics: pro-natalism, concerns of a birth crisis and calling for population growth were prominent on the agendas of early eugenicists such as Alfred Ploetz and Wilhelm Schallmayer in the early 20th century.[68] Thus, leading eugenicists and eugenics lobbyists including Pauls Stadiņš made it clear that "rising quality" and the "improvement" of genetic value was the main goal of Latvian eugenics – more important than rising "quantity" and birth rates. Additionally, Latvian psychiatrist Verners Kraulis, who was member of the eugenics commission that ruled on eugenics-based abortions and sterilisations, and who headed the TDSPI's eugenics department, asserted the importance of supporting families with many children. He emphasised that there might be no government assistance for "inferior families". Mothers in such families could only be offered sterilisations.[69] Professor Pēteris Starcs, secretary of the TDSPI, who focused on the government support for families with many children, stressed that marriages should only be held following eugenics counselling, and that only "healthy families" should receive special funds for children.[70]

The main task of the eugenics institute – and the eugenics project as a whole – was the "rise of the quality" of the nation. This was also intention of health official Oskars Alks when he spoke of "racial selection".[71] Prīmanis referred to this as the "rise of the biological value" of the nation. In the preface to the published eugenics lectures by Gustavs Reinhards, Prīmanis declared the meaning of "race" to be not an anthropological category, but "inherited genetic features". Race was frequently used as a synonym for genes at the time.[72] The Latvian nation therefore should be biologically shaped in order to "improve" the national health and the national body.

Applied eugenics – the work of the eugenics commission – highlights this approach: about half of the approved abortions were conducted on women with mental diseases such as schizophrenia and epilepsy. The rest of the victims were women with tuberculosis. The aim was to eliminate supposed hereditary diseases, which was considered to include many mental conditions. In the case of tuberculosis, there was a widespread belief among physicians that not the disease, but the biological precondition to suffer from it, was inherited.

The sterilisations of 1938 and 1939 also document the agenda to form the nation biologically. Two-thirds of the victims had been labelled "feeble-minded", a category that encompassed biologically and socially "inferior" people including the mentally disabled, uneducated and socially weak: somehow a code for the underclass and the "a-social". The other third consisted of the mentally ill, with the exception of one castrated homosexual criminal.[73] The perceived bottom layer of Latvian society was to be "purged" by sterilisation. This was no "humanitarian" act motivated by the will to make life easier for the poor and the ill. Latvian eugenics was far from the so-called "reform eugenics" of the post-World War II years. This fact is outlined by the debate over "inferior" and "superior" individuals – persons of "high" and "low" biological value. To ensure the national health, biologically "inferior" individuals should be eliminated from society, or at least be stripped of the ability to reproduce. This was "racial selection" as practiced by the Latvian government. "Inferior" people were those categorised as mentally ill or disabled, criminals and those with other alleged diseases. Prīmanis also placed alcoholics and members of the underclass in this category.[74] Following Michel Foucault's definition, this could be seen as a form of racism.[75]

The importance of genetic cleansing to Latvian eugenicists illustrates the fact that when the TDSPI was being planned, it should have had four, rather than three departments. The unrealised department was called the "genetic department"; its proposed research areas were later distributed to the anthropological and to the eugenics department.[76] The eugenics project sought to "improve" the nation "racially" as well as genetically, "reducing the inferior" to ensure the survival of the nation. The dawn of applied eugenics not only involved biologically shaping the nation by eliminating the "inferior", but also forming a "superior" race. This would lead to a biologically formed nation – a special "Latvian race", as called for by Ernsts Brastiņš in 1934.[77]

The eugenics debate over biological value and "inferiority" was also connected to anthropological categories of race. Prīmanis became a central figure in the formation of the ideas of a Latvian race and ethnic eugenics. In the mid-1930s he laid the foundation for later biological selection of the nation. By including genetic anthropology to his research, he began collecting anthropometric data connected to social and medical areas, such as psychopathology. The questionnaires included the usual blanks for the results of the anthropometric measurements, though the person's profession and their parents' professions were requested. The individual had to provide information on their own and their relatives' health. Genealogical questionnaires had been used by eugenicists as tools since the beginning of the century for hereditary research. Further, demographic data on children and parenthood were important for the TDSPI's pro-natalistic research. Considering Prīmanis' belief in eugenics, the material he collected on genetic anthropology was meant to be useful for eugenicists in forming the Latvian nation. The genetic information could provide vast data for eugenics-based selection. For example, the data on health and the information on diseases could have been used in a card index on hereditary diseases. As the eugenics law was being enacted, the Latvian health administration began building up a record of mental diseases similar to initiatives in Nazi Germany. In December 1939, Latvian psychiatric and medical hospitals were informed that they had to register all cases of mental illness and hereditary physical disease for "further research of hereditary mental disorders in Latvia".[78]

With Prīmanis speaking of the importance of fertility within the Latvian middle class, and especially for academics to have more children, the anthropological data could provide important information related to social class. Scientists could be able to make connections between data on race (using the cephalic index), class, education, fertility, hereditary health and ethnicity. Because these anthropological data were collected for the Latvian eugenics institute, they easily could have been used as a basis for a genetic record similar to Charles Davenport's record in Cold Spring Harbor.[79] Beyond scientific interests, the data also could have been abused for eugenic breeding purposes.

This was not only a hypothesis but the future agenda of Latvian eugenicists is illustrated by further eugenics propaganda of Prīmanis. Beginning in the early 1930s he was obsessed with "purity", not only in terms of the supranational anthropological categories of "race", but also in terms of a biologically defined Latvian nation. Through palaeoanthropological as well as racial anthropological research, he searched for the "pure" Latvians who carried much of the genetic material of ancient Latvians. Further, his obsession not only concerned racial characteristics but also the tangible form of "blood mixing" – the intermarriage of Latvians and non-Latvians. In the editorial of the first issue of the TDSPI journal, which presented its current research, Prīmanis pointed out the "unpleasant" racial influence of Russians and Poles on the Latvian national body during the previous 200 years.[80] He also condemned "blood mixing" between Latvians and Russians, Poles and even Lithuanians, who were actually of "Baltic" origin, because this would lead to rise of "biologically inferior" people and biological decay.[81] Foreign nations posed dangers to Latvia's biological value. Other dangers from Latgallians, ethnic Latvians from the country's eastern region. Prīmanis called the rural migration of Latgallians to western Latvia an "inner colonisation" that he said could cause biological decay, because Latgallians were racially "different" from the typical Latvian – who was dominated by "Nordic race" following to Primanis' categories. Regarding Prīmanis' anthropological research, Latvians in the East lost their "Nordic" feature and took on more characteristics of the "Eastern Baltic race".[82] According to Backman and Prīmanis, Latgallians had been strongly influenced by their Russian neighbours in terms of biology and race. Again, Prīmanis feared for the "purity" of Latvians. He believed that if this process of migration from the East to the West did not stop, Latvians would change and eventually "become totally different from the current population". Suggesting that Latvian eugenicists would prevent this, Prīmanis quoted the dictator Ulmanis: "Who owns the land should also sow".[83] The ruling power and titular nation should have the power to decide on the biological shape of the nation as well as on the question of who should have the right to procreate.

The genetic approach to anthropology and the definition of Latvians as a biological group was also emphasised in the TDSPI's research. In 1939 the demographic department searched for people who voted for German nationality but who had Latvian family names according to the 1935 census.

The eugenicists were looking for biological or ethnic Latvians who had become Germanised or Polonised over time, to re-Latvianise them. Demographers were obsessed with this idea. Magers Skujeniks was convinced, for example, that a large portion of the population in the Abrene district near the Russian border, which consisted of half Latvian and half Russian people, was biologically of Latvian origin.[84]

Given the search for purity in terms of race as well as a biologically defined nation, the objective of Latvian ethnic eugenics, racial hygiene and ethnic purity became more concrete. Following the German invasion, the eugenics project was relaunched in 1942 by precisely the same protagonists as before, including Prīmanis and other eugenicists and health officials. It is little wonder that Oskars Dankers, the head of the Latvian administration under Nazi occupation, enacted different laws that banned intermarriage of Latvians and Germans, Russian prisoners of war and other non-Latvians.[85]

7. Conclusions

Anthropology played a crucial role not only in creating a racial identity but also providing a basis for biologically defining the Latvian nation. Physical anthropology, with its ancillary sciences of palaeoanthropology, craniology and archaeology, created a narrative of the biological and historical origin of Latvians that was also significant in legitimating the nation. Identifying themselves as "originally Nordic", Latvians tried to achieve biological value in accordance with the racial paradigm of the Western European value system and its domination by the "Nordic" approach. High biological value had evolved in the context of social Darwinism. Prīmanis asserted that Latvians had survived the biological struggle with other nations through the century, preserving their "physical and mental strength" during his day.[86]

In creating a biological history of the nation, racial identity also implied a biological definition of the nation. Hence, it fuelled ethno-nationalism that spread further after World War I. Racial anthropology and ethno-nationalism were linked to eugenics in order to biologically shape the nation. After Ulmains' *coup d'état* in 1934, the new regime followed the paradigm of the biological re-creation of a nation.[87]Anthropology assumed a new significance as a national science with the duty to describe and measure the entire national body. The new regime shifted in its official propaganda toward ethno-nationalism, race and eugenics. Further, it declared bio-politics as a national goal, while anthropology played an important role in connecting eugenics with race in the process of social renewal.

With the idea of a total measurement of the population, the enlargement to demography and genetics, anthropology became a potential tool for eugenic engineering. Consequently, anthropology became the most important research field after genetics at the Latvian eugenics institute beginning in 1938.

Following the eugenics law of December 1937, applied eugenics in Latvia had not openly been connected to race. With the tools of abortion and sterilisation, government-organised eugenics aimed at "hereditary hygiene", genetic hygiene and "purity". People with mental diseases, physical and mental disabilities, and social deviant behaviour – commonly described as "feeble-minded" – allegedly suffered from hereditary diseases. Eugenicists attempted to "improve" the "biological value" of the nation.[88]

The emerging ethno-nationalism was fostered by the Ulmanis regime beginning in 1934, being heavily influenced by racial biology. The "Nordic" character of Latvians was emphasised. The late 1930s saw an increase of xenophobia and ethno-nationalism, with an exclusive approach towards non-Latvians. "Latvian" was even seen as a biological group that could be described as a "race". The idea of a "pure" Latvian nation influenced the eugenics agenda, not in terms of applied eugenics but concerning eugenics theory and research. Eugenicists such as Prīmanis warned the public of "blood mixing", not only in terms of racial anthropological categories, but also in ethnic terms. Intermarriage between Latvians and non-Latvians was to be be banned, because it was seen as a form of biological decay. The inclusion of "ethnic eugenics" in the Latvian national agenda becomes obvious by the writings of Prīmanis. Applied eugenics after 1942 under Nazi occupation drew on the ethnic eugenics approach in Latvia.

It would be too simple to call Prīmanis and other eugenicists merely racist or xenophobic. Prīmanis was a scientist who strictly followed international scientific rules and methods. He and others were true believers in scientific positivism who shared an eschatological conviction that science would heal the world, or at least his nation. Prīmanis also believed in strict genetic determinism. In his speech to athletes in 1938, he stressed that sports could promote health, but if an individual carries a defective gene, he never will become genetically healthy.[89]

Latvian eugenicists not only were Darwinist, but they also had a secular religious belief in "nature" as an omnipotent power, which fuelled their biologistic world view: while "natural selection" used to form humans in ancient times, by the 1930s "materialism", "liberalism" and social care hindered nature from the formative processes of evolutionary selection resulting in the degeneration of society.[90] There was a duty to restart this "selection", at least as "artificial" selection, to stop degeneration. "God forgives – but nature never will": this statement by Prīmanis implied that without eugenic intervention in human genetics, the nation would become weakened in the "natural" Darwinist struggle. Prīmanis and others eugenicists developed their own secular religion – as Francis Galton had suggested some 50 years earlier. Prīmanis spoke of a "sin" against nation and nature by ignoring the signs of degeneration and not following the path of eugenics. The health official Oskar Alks called eugenics a "holy mission".[91] The agenda of the racial state was a secular religious act.

Comparing Latvia's eugenics project to the Swedish welfare-state model it is quite obvious, that the Latvian project was quite differed even if it also had a strong collective approach: for Prīmanis, a person's achievements and strength were not as important as the "gene he left for his nation".[92] With its focus on race, biology and "genetic improvement", Latvian eugenics was far from Sweden's model. Focusing on applied eugenics, the Latvian project was actually quite near Nazi racial hygiene, with the exception that there were no mandatory sterilisations and certainly no euthanasia-killings of the mental ill. But similar to German racial hygiene, the Latvian "hereditary hygiene" targeted primarily the "feeble-minded" as well as other mentally ill people.[93] An additional parallel to the German model was the rhetoric of ethnic and racial purity, and the concept of the racial state. It became clear how deeply racial anthropology and its creation of Latvian racial identity influenced the debate on inter-war Latvian identity in general, and the debate on and practice of the biological formation of the Latvian nation in the late 1930s.

Notes

---

[1] Prīmanis, 1938a, 2-4.

[2] For a theoretical approach on the "racial state" see: Burleigh and Wippermann, 1991; Turda, 2010.

[3] Baehr, 1850.

[4] For the anthropological research at Dorpat see: Käbin, 1986.

[5] Waeber, 1879; Weinberg, 1897.

[6] Virchow, 1877a, 255; Virchow, 1877b.

[7] Ripley, 1915, 341, 351.

[8] Grant, 1916,188, 203.

[9] Deniker, 1900, 327, 346-347.

[10] Günther, 1925, 86.

[11] Grāvere, 2008, 110-121.

[12] Ezergailis, 2004, 42-53.

[13] Felder, 2005, 16-17; Felder, 2009, 275-296.

[14] Zelče, 2006, 94-137.

[15] Stranga, 2005, 2; Stranga, 2008.

[16] On the number of students see: Skujenieks, 1938, 20-21.

[17] An example for Latvian historiography might be Francis Balodis, an archaeologist, who claimed that the ancient Samgallians, who were alleged to be the forefathers of the Latvians, were so strong that they controlled not only the river Daugava (Düna) but also repelled the invading Swedish

Vikings. Only because of the Semgallian strongholds did the Vikings follow the way via the lake Ladoga to inner Russia instead of the river Daugava in 9th/10th century: Balodis, 1948, 317-365.
[18] On the Baltic German historiography see for example: Rauch, 1986.
[19] See for example: Vīķe-Freiberga, 2002.
[20] An example of the phenomenon of creating bio-history in 21th century is provided by: Sommer, 2008, 473-528; Sommer, 2010, 366-390.
[21] For Backman see his university file at the University of Latvia: Latvijas Valsts Vēstures Arhīvs (Latvian Historical State Archive – further: LVVA) 7427/13/114; See also: Lindberga, 2003, 42-44; Kott, 2009, 59-82.
[22] Backman,1919.
[23] See: Kott, 2009, 68-73.
[24] Backman, 1915, 330-350; Hedin also supported the German war effort: Hedin, 1915.
[25] Backman, 1915, 330-338, 350; Backman, 1937, 1-64.
[26] Vilde, 1924, 93-181; Prīmanis, 1925, 429-485; Jerums and Vītols, 1928, 273-386; Knorre, 1930, 256-312.
[27] Backman, 1925, 367-379.
[28] Backman, 1937, 48.
[29] For the numbers of minority members see: Skujenieks, 1938, 13-17.
[30] On the Rainis case see: Grāvere, 2004, 153.
[31] Arājs, 2005, 90, 99-100.
[32] See: Lindberg, 2003; See also Prīmanis' student file at the University of Dorpat: Eesti Ajalooarhiiv (Estonian Historical Archive – further: EAA) 402/1/21450, and his personal file at the University of Latvia: LVVA 7427/13/1367.
[33] See: Arājs, 2005, 126-139; and also Prīmanis' personal file at the University of Latvia: LVVA 7427/13/1367.
[34] Arājs, 2005, 42, 186-192.
[35] See: Jeruma, 1935.
[36] On the governmental demand for anthropology see: "Ministru Kabinets [The Cabinet]" *Valdības Vēstnesis*, no. 156 (15 July 1936), 1; Arājs, 2005, 148.
[37] Prīmanis, 1937a, 254-259.
[38] Arājs, 2005, 142; "L.U. Medicīnas Fakultātes Anatomijas Institutā 20 darba gadi [The 20 years old work of the Institute of Anatomy at the Medical Faculty at the University of Latvia]" *Ārsts*, vol. 3, (1940), 179-180.
[39] See the Report of the TDSPI from 1938 to 1939: LVVA 3112/1/19, 1-23; LVVA 3112/2/21, 194-202; See: Arājs, 2005, 144-186.
[40] Cauna, 1943, 260-265.
[41] Prīmanis, 1929, 46.
[42] Prīmanis, 1937b.

[43] See: Arājs, 2005, 156. The Archive of the Museum of Anatomy in Riga still keeps in several fonds the materials of the anthropological excursions under Prīmanis, see: Grāvere, 2002, 111-130; Grāvere, 2004.
[44] Prīmanis mentioned this further meaning of race in the preface of the publication of the eugenic lessions of Gustavs Reinhards: Reinhards, 1940, 5.
[45] Ducmanis, 1936, 111-128.
[46] Ibid., 124.
[47] Brastinsch, 1937.
[48] Lapiņš, 1936, 5-19.
[49] Vecozols, 1935, 2-3.
[50] For an overview on the Latvian eugenic project: Felder, 2009, 275-285.
[51] Altogether 63 sterilisations have been documented from 1937-1939. For the first half of 1940 there are no extant documents but indications that the commission worked until June 1940; Report on the work of the eugenic commission in 1938: LVVA 4578/1/204, 117. For the report of 1939 see: *Ārsts*, vol. 2 (1940), 127.
[52] For Latvian eugenics under Nazi occupation see: Felder, 2013.
[53] Arājs, 2005, 155.
[54] For the early eugenic debates in Dorpat see: Felder, 2011, 335-365.
[55] Arājs, 2005, 205.
[56] Buduls, 1909; On Buduls' role in establishing eugenics in Latvia see: Felder, 2009, 281-284; and the chapter by Vladimirs Kuznecovs in this volume.
[57] Shepilevskii, 1914; Gamaleja, 1912.
[58] See eugenic writings by Backman as: Backman, 1919.
[59] Arājs, 2005, 177-178, 205; Reinhards' lectures were also announced in the daily press: "galvas pilsēta [The Capital]" *Rits*, no. 35 (1 February 1935), p. 8; and were later published: Reinhards, 1940.
[60] This is still only an assumption by Arājs, but there is no documentary evidence found yet: Arājs, 2005, 178. See also the new civil law with paragraph 34 on the eugenic marriage ban: *Civillikums [Civil law] - Kodifikācijas nodaļas 1937 gada izdevums. 3. iespediums, 2. Nachdruck der Ausgabe Riga, 1938* (1990[1938]). Hamburg: Loeber, 10.
[61] See the annual report of the TDSPI for 1938: LVVA 3112/1/19, 8-11.
[62] For the radio lectures see for example: "Ko raidīs mūsu radiofoni? [What will the radio broadcast]" *Latvijas Kareivis*, no. 59 (13 March 1938), 6; For Jēkapils see: "Ārstu sanāksme Jēkabpilī [Meeting of physicians in Jekabpils]" *Jēkabpils Vēstnesis*, no. 37 (16 September 1937), 3; On Alks' lectures in Riga see: Alks, 1938, 7-12; On the eugenic lessons for health personnel see: Stradiņš, 1937, 20.
[63] *Rits*, no. 55 (24 February 1939).

[64] The Ministry announced the popularisation of eugenics in 1936: "Ministru Kabinets [The Cabinet]" *Valdības Vēstnesis*, no. 156 (15 July 1936), 1; "Valsts plānveidīgā celtniecība [Plan for the development - of the State]" *Rits*, no. 199 (21 July 1936), 2; "Valdības darba plāns [The State working plan]" *Ventas Balss*, no. 78 (16 July 1936), 2; A first overview on the governmental eugenic agenda is given by: "Projekts dzimstības pacelšanai un ģimenes nasti izlidzināšanai [A project to rise the birthrate and to ease familiy's burden]" *Brivā Zēme*, no. 174 (5 August 1936), 2.
[65] Apsīts, 1938, 2; Stradiņš, 1938, 1-2; Prīmanis, 1938b, 7.
[66] Stradiņš, 1938; Prīmanis, 1938b.
[67] See the annual Report of the VVB concerning the institute for 1939: LVVA 3112/2/19, 9; also: LVVA 3112/2/21, 194; Salnītis, 1939, Prīmanis, 1938b, 7; See also: "Sāks darboties tautas dzīvā spēka pētīšanas institūts [The Research Institute on the National Vitality took up its work]" *Latvijas Kareivis*, no. 20 (26 January 1938), 1.
[68] Ploetz, 1895, 54-90; Schallmayer, 1910, 280-303.
[69] Kraulis, 1939, 167; On Kraulis see: Felder, 2009, 281-285.
[70] Starcs, 1939, 360-362.
[71] See: "Mūsu fiziskās kultūras un sporta jaunie ceļi [The new Ways for our Physical Culture and Sport]" *Jaunakās Ziņas*, no. 91 (25 April 1938), 6.
[72] Prīmanis, 1938a, 2; Reinhards, 1940, 5.
[73] See the files of the eugenic commission: LVVA 4578/1/202, 4578/1/204, 4578/1/205.
[74] Prīmanis, 1939, 6.
[75] On Foucault's concept of racism see for example: Sarasin, 2003, 55-79.
[76] For the blueprint of the TDSPI of Prīmanis sent to the Ministry of Public Welfare in 1937 see: LVVA 1642/1/38, 2-4; On the internal structure of TSDPI see the annual report of the VVB of 1938: LVVA 3112/1/19, 12-23; "Tautas dzīvā spēka pētīšanas istitūta turpmākā darbībā [The Future Work of the Research Institute for National Vitality]" *Latvijas Kareivis*, no. 35 (11 February 1939), 5.
[77] See: Brastinsch, 1937, 45.
[78] "Pārskats par veselības departamenta darbību [Overview on the Work of the Health Department]" *Tautas labklājības ministrijas mēnešraksts*, no. 3 (1939), 224.
[79] On Davenport see: Black, 2003.
[80] Prīmanis, 1937b, 4-5.
[81] Prīmanis, 1939, 5-6.
[82] Ibid., 6; This picture of "inferior" Latgallians was also mentioned in Prīmanis' post-war works: Prīmanis, 1952, 461-468.
[83] Prīmanis, 1939, 9.

[84] For the research of the TDSPI see the annual report of the VVB for 1939: LVVA 3112/2/19, 9; also: LVVA 3112/2/21, 194; For Skujenieks see: Skujenieks, 1938, 14.
[85] Felder, 2009, 286-297.
[86] Prīmanis, 1938a, 2.
[87] Turda, 2010.
[88] Prīmanis, 1938a, 2.
[89] Ibid., 3.
[90] The support of "natural selection" was often claimed in the context of the eugenic project: "Jaunnedēļ sāks daboties pirmā eigeniskā komisija [The first Eugenic Commission will take up its Work in the Next Week]" *Rits*, no. 7 (7 January 1938), 7; Materialism, Socialism and Liberalism were often alleged of be responsible for the biological decay and elimination of the nation, see for example: Alksnis, 1938, 13, 20.
[91] Prīmanis, 1938a, 3; Alks, 1938, 12.
[92] Prīmanis, 1938a, 3.
[93] Schmuhl, 2005, 381; For a more detailed analysis: Bock, 2010.

References

Alks, O. (1938), "Arstu uzdevumi atjaunotā Latvijā [The assignment of physicians in the new Latvia]" *Tautas Labklājības Minstrijas Mēnešraksts*, no. 1, 7-12.

Alksnis, J. (1938), "Vārdi un Darbi. Pārunām par tautas dzīvo spēku [Words and Work. Some words on the National Vitality]" *Brīvā Zeme*, no. 114 (21 May), 20.

Apsīts, H. (1938), "Kas vēl vajadzīgs [What else is needed]" *Briva Zeme*, no. 18 (24 January), 2.

Arājs, K. (2005), *Latviešu kauli...Atmiņas par anatomikunu un antropoloģijas ziedu laikiem Latviā* [Latvian Bones...Memories on the Blooming Years of Anatomy and Anthropology in Latvia]. Riga: Rigas Stradiņa Universitāte.

Backman, G. (1915), "Den europeiska rasfrågan ur antropologiska och sociala synpunkter [The European racial question from an anthroplogical and social perspective]" *Ymer*, no. 4, 330-350.

Backman, G. (1919), "Det svenska folkets urspung och rasbeskaffenhet enligt anropologiens vittensbröd [The origins and racial character of the Swedish people according to the results of anthropological research]" in:

Lundborg, H. (ed.): *Rasfrågor i modern belysning, med särskild hansyn till det svesnka folket. Populär handledning* [Racial questions in a modern view, especially the Swedish people]. Stockholm: P.A.Norstedt.

Backman, G. (1925), "Anthropologische Beiträge zur Kenntnis der Bevölkerung Lettlands" *Latvijas Universitātes Raksti*, no. 12, 367-379.

Backman, G. (1937), "Die Augenfarbe der Letten" *Lunds Universitets Årsskrift*, vol. 33, no. 3, 1-64.

Baehr, J. C. (1850), *Die Gräber der alten Liven*. Dresden: Rudolph Kunze.

Balodis, F. (1948), "Handelswege nach dem Osten und die Wikinger in Russland" *Antikvarisk Studier*, no. III, 317-365.

Black, E. (2003), *War against the Weak. Eugenics and America's Campaign to create a Master Race.* New York: Four Walls Eight Windows.

Bock, G. (2010), *Zwangssterilisation im Nationalsozialismus. Studien zur Rassenpolitik und Geschlechterpolitik*. Münster: MV-Wissenschaft.

Brastinsch, E. (1937), *Volkstumslehre (Dienstliche Übersetzung der Publikationsstelle des Preussischen Geheimen Staatsarchivs in Berlin-Dahlem ausgeführt von Dipl. Volkswirt Paul Sender im Auftrage des Instituts für Osteuropäische Wirtschaft)*. Berlin: *Publikationsstelle des Preussischen Geheimen Staatsarchivs* .

Buduls, H. (1909), *Lauliba un zilweka dsihwes mehrķis. Bioloģisks un etisks apzerejums* [Marriage and the Goal of Human Life. Biological and Ethical Essay]. Riga: Ed.Sirgela.

Burleigh, M. and Wippermann W. (1991), *The Racial State: Germany 1933-1945*. Cambridge: Cambridge University Press.

Cauna, N. (1943), "Über Blutgruppen der Letten" *Latvijas Ārstu Žurnāls*, no. 4, 260-265.

Deniker, J. (1900), *The Races of Men. An Outline of Anthropology and Ethnography*. London: Walter Scott.

Ducmanis, K. (1936), "Mūsu nācionālo centienu garīgie pamati [Spiritual foundations for our national efforts]" in: Lapiņš, J. (ed.), *Jaunais nācionālisms*. Riga: Valters un Rapa, 111-128.

Ezergailis, A. (2004), "Kolaborācija vācu okupētajā Latvijā: piedāvātā un atraidītā [Collaboration in German occupied Latvia: offered and rejected]" *Latvijas Vēsture*, vol. 54, no. 2, 42-53.

Felder, B. (2005), "Mazvērtīgo samazināšana- eigēnika Latvijā [Reducing the Inferior - Eugenics in Latvia]" *Kultūras Diena*, no.95 (23 April), 16-17.

Felder, B. (2009), *Lettland im Zweiten Weltkrieg. Zwischen sowjetischen und deutschen Besatzern 1940-46*. Paderborn: Schöningh.

Felder, B. (2011), "Rassenhygiene in Russland. Der Hygieniker und Bakteriologe Evgenij A. Šepilevskij (1857-1920) und die Anfänge der Eugenik in Russischen Zarenreich. The Hygienist and Bacteriologist Evgenii A. Shepilevsky and the Beginning of Eugenics in Russia" in: Riha, O. and Fischer, M. (eds.), *Naturwissenschaft als Kommunikationsraum zwischen Deutschland und Russland im 19. Jahrhundert. Internationale Tagung. Leipzig, 29.2. - 1.10.2010*. Aachen: Shaker, 335-365.

Felder, B. (2013), "Eugenics and racial paradigms in Latvia under Nazi Occupation 1941-1945", in: Weiss-Wendt, A. and Yeomans, R. (eds.), *Racial Science in Hitler's "New Europe"*. (forthcoming)

Gamaleja, N. (1912), "Tõutervenduse-õpetuse, eugeenika põhjusmõtteist ja ülesannetest [Basic principles and tasks of racial hygiene/eugenics]" *Postimees*, no. 257-259 (5-9 November).

Grant, M. (1916), *The Passing of the Great Race or the racial Basis of European History*. New York: Scribner.

Grāvere, R. (2002), "Daži zemgales dzimtu identitātes un demogrāfijas dati [Some informations on the herediditary identity and demography in Semgale]" *Latvijas Vēstures Institūta Žurnāls*, no. 4, 111-130.

Grāvere, R. (2004), "Dažas LU Anatomijas muzeja 20.gs. 30. gadu antropoloģiskās kolekcijas mūsdienīgas izpētes iespējas [Possible research on the anthropological collection of the Museum of Anatomy of the University of Latvia]" *Latvijas Universitātes Raksti*, vol. 661, no. Zinātņu vēsture un muzejnība, 153-161.

Grāvere, R. (2008), "Vācbaltu anatomu devums Latvijas antropoloģijā. Balt-German Anatomists Contribution to Anthropological Rearch in Latvia" *Latvijas Universitātes Raksti*, vol. 716, no. Zinātņu vēsture un muzejniecība: 110-121.

Günther, H. F. K. (1925), *Kleine Rassenkunde Europas*. München: J.F-Lehmann.

Hedin, S. (1915), *Kriget mot Russland. Minnen från fronten i öster. Mars-Augusti 1915* [The war against Russia. Memories from the front in the East]. Stockholm: Bonnier.

Jeruma, L. (1935), *Die Lettin vom anthropologischen Standpunkt*. Riga: Anatomisches Institut der Universität Lettlands.

Jerums, N. and Vītols, T. (1928), "Beiträge zu Anthropologie der Letten" *Latvijas Universitātes Raksti*, no. XVIII, 273-386.

Käbin, I. (1986), *Die medizinische Forschung und Lehre an der Universität Dorpat/Tartu 1802-1940. Ergebnisse und Bedeutung für die Entwicklung der Medizin*. Lüneburg: Nordostdeutsches Kulturwerk.

Knorre, G. (1930), "Kranologische Untersuchungen an Schädeln aus Skelettgräbern Lettgallens" *Zeitschrift für Morphologie und Anthropologie*, vol. 28, no. 3, 256-312.

Kott, M. (2009), "Anthropologen Gaston Backman och den uppsaliensiska rasbiologins spridning i tid och rum" in: Müssener, H. and Jegebäck, P. (eds.), *Rasen och vetenskapen*. Uppsala: Centrum för multietnisk forskning. Programmet för studier kring Förintelsen och folkmord, 59-82.

Kraulis, Verners (1939), "Eigenikas ceļi Latvijā [The Way of Eugenics in Latvia]" *Tautas Labklājības Minstrijas Mēnešraksts*, no. 3, 161-168.

Lapiņš, J. (1936), "Ievadam: Jaunais nācionālisms [Introduction: New Nationalism]" in: Lapiņš, J. (ed.), *Jaunais nācionālisms*. Riga: Valters un Rapa, 5-19.

Lindberga, R. (2003), "Medicīnas Doktors, Professors Gastons, Viktors Bakmanis - Leģendāra personība Eiropas un Latvijas zinātnē [Medical Doktor, Professor Gaston Viktor Backman - Legendary Person of European and Latvian Science]" in: *The Professor Gaston Victor Backmann scientific conference*. Riga: Anatomicum, 42-44.

Ploetz, A. (1895), *Die Tüchtigkeit unsrer Rasse und der Schutz der Schwachen. Ein Versuch über Rassenhygiene und ihr Verhältniss zu den humanen Idealen, besonders zum Socialismus*. Berlin: S. Fischer.

Prīmanis, J. (1925), "Pales galvas kausi [The Skulls from Pale]" *Acta unversitates Latviensis*, no. 12, 429-485.

Prīmanis, J. (1929), *Latviešu ķermena uzbūve Latvju Daiņās* [The Shape of the Latvian Body in the Latvian Dainas]. Riga: AS Golts un Jurams.

Prīmanis, J. (1937), "Antrōpoloģiskais darbs Latvijā [Anthropological Work in Latvia]" *Sējējs*, no. 3, 254-259.

Prīmanis, J. (1937b), *Latviešu antropoloģiskais raksturojums* [The Anthropological Character of the Latvians]. Riga: Valtera un Rapas akc. sab. apgads.

Prīmanis, J. (1938a), "Dievs piedod, bet daba nekad! [God forgives - but Nature never will!]" *Latvijas Kareivis*, no.91 (24 April), 2-4.

Prīmanis, J. (1938b), "Kā daboies tautas dzīvā spēka pētišanas institūts" *Jaunakās Zīnas*, no. 20 (26 January), 7.

Prīmanis, J. (1939), "Tautas dzīvā spēka pētīšanas uzdevumi [Tasks of Research into the National Vitality]" *Tautas dzīvā spēka pētīšanas institūta raksti*, no. 1, 3-9.

Prīmanis, J. (1952), "Latviešu ķermena pazīmes [The characteristics of the Latvian body]" *Cela zīmes: mēnešraksts svešuma*, no. 10, 461-468.

Rauch, G. von (ed.) (1986), *Geschichte der deutschbaltischen Geschichtsschreibung*. Köln, Wien: Böhlau.

Reinhards, G. (1940), *Eugenika*. Riga: Valters un Rapa.

Ripley, W. Z. (1915), *The Races of Europe*. London: Paul.

Salnītis, V. (1939), "Mūsu dzīvais spēks", *Tautas dzīvā pētīšanas instituta raksti*, no.1, 10-34.

Sarasin, P. (2003), "Zweierlei Rassismus? Die Selektion des Femden als Problem bei Michel Foucault's Verbindung von Biopolitik und Rassismus" in: Stingelin, M. (ed.), *Biopolitik und Rassismus*. Frankfurt/M: Suhrkamp, 55-79.

Schallmayer, W. (1910), *Vererbung und Auslese in ihrer soziologischen und politischen Bedeutung, zweite Auflage*. Jena: Gustav Fischer.

Schmuhl, H.-W. (2005), *Grenzüberschreitungen. Das Kaiser-Wilhelm-Institut für Anthropologie, menschliche Erblehre und Eugenik 1927-1945, Geschichte der Kaiser-Wilhelm-Gesellschaft im Nationalsozialismus, Bd.9*. Göttingen: Wallstein.

Shepilevskii, E. A. (1914), *Osnovy i sredstva rasovoj gigieny (gigiena razmnoženija)*[Basic Principles and Means of Racial Hygiene (hygiene of procreation)]. Jur'yev': Mattisen.

Skujenieks, M. (1938), *Latvijas Statistikas atlass* [Statistical Atlas of Latvia]. Rīga: Valsts Statistikā Pārvalde.

Sommer, M. (2008), "History in the Gene: Negotiations between Molecular and Organismal Anthropology" *Journal for the History of Biology*, vol. 41, no. 3, 473-528.

Sommer, M. (2010), "DNA and Cultures of Remembrance: Anthropological Genetics, Biohistories, and Biosocialities" *Bio Societies*, vol. 5, no. 3, 366-390.

Starcs, P. (1939), "Kas darīts, kas vēl jādara tautas dzīvā spēka pieauguma veicināšanai [What is done and what has to be done for the Advancement of the Growth of the National Vitality]" *Tautas Labklājības Minstrijas Mēnešraksts*, no. 5, 351-362.

Stradiņš, P. (1937), "Par tautas veselību un viņas možo garu [On the national health und its vivid spirit]" *Brīvā Zeme*, no. 259 (13 November), 20.

Stradiņš, P. (1938), "Vienots darbs tautas veselības veicināšanai [United efforts on the advancement of National Vitality]" *Jaunākās Ziņas*, no. 22 (22 January), 1-2.

Stranga, A. (2005), "Replika [The Remark]" *Diena*, no. 101 (30 April), 2.

Stranga, A. (2008), *Ebreji Baltijā. No ienākšanas pirmsākumiem līdz holokaustam 14.gadsimts - 1945 gads* [Jews in the Baltics. From the beginnings of their arrival to the holocaust, 14th century to 1945]. Riga: Latvijas Vēsture.

Turda, M. (2010), *Modernism and Eugenics*. Houndsmill: Palgrave Macmillan.

Vecozols, J. (1935) "Latvijas tautas asiņu kopība [The blood community of the Latvian nation]" *Latvijas Kareivis*, no. 235 (19 October), 2-3.

Vīķe-Freiberga, V. (2002), *Trejādas saules. Meteroloģiskā saule, Siltā saule* [The Threefold Sun. The Meterological Sun. The Warm Sun]. Riga: Karogs.

Vilde, J. (1924), "Materiāli par lībiešu antropoloģiju [Materials on the anthropology of Livonians]" *Latvijas Universitātes Raksti*, no. XI , 93-181.

Virchow, R. (1877a), "Ältere Gräber in Livland" *Verhandlungen der Berliner Gesellschaft für Anthropologie, Ethnologie und Urgeschichte*, vol. 9. no. 6, 255-259.

Virchow, R. (1877b), "Archäologische Reise nach Livland" *Verhandlungen der Berliner Gesellschaft für Anthropologie, Ethnologie und Urgeschichte*, vol. 9, no.6, 365-437.

Waeber, O. (1879), *Beiträge zur Anthropologie der Letten*. Dorpat: Schnakenburg.

Weinberg, R. (1897), "Über einige Schädel aus älteren Liven-, Letten- und Estengräbern" *Sitzungsberichte der Gelehrten Estländischen Gesellschaft 1896*.

Zelče, V. (2006), "Vara, zinātne, veslība, un cilvēci. Eigēnika Latvijā [Power, Science, Health and Men. Eugenics in Latvia]" *Latvijas Arhīvi*, no. 3, 94-137.

Latvian Psychiatry and Medical Legislation of the 1930s and the German Sterilisation Law

Vladimirs Kuznecovs

Abstract

The founder of Latvian psychiatry Hermanis Buduls and his pupil Verners Kraulis played an important role in the development of Latvian eugenics. In 1933 Buduls' expert opinion was critical in resisting Nazi Germany's eugenic law of 1933 (in its Danzig version) as a possible model for a national eugenic law. Settling on Mendel's genetics principles Buduls proposed conceptual narrowing of the hereditary indications for mass sterilization rejecting biosocial determinism and paternalism of the Nazi sterilization law. Viewing survival interests of a "small nation" in pro-natalistic perspective, Kraulis rejected radical sterilization policies of some other European nations headed by Germany. He stressed, by contrast with the latter, the formation of eugenic commissions with only counseling/administrative functions, an absence of the list of specific "hereditary diseases" for mandatory sterilization, and the principle of voluntariness as distinctive features of 1937 Latvian eugenic law.

The aim of this paper is to elucidate one aspect of the history of eugenics in Latvia, namely, the role of psychiatry and psychiatrists in the local eugenics discourse during the 1930s. In particular, it will compare the theory and practice of the eugenics portion of Latvia's medical legislation to the most radical eugenics law of Europe – the Nazi German "sterilisation law" (*Gesetz zur Verhütung erbkranken Nachwuchses*) of 1933. As an interdisciplinary science, eugenics attracted loose groups of participants from different institutions, each developing in its own historical national context. Hereditarian-minded Latvian physicians, including eminent psychiatrists Hermanis Buduls and Verners Kraulis, had a keen professional interest in eugenics as a working tool at all levels of individual and social health prophylaxis, being deeply interested at the same time in its biological, genetic and ethical aspects. Psychiatrists played an important role in developing applied programmes that implemented practices of so-called "positive" and "negative" eugenics. According to the "prescriptions" of eugenics, ineffective natural selection had to be changed artificially, encouraging fertility among the racially robust (positive eugenics) and limiting the fertility of "inferior" people through the life-long institutionalisation of chronic psychotic patients and the mentally disabled, legalising abortion and sterilisation, and using euthanasia (negative eugenics). Psychiatrists' support of the norms of negative eugenics assumed its most extreme form in the case of Nazi Germany's "sterilisation law" of 1933, and through additional mass actions against "unfit", first of all mentally disabled and other persons with "hereditary diseases". The impact of psychiatrists in less dark, pro-natalist forms is evident in other countries such as Latvia.

Different theoretical and applied aspects of Latvian eugenics during the interwar period were studied mainly recently in Latvia and outside Latvia. As a representative of (very rare) Soviet-era research of racial hygiene in Latvia, V. J. Rikshe in 1978 utilised strong ideological clichés in analysing the views of the main figures, including J. Prīmanis, E. Fērmanis and G. Reinhards (none of whom were psychiatrists).[1] In 2005, Björn Felder first focussed on the issue with a scientific approach. His first groundbreaking article and his fundamental book on Latvia in World War II contain a detailed history of Latvian eugenics between the wars and during the 1940s, presented in a broad historical and cultural context. This research on Latvian racial hygiene – its programmes, people and institutions – is based on a thorough analysis of sources and literature. Felder also analysed the eugenics component of Latvia's laws of 1933 (Criminal Code) and 1937 (Medical Law), and focused on the activity of the eugenics commission in the late 1930s and 1940s. Felder draws a plausible conclusion about the dominant philosophy of the survival of a "small people" possibly being a main cause of the refusal by politicians to institute forced sterilisation in Latvia. Felder also focuses on the development of Latvian psychiatry in the 1930s-1940s in the context of eugenics. Felder's interpretation of a typical psychiatrist during the interwar period, mercilessly treating his patients as unfit persons by shock therapy, is a moot question.[2] In an answer to Felder, Vita Zelče published an article on the history of Latvian eugenics in the 1920-1930s. Among other topics, she analysed the law on abortion within the Latvian Criminal Code of 1933 and the 1937 Health Law's provision on eugenics and its application via the Eugenics Commission in the late 1930s. She cited V. Kraulis' brief evaluation of the local eugenics law. However, she did not analyse systematically Kraulis' eugenic views that contrasted with the principles of Germany's sterilisation law of 1933. The diagnostic aspect of sterilisation cases in 1938-39 was also not analysed by Zelče.[3] The main trends of the institutionalisation of eugenics during the 1930s, including V. Kraulis' role in this process, and his attitude towards sterilising the mentally ill regardless of their ethnicity, are mentioned by A. Stranga in his history of the Jewish minority in Latvia (but only in one footnote).[4] Some aspects of the history of Latvian eugenics have been treated by other Latvian researchers from different scientific and ideological positions. Hermanis Buduls' eugenic publications are mentioned in his bibliography edited by Arnis Vīksna, who also wrote Buduls' biography for this collection. However, this booklet does not provide any special comment concerning Buduls' interest in eugenics.[5] Viskna also talks about Buduls' ethical activities in saving the lives of hospitalised chronic psychotics and mentally disabled people during the Nazi occupation.[6] R. Vīksne wrote a detailed article on the extermination of patients in Latvian psychiatric institutions during the Holocaust and the positions of psychiatrists regarding these actions during the 1940s (which is beyond the chronological boundary of this article).[7]

1. Hermanis Buduls and the Problems of Eugenics during the Interwar Period

Latvian psychiatry and its representatives played an important role in the development of national eugenics or racial hygiene, including its legal aspect. This contribution was largely associated with two major psychiatric figures, Hermanis Buduls and his pupil, Verners Kraulis.[8] As one of the most important personalities in the history of Latvian psychiatry, Buduls (1882-1954) was the founding father of Latvian psychiatry. Beginning in 1919, he was the chair the Department of Psychiatry of the School of Medicine of the only indigenous university, the University of Latvia in Riga. Beginning in 1924 he also served as the chair of the Latvian Society of Psychiatrists and Neurologists, and he was the medical director of the University of Latvia's psychiatric clinic Sarkankalns. He prepared his doctoral thesis, "On Comparative Racial Psychiatry," in the psychiatric clinic of Yur'yev' University in 1914. Buduls was an author of the first Latvian textbook on psychiatry, published in two parts in 1924 and 1929. Eugenics issues themselves were mostly covered in popular scientific publications, such as his early book *Marriage and the Goal of Human Life* (1909).[9] Among other topics, Buduls' book follows German biologist and early geneticist August Weismann's strict biological determinism in polemics with Lamarckism (the possibility of inheriting acquired characteristics).[10] Though without using the terms "genetics" and "eugenics," he appealed to the "weak individuals" to abstain from intensified procreation. Buduls did not mention concrete methods or support any forced means. He wrote: "The society should not allow to barrier the life ways of its weak members through various means of force." Calling the intention of such people to create a family "to produce cripple offspring" as "a dark work", Buduls suggested to "bring the light in" through education or in later terminology, positive eugenics.[11] However, Buduls' advice that "weak individuals" abstain from procreation was not forgotten. Later, it was cited with appreciation by neurologist T. Upners, author of university courses on eugenics. In 1942-1944 he was an expert for the Eugenics Commission and an advocate of Nazi Germany's eugenics law[12].

Since the beginning of 20th century, and particularly by the 1930s, the entire range of scientific, demographic, economic and cultural problems in Europe contributed to the appearance and spread of an atmosphere of growing pessimism. This featured "etiological" factors, according to Latvian psychiatrists and other specialists: "culture", modernisation and its tempo, urbanisation with its social contrasts, progress in social politics and medicine, both of which enhanced the survival of the "unfit"; and biological, psychological and moral "degeneration" of individuals and nations with high rates of mental disease and suicide. There was a growing fear of popular "degeneration" due to excessive procreation by certain groups of people – the mentally ill and minorities – and decreasing birth rates among the "valuable" population. In addition to the traditional fear of the negative influence of mental disorders on healthy people, eugenicists (psychiatrists, physicians and representatives of other sciences) argued that caring for institutionalised people was a growing economic burden. In particular, "feeble-mindedness", the most significant and widespread "hereditary disease", was frequently linked by eugenicists to delinquency, promiscuity and social dependence.[13] In this regard, the interest in eugenics as an instrument of social and biological regulation of societal development was growing in the Western world. Certain European countries enacted eugenics laws dealing with sterilisation. This development began in the US in 1907, moved to the Swiss canton of Vaud in 1928, and ended in Iceland in 1938.[14] Latvia fell toward the end of this period, enacting its eugenics law in 1937. The small Baltic state had before it a panorama of legal examples that were studied in university eugenics courses and discussed in the press.[15] The most radical of these, Nazi Germany's sterilisation law (whose authors included world-famous psychiatrists such as Ernst Rüdin) was enacted on 14 July 1933 and took effect on 1 January 1934. It included nine conditions that were assumed to be hereditary as criteria for mandatory sterilisation. The law did not exclude voluntary sterilisation. Considering them to be "hereditary diseases", the law included mental and other disorders such as schizophrenia, manic depression, inherited epilepsy, Huntington's chorea, vague categories such as "alcoholism" and "congenital feeble-mindedness", "hereditary" blindness and deafness, and serious bodily deformities. After the law's passing, its ambiguous formulation allowed including onto the list of non-pathological forms of behaviour such as mental disablement and a category called "moral feeble-mindedness", which greatly enlarged those subject to the law. Decisions by special hereditary health courts to sterilise people were routine. Vasectomy was the typical sterilisation method for men and tubal ligation for women, the latter of which was an invasive procedure that caused the death of hundreds of women. During the first three years of the law, the number of sterilisations totalled 169,089: 32,368 in 1934, 73,174 in 1935, and 63,547 in 1936 [16].

In independent Latvia prior to 1937, the legal regulation of birth rates manifested itself only in legislation on abortions as part of the Criminal Code. This underwent substantial changes on 13 January 1933, when amendments to the Criminal Code significantly enlarged indications to abortion (including, in particular, the eugenics component) compared to those in the previous version, which was based on the Penal Code of the Russian Empire. The latter was in force until 1 August 1933, when the new Latvian Criminal Code took effect. After 13 January 1933, the Code's Article 400, though punishing illegal abortion with imprisonment, provided for an entire range of eugenic, medical, legal and social indications for abortion. It justified an operation performed by a physician during the first three months of pregnancy for reasons not only associated with eugenics and terminating pregnancy (to prevent the birth of a child with a severe mental or physical defect, or if the mother's life or health was in danger), but also based on moral/forensic (rape) causes and "social reasons" (when to prevent the birth of a child was seen as a heavy burden by a pregnant woman and her family). On 22 March 1935, a decision by the Cabinet of Ministers on Article 440 provides an exception for the above-mentioned social indication for abortion.[17] Simultaneously, a new Article 440 included a Medical Regulation (points 55-2) that tried to define more precisely the eugenics indications for abortions: if the parents suffered from 1) a "non-treatable (or hardly treatable) mental disease that is transmitted to a child, as rule, hereditably"; 2) "other severe hereditable mental or physical defects", or 3) "active syphilis" [18]. The Latvian medical press criticised the new regulation (including the sections concerning mental illness) due to the eugenic and medical incompleteness of its inclusion criteria. For the Riga-based forensic doctor Miķelis Veidemanis, the regulation did not foresee the possibility of a split between pheno- and genotype in parents and offspring with recessive heredity (one or both parents could be formally ''healthy" but the offspring ill, especially if both parents had recessive genes of an illness). Second, the phrasing "other mental or physical hereditable defects" also raised questions over the inclusion of many other non-mental hereditary – i.e. neurological – diseases such as Huntington's chorea, hereditary ataxia and eye diseases such as retinitis pigmentoza. Third, citing "active syphilis" as an indication for eugenic abortion raised questions concerning, for example, the degree of eugenic risk for the foetus in case of intrauterine infection and the possibility of successfully treating it. This group of infectious diseases, according to Veidemanis, was also marked by incompleteness (tuberculosis was not mentioned). All in all, according to Dr. Veidemanis, the indications for abortion could have been expanded through elaboration of eugenics criteria.[19]

The procedure of sterilisation, either in parallel with or independent of abortion, was not stipulated by the 1935 law, nor was it part of scientific or public debates. However, the quest for a list of concrete "hereditary" diseases might be seen as one of the early signs of influence by Germany's eugenics law on eugenics thought in Latvia during the 1930s. Loud praise for sterilisation law, in comparison to Latvian legal situation in the first part of

1930s, came from Professor Ernsts Fērmanis (1872-1947), the University of Latvia's chair of hygiene.[20] For Fērmanis, in his concerns for the qualitative improvement of "blood and race"[21] described Germany's law with its list of "hereditary" diseases and defects as a reason for forced sterilisation as an adequate means for national liberation, firstly from the mentally ill, whose diseases were mentioned in the first seven points from the nine cited in the list of "hereditary diseases" (a modest figure compare to the variety of such diseases in the population, according to Fērmanis) by the German law. Fermanis shared the fears of other eugenicists that "mental defects and mental diseases are much more important than the physical defects, because the former ones often make individuals socially unworthy, and even into socially dangerous person compared to the latter ones".[22] Fērmanis blamed "our democratic era", with its attention to individual freedom, for preventing the enacting of a eugenics law in Latvia similar to that in Germany.[23]

Among other things, it is possible that due to circumstances that are unclear today – specifically, the incomplete expression of eugenics principles in the law of 1933 – the head of the Department of Public Health under the Ministry of Welfare of Latvia, A. Peterson, made a request to Buduls for an expert review of the Danzig eugenics law on 20 December 1933.[24] At the time the city Danzig was under the aegis of the League of Nations, and was a city-state ruled by Senate of the Free City of Danzig. Passed on 24 November and published on 29 November 1933, the Danzig law was a precise copy of Nazi Germany's eugenics law of 1933.[25] Both laws easily could have been misused due to their vague terminology, and they served as "sanitary measures" that required curing not individuals but the collective population. It is not clear today if these close links were known to both sides, or how a copy of the Danzig sterilisation law came into the hands of Peterson, and who else he might have been able to approach about an expert review.

The draft was likely only sent to Buduls as the most authoritative expert in the field of racial medicine and heredity in Latvia, at least until the mid-1930s. Buduls' short response followed immediately (his pencilled draft is dated 23 December). This could be indirect proof of his previous knowledge of the German original serving as an example for the Danzig law. Buduls' draft gives a concise but sufficiently representative opinion regarding a number of fundamental principles of eugenics due to its close connection with genetics and standards for appropriate legislation. The response, first, indicates that its author was not fundamentally opposed to negative eugenics and recognised the need in the long term to introduce a law in Latvia similar to the one in Danzig: "in principle there are no objections against the law also for our Situation". Second, Buduls was cautious concerning the use of coercive methods to decrease the frequency of certain hereditary traits being passed on to the next generation. Buduls pointed out the difficulty of identifying and interpreting genetic aspects of the mental state of patients' offspring, due to the possible incongruence between geno- and phenotypical issues, as well as the variability of risk that any genetic factors might cause abnormalities. He warned of "serious dangers" in the law's implementation with regard to the destiny of the offspring of the mentally ill, as "very often in genetic questions the collected material in concrete cases does not gives any foot-hold to foresee the mentally ill offspring's health condition". This is why, third, he insisted on the conceptual narrowing of targeted hereditary diseases in the interests of "racial improvement". Fourth, as an additional technical obstacle to adapting the Danzig law, Buduls mentioned difficulties in establishing an executive institution in applying the eugenic law (supposedly he speaks about system of hereditary courts in Germany) in the form of mass sterilisation in Latvia.[26] Actually, Buduls' professional statement meant "no" to attempting to adapt in Latvia the German sterilisation law disguised as the Danzig law. Later, this Latvian psychiatrist followed liberal eugenics philosophy in his polemics against the totalitarian *Rassenhygiene* of the 1930s.

In fact, Buduls could well agree with the opinion of prominent Munich psychiatrist and geneticist Oswald Bumke on castrating "psychopathic persons", as expressed in his book *Culture and Degeneration* (1912/1922). For Bumke, who definitely rejected such an idea, the operation could affect the immense spectrum of people, including phenotypically healthy ones: "If that measure would have had practical results for all the society, then the forced surgery should be performed not only on the numerous inmates of mental asylums and prisons, but also on much more numerous quantity of light nervous psychopaths and, probably, on the same quantity of healthy persons. Those that suggest something else fail to take into account facts of covered (recessive) hereditary transmission"[27]. O. Bumke lectured in Riga at least twice – 20-23 October 1933 and 7 October 1934 – and a similar passage (the text of lecture was not preserved) could be part of his first lecture on 21 October, named "Civilisation and regressive changes". All members of the Latvian Medical Society were invited by the local Herder Society and Society for Practicing Physicians to attend Bumke's lectures free of charge.[28] The programme of the second lecture series in 1934 is not known. Another racial hygiene advocate, eminent Austrian psychiatrist and geneticist Erwin Stransky, visited Riga earlier, in 1931, and gave two lectures. One of them, titled "Heredity and racial hygiene", was given in one of the most prestigious buildings in Old Riga, *Melngalvju nams*, on 24 March 1931.[29]

Buduls returned to the legal questions of eugenics in the article "Eugenics of Race" in 1936.[30] The article represented a complete expression of Buduls' eugenics credo in the public press, attempting to give a Latvian equivalent of the term (close to both the German and Anglo-Saxon version), and at the same time a polemic on Nazi Germany's sterilisation law. Buduls recognised the popularity of racial hygiene issues in the public discourse of many developed countries, only comparable, in his opinion, to the popularity of the ideology of nationalism. In analysing sections of Germany's law, he noted how radical they were as compared to analogous laws elsewhere, such as in Scandinavia, where they "closely observe the sterilised person's absolute free-will and consent".[31] Buduls thought the most important feature of Germany's eugenics law was its attempt to totally subordinate the interests of the individual to the needs of the collective, as "Germany asks of its corresponding individuals a much greater degree of subordination to the overall goals of eugenics for the public as a whole."[32] In defining the main purpose of sterilisation law as the struggle against the spread of mental illnesses and degeneration of the human race, Buduls considered it essential to elaborate further questions of mental inheritance and eugenics. Moreover, as he noted, the problem of sterilisation had not been discussed in the scientific community, and only brief statements about it had appeared in the general press. Hopes for supposed benefits should a sterilisation law be adopted had been exaggerated. In his reasoning – in the spirit of classical genetics, and in accordance with his previous report to the Department of Health – he referred to the dominant-recessive mechanism of heredity and the absence of any strict correlation with phenotype and genotype, thus leading to the appearance of practically healthy offspring of patients. Emphasising the great complexity of the problem of the inheritance of mental features, Buduls referred to research by another famous Munich psychiatrist and geneticist, Hans Luxemburger. Luxemburger also underscored the probability structure of the expression of psychopathological traits, capable, in his opinion, of leading even to the normalisation of the genotype of offspring within four generations. Both authors spoke about regeneration as opposed to degeneration in cases of many recessive genetic disorders with an unknown genetic cause, where the proportion of heterozygous carriers has a tendency to vastly exceed the affected homozygotes.

Regeneration as a sort of natural law, as opposed to the proclaimed biologically determined degeneration of individuals and nations, was also mentioned by V. Kraulis in one of his early popular articles in 1928.[33] The laws of genetics, in Buduls' opinion, might confirm the postulates of the laws of Moses, which says God punishes children for the sins of their fathers to the third and fourth generation.[34] Referring to the calculations of Luxemburger (ironically, himself a vigorous supporter of Germany's sterilisation law), Buduls observed that in the general population, the number of patients with psychopathologically afflicted inheritance was 15.6 %, while the remainder was phenotypically healthy. This is why Buduls considered concerns about the steady degeneration of "the race" greatly exaggerated, as well as fears of excessive fertility among the mentally ill, especially for people living in special institutions where they are deprived of the opportunity to have a family. But even in less severe cases of people not living in such institutions, the patients usually remained unmarried and did not have children. This is why mental disorders, to the extent that they are inherited, cannot be dangerous to humanity even if there is no attempt to sterilise mentally ill people. Buduls was worried that the consequences of mass sterilisation might be "radical changes in the natural development of mankind", and he called for abstaining from interfering with natural selection. Again he made reference to the Bible, comparing the consequences of widespread application of any means of artificial selection with an attempt to separate the good wheat from the chaff – in which case the good part of the good wheat would be thrown into the wind, along with the chaff and the bad wheat.[35] The repeated use of biblical analogies suggests that moral and religious principles played a significant role in the moderate position of Buduls. Later, in his last book, *Man in the Mirror of Life* of 1954, he wrote: "Both European and the New World culture are closely associated with Christianity. The end of the latter would have meant the end of Western Europe's culture, too... The modern life would be much more chaotic, merciless and full of hopelessness if it would not have been more impressed by religion".[36] However, it is possible that the Latvian professor used such analogies to make his scientific principles easily understandable for a wide audience. His article's conclusion is significant for its seemingly mild eugenic views, as Buduls once again mentioned the benefit of adopting a reasonable sterilisation law that first and foremost would be implemented among people with hereditary mental disabilities.

Buduls touched again upon the eugenic aspect of psychiatry in one of his last articles written in independent Latvia, a year before the Soviet invasion of 1940, leading to the eclipse of the discussion of eugenics. The 1939 article "Are Mental Diseases Spreading?" was written almost two years after the Medical Law was enacted in 1937. The law was intended to dispel prevailing public fears about excessive growth in the number of mentally ill, the related threat of biological degeneration of the population, and the high cost of caring for institutionalised individuals.[37] Latvian newspapers reported that huge sums of money were paid for the care of untreatable patients at the Sarkankalns clinic, where Buduls had been director since 1919. Noting the rapid growth of hospitalised mental patients, which exceeded all projections, he emphasised the social and medical reasons for this phenomenon: a new law on rental housing that often forced people to send relatives with overt signs of behavioural deterioration to hospitals due to the threat of eviction; the accumulation of many mentally ill people (especially those with alcoholism and syphilis) in large cities; and expanded and improved social services for patients whose treatment was increasingly paid by social organisations (82% of patients in Riga's clinics). While in agreement with the principles of eugenics, Buduls was ambivalent toward the achievements of modern medicine that increased survival rates of biologically "inferior" individuals. He pointed out these increasing rates not only for mental, but also for somatic patients. On the other hand, Buduls, as director of a clinic and author of publications on treating the mentally ill,[38] could not help but notice the fact that new treatment methods (malaria therapy for progressive paralysis; insulin and cardiazol treatment for schizophrenia) were returning to society people who previously were considered to be as incurable as schizophrenics, which made such patients if not equal then close to somatic patients. According to Buduls, about 60% of 200 schizophrenics treated in Sarkankalns during a two-year period had their capacities restored after being treated with insulin and cardiazol. So along with attention to early treatment, Buduls again questioned the advisability of massively applying legislative measures for negative eugenics with regard to so-called hereditary diseases. He regarded eugenics basically as a special area of preventive medicine designed to restrain the growth of mental illness. At the time, Buduls was sceptical of the success of eugenics, which in a number of countries was respected as a major pillar in efforts to improve "race" quality and to prevent it from further degeneration. "Nevertheless, one of the basic problems of eugenics, the doctrine of the origin of diseases and inheritance, still contains too many mysterious circumstances. Therefore, at the present time, the use of eugenics in resolving practical questions is sufficiently limited".[39]

On the whole, Buduls proved to be a supporter of a milder form of eugenics. He only theoretically considered more harsh eugenics measures. Buduls was not formally engaged in the activities of any local eugenics institution. His ethical position that respected patients' right to consent to sterilisation strongly influenced the Latvian eugenics law of 1937. According to some testimonies, he tried to help at least some categories of mentally ill escape mass executions by the Nazis, discharging them from his clinic "Sarkankalns".[40]

2. Verners Kraulis and Eugenic Legislation of 1937

The next step in the development of eugenics and eugenics legislation in Latvia was closely associated with the activity of another psychiatrist, doctor of medicine and docent – Verners Kraulis of the University of Latvia.[41] After graduating *cum laude* from the University of Latvia Medical School, in 1932 Kraulis defended his doctoral thesis on questions of the heredity of hysteria. While working on his thesis, Kraulis was on an assignment from the University of Latvia in 1929 to spend six months at the universities of Berlin and Munich. He was advised by well-known German psychiatrist and geneticist Ernst Rüdin, one of the authors of the German sterilisation law and director of the Genealogical–Demographic Department at the German Institute for Psychiatry in Munich. In 1929 Kraulis was on assignment in Vienna and Rome, and in 1936 he was assigned to London's Galton Institute to learn methods of twin studies. In 1939 he was again on assignment in Germany. He also visited Vienna and Rome for scientific purposes in 1931. Kraulis participated actively in the founding and activities of the Institute for the Research of the National Vitality, headed by the celebrated anatomist and anthropologist, Professor E. Prīmanis. Kraulis was a member of its governing board, along with prominent figures in Latvian medical science Professors P. Stradiņš and P. Sniķers. He was also the head of the Institute's eugenics section. He was one of the three members of the eugenics commission that was engaged in the practical implementation of the law until the autumn of 1941. A course in eugenics, led by privat-docent Kraulis at the University of Latvia, was made obligatory for 1940-41 on 22 May 1940 by the University of Latvia Academic Board. It was rapidly banned by the board on 29 May 1940 not long before the Red Army entered Latvia, on 17 June 1940. During this "first" Soviet presence in 1940-41, Kraulis was appointed as University of Latvia lecturer by Soviet officials. On 2 September 1941, during the German occupation, Kraulis was dismissed from the university by the Academic Board along with the statement "for ever and ever", presumably due to his alleged "leftist" political views.[42] He went to Liepaja, where he continued to work as a psychiatrist until his death from pneumonia in 1944.[43]

Statements by Kraulis regarding questions of eugenics are particularly interesting from the standpoint of comparing Latvia's eugenics law to Germany's "sterilization law". These statements are also directly related to the former practical implementation, while Kraulis' participation in the work of the Eugenics Commission was as one of its two experts. Kraulis' article "Ways of Eugenics in Latvia"[44] provided an analysis of Latvian eugenics legislation and its practical implementation during the first year of its existence. (He also presented a paper to the Latvian Medical Association on 19 January 1939 with the same title. The text is not preserved.[45]) Eugenics was characterised as a science, which was based on the biology of heredity that had made considerable progress since the discovery of the laws of classical genetics. In his evaluation of this progress, Kraulis was overly optimistic: "Human hereditary diseases already are so well known that it is possible to wage a practical battle against them. The laws of heredity are the same everywhere: they are the same for midges, snapdragons, and human beings. ... Circumstances require us to immediately care about human hereditary characteristics and increase the birth rate in civilised countries".[46] Kraulis viewed the main objectives of Latvian eugenics from a moderate perspective: intervening early to halt the drop in the birth rate and improving the hereditary qualities of children. Coming out against the widespread use of negative eugenics measures, he cited advanced cultural, historical, social and demographic arguments associated with the survival of a "small nation": traditions; a population with a declining proportion of ethnic Latvians; being surrounded by powerful neighbours; and losing one-third of the population as a result of World War I. Kraulis summarised his aims: "Therefore, eugenics in Latvia must follow a different pathway. We cannot simply copy the laws of other nations in the area of eugenics".[47] The first country he named was Germany, implying that its sterilisation law was radical.[48] It is interesting that in the same collection of papers, the director of the Department of Public Health, Oskars Alks (who succeeded A. Peterson), invoked similar arguments in favour of a special pathway for Latvian eugenics. Alks was also president of the Eugenics Commission and assumed a key role along with V. Kraulis in forming the eugenics section of the Medical Law of 1937.[49] This proves that the eugenics philosophy regarding the survival of the "small nation" was shared by representatives of the local medical elite during the 1930s and later, regardless of their eugenics or political orientation (e.g. "liberal" eugenicist Kraulis and "hard-line" eugenicist, the neurologist T. Upners).[50] Upners taught a course on eugenics at the Latvian University and was a member of the Eugenics Commission after Kraulis' dismissal under the Nazi occupation.

Kraulis considered the four most important legislative features of this special pathway. First was the establishing of eugenics commissions. In organising the work of the commissions, Kraulis viewed this as something new that distinguished Latvian eugenics legislation from the laws of other countries. Kraulis focused primarily on concentrating in a single institution (e.g. a eugenics commission) issues related to the termination of pregnancy (which was not rare in eugenics matters) and sterilisation. Kraulis did not comment on the significant differences between the administrative activities of such a commission and decisions by German eugenics courts, which possessed legal force. However, it was expected, considering the voluntary nature of sterilisation, that the number of applicants for surgery would not be large. Hence, a purely "sterilisation" commission would be only occasionally used and have little effect.[51]

Second was the ability to permit abortion and sterilisation based on medical indications in the case of acquired diseases such as syphilis in the case of its threat to posterity. On this point, there was agreement that only cases of active syphilis symptoms would be under consideration, since "treated syphilis poses no threat to posterity". The criteria for curing were considered to be the absence of active luetic symptoms, confirmed by a negative Wassermann blood test, as well as other negative tests, such as Lange's colloidal gold test from a spinal puncture. At the same time, Kraulis acknowledged the possibility of the appearance of signs of neurosyphilis among offspring even with these measures, identifying, in agreement with German venereologist Georg Loewenstein, the proportion of these newborns as 16%.[52] This, in his eyes, justified the decision to abort. Kraulis also justified sterilisation based on purely medical indications such as severe tuberculosis or tumours, in which case multiple births might threaten the life of a woman in labour. By way of contrast, Sweden's law with its "social engineering", and Nazi Germany's law, Latvia's law provided abortion and sterilisation based on individual free choice. This fact was evaluated by Kraulis positively from a moderate perspective.

The third essential feature of Latvian eugenic law, according to Kraulis, was the absence of a list of specific hereditary diseases as a basis for sterilisation. The law left the resolving of these issues in specific cases to the discretion of expert physicians. This situation once again radically distinguished Latvia's law from Nazi Germany's, with its specified nine conditions denotated as hereditary. Kraulis saw the advantages of such a non-inclusive position, since scientific research in the area of hereditary diseases was far from complete. Therefore it would be inappropriate to draft legislation each time new indications for sterilisation emerged. This clearly liberal attitude toward applying the law drew Latvia's eugenics law close to the most liberal eugenics laws of the time

Fourth, bearing in mind the incompleteness of scientific research on hereditary diseases, Kraulis recognised the principle of voluntary procedures with its ambiguity as the most appropriate. This ultimately differentiated Latvia's eugenics law from the harsh biological determinism and paternalism of Germany's "sterilization law" in that Latvia formally preserved respect for principle of voluntariness. At the same time, as in other European countries, Kraulis considered the issue of possibly sterilising the mentally disabled as being the most pressing. Referring to data from genetic studies of twins and from Kraulis' own research data (see below), he believed these data proved that the paramount influence on severe mental disability was genetic factors, while minimising the impact of external factors. In this regard, he insisted that mental disability was incurable and that the mentally disabled themselves "multiply several times more intensively than healthy people".[53] This notwithstanding, the massive application of radical eugenics measures appeared unacceptable to Kraulis. To resolve the issue of (highly desirable but still voluntary) sterilisation, it was necessary, in his opinion, to attract special attention particularly among the general public and especially among physicians, teachers and parents to the problem of voluntary consent for this operation. If it was not possible to obtain consent for sterilisation from the patients themselves, permission would have to be provided by their legal representatives in cases of incompetence due to mental illness. Kraulis referred to the positive experience in Switzerland in this regard.[54] As was known (with the exception of the canton of Vaud, which in 1928 became the first in Europe to legislate legal standards for compulsory sterilisation related to chronic mental illness and mental disability), in other Swiss cantons sterilisation legislation formally did not exist (as differentiated from "soft" sterilisation practice, e.g. in the canton of Zürich. This practice has not been proven to exist in Latvia in the 1930s).[55] This said, Kraulis made no mention of eugenics laws in northern European countries, with which Latvian sterilisation law also had much in common, and which at times was less harsh (e.g. the Danish Law of 1934, which provided for sterilisation of the mentally disabled).[56]

The principle of voluntary sterilisation, firmly established by Kraulis, was as previously by Buduls, based on classical genetics. Kraulis cited estimates indicating that only 10-20% of children suffer from the same hereditary disease as their parent. The sterilisation of such parents would deprive five to nine children who would be born healthy of the right to life: "With a calm heart we can sterilise patients with hereditary diseases only when healthy families will be able to multiple more intensively".[57] In anticipation of this, Kraulis thought that one should not sit idly by. Therefore he proposed a system of pro-natalist measures: marrying at an earlier age, improving social conditions that in turn would improve the psychological makeup of healthy families, a system of taxes and payment of child welfare benefits, and financial support for families with many children. Similar suggestions were made by other representatives of the Latvian eugenics.[58] However, Kraulis viewed voluntary sterilisation of at least one parent as a preliminary condition for relief measures for those suffering from hereditary diseases and for handicapped families. He cited a system of positive eugenics measures adopted in Germany in 1933 that resulted in a significant increase in the birth rate as an example of the success of positive eugenics. Nevertheless, Kraulis considered Nazi Germany's pro-natalist experience as entirely unacceptable in view of the structure of the population of Latvia. Local eugenicists were concerned about the comparatively large increase in the number of ethnic minorities.[59] However, no special eugenics actions were taken against minorities in independent Latvia.[60] Later, in 1939, when speaking about ethnicity and mental illness, Kraulis noted that diseases of an endogenous nature occurred among people of Jewish and German ethnicity five to six times more frequently as compared to the proportion of these ethnic groups in the general population. Regarding those suffering from such psychoses, Kraulis did not propose implementing negative eugenics measures similar to those implemented by German radical eugenicists.[61]

It is important to trace Kraulis' attitude on applying measures of negative eugenics to the "congenitally feeble-minded". In 1930 Kraulis presented a report on "Criminality and Heredity" for the Society of Neurologists and Psychiatrists. The text has not been preserved, but it is possible that he did discuss the problem of asocial behaviour of the mentally disabled.[62] At the end of the 1930s, Kraulis returned to the issue of studying the mentally disabled within the framework of the research project completed by the Institute for the Research in National Vitality's eugenics section. He was helped in his work by medical student Rudolfs Krūmiņš (1904-1982), who later became a Latvian psychiatrist.[63] Kraulis referred to data from his genealogical analysis of 62 mentally disabled people from the Riga Home for Handicapped Children, who came from various areas of Latvia, as well as members of their families.[64] The results of this study, which according to the researcher reflected a representative sample, allowed the calculation of the percentage of people suffering from varying degrees of severity of mental disability in the population. Among the probands, 51% suffered from severe "feeble-mindedness" (*idiocy*), 41% from moderate one (*imbecility*), and 8% from mild one (*debility*). Only 22% of probands were found to have external causes of mental disability, and 78% were determined to be endogenous. Parental alcoholism, syphilis, tuberculosis and nutritional deficiency were mentioned among the main exogenous causes. Members of 41 proband families were surveyed, totalling about 1,200 people living in various areas of Latvia. Quantitatively, families of the mentally disabled had twice as many children as healthy families, and often both parents were mentally disabled. Among the parents, only 38.3% were mentally healthy, and of those who were mentally unhealthy, 21.4% were also found to have "feeble-mindedness". Kraulis' eclectic classification of the other nearly two-thirds of unhealthy people from the parental group is typical for biases toward "feeble-minded": only 1.9% were marked as mentally ill, and others were classified as "socially degraded drunkards", "psychopaths" or accentuated persons, and criminals.[65] A significant prevalence of mental disability found among probands' siblings – 34% of them were "somewhat abnormal individuals", of whom 22% were persons with "feeble-mindedness": 10% severe, 20% moderate, 45% mild, and 25% borderline (*mazapdāvinātie*).[66]

The results seemed to support the views of the world's eugenicists toward the "feeble-minded", justifying mandatory sterilisation. However, the study allowed Kraulis to elaborate more differentiated indications for sterilisation of the mentally disabled. This was more needed for people with moderate mental disability, given the obvious biological origins and the grossly inappropriate behaviour, including promiscuity. Sterilisation, in his opinion, was not as urgent for those with severe mental disabilities, as the majority of them were kept in special institutions and were unable to have children. With great caution, Kraulis recommended sterilisation for people with developmental disorders (the mentally disabled), as well as those with mild mental disability (debility). For the latter, he specified particular cases in which children from healthy parents (perhaps as carriers of a recessive gene of a disease) or in which there was information about an exogenous cause of mental deficiency (e.g. clearly expressed external factors in its genesis) and social causes. In this case, sterilisation was considered undesirable because healthy children still could be produced. At the same time, in the spirit of eugenics legislation that had been adopted in Latvia, even in these cases sterilisation required the expressed consent of the person to be sterilised. If someone did not agree to the operation, according to the law, the decision could be made by the person's guardian or trustee. This formal voluntary arrangement (strictly observed by experts of the eugenics commission in the case of sterilisation, at least in independent Latvia),[67] of course differed greatly with the letter and spirit of Nazi Germany's sterilisation law, and drew Latvia's eugenics law closer to most liberal laws of this type in Protestant countries in Europe.[68] The total of 63 people who were surgically sterilised in Latvia in 1938-1939 contrasted hugely not only with Germany's statistics during this period but also with Scandinavia.[69] However, in the opinion of a "liberal eugenicist" as Kraulis appears to have been, it was not difficult to obtain consent from "imbeciles", since they were "easily persuaded".[70] Kraulis addressed his appeal to parents, teachers, physicians and those representatives of state and local authorities for whom such "imbeciles" were, in the first hand, a public nuisance. In his earlier appeal to physicians and priests he expressed hope that the efforts to persuade the "hereditary mentally ill" to agree to surgery would help society to "once liberate itself from individuals of little value".[71] His position demonstrates how delicate the balance was between free choice and coercion in sterilisation issues in Latvia during the 1930s.

3. Conclusions

The development of eugenics in Latvia, including the role of psychiatry, is still insufficiently understood.. The impact of physicians, and especially key psychiatrists such as Hermanis Buduls and Verners Kraulis, was significant – both in terms of forming Latvian eugenics priorities (including pro-natalist legal issues), and the institutionalisation of Latvian eugenics in its alliance with Mendelian genetics. The pro-natalistic orientation, the cautious attitude toward "hereditary" indications for applying negative eugenics measures, the lack of individualistic and social indications for abortion and sterilisation, and ethical standards regarding the expression of will by even seriously ill individuals in matters of sterilisation, were important parts of local eugenics, one of the poles of its continuum. Latvian eugenics also provided further features (sometimes expressed by the same people, presumably modest eugenicists) concerning the amelioration of the nation's demographic, social and economic standards on account of certain groups of people, especially the mentally disabled and "asocial" – either for the best of the nation's biological make-up or for the best of the "unfit". The influence exerted throughout Europe by Germany's sterilisation law of 1933 was, in the case of Latvia, early and at times indirect (through the 1933 Danzig eugenics law), and with very modest results with regard to assimilation of its norms by the Latvian eugenics legislation. Latvia's law was more influenced by the ideas of modest eugenics in some Swiss cantons than by the ideas of forced sterilisation of the "unfit" and the Nazi German model of practical implementation. The impact of coercive negative eugenics measures such as sterilisation in Latvia's eugenics law was insignificant due to the distinct establishment of a voluntary arrangement and the absence of a list of mandatory eugenic indications. In conjunction with the administrative nature of sterilisation decisions made by the eugenics committee, local eugenics legislation contrasted sharply with Germany and, less dramatically, with the legislation of neighbouring democratic countries in northern Europe. Like applied eugenics in other European states, the practical implementation of the local eugenics law was not free from divergences between stated intentions and practices. Interestingly, to support their liberal eugenics positions, Latvian psychiatrists widely used the statistical data of German psychiatrists who also supported the compulsory sterilisation programme.

Notes

[1] Rikshe, 1978, 139-142.
[2] Felder, 2005; Felder, 2009.
[3] Zelče, 2006, 103-104.
[4] Stranga, 2008.
[5] Vīksna, 2002.
[6] Vīksna, 1990.
[7] Vīksne, 2007, 323-350.
[8] For his biography see: Vīksna, 2002; Felder, 2009, 283.
[9] Buduls, 1909.
[10] Ibid., 72.
[11] Ibid., 150-151.
[12] Upners, 1943, 52.
[13] Kraulis, 1928, 78-83; Buduls, 1926; Fērmanis, 1935, 3-8.
[14] Broberg and Roll-Hansen, 2005.
[15] Reinhards, 1940. To name only a few newspaper articles: "Igaunijā pieņemts sterilizācijas likums [Sterilization law is accepted in Estonia]" *Jaunākās ziņas*, no. 271 (28 November 1936), 7; "Igaunija grib sterilizēt 1000 cilvēku [Estonia wants to sterilise]" *Jaunākās ziņas*, no. 276 (4 December 1936), 7; "Igaunijā stājas spēkā sterilizācijas likums [Sterilisation law comes into effect in Estonia]" *Jaunākās Ziņas*, no. 71 (1 April 1937), 10.
[16] Bleuler, 1937; Schmuhl, 2008; Friedlander, 1995.
[17] Zelče, 2006, 103-104.
[18] Veidemanis, 1935, 571-578.
[19] Ibid., 573-575.
[20] Fērmanis, 1935, 3-8.
[21] Ibid., 4.
[22] Ibid., 7.

[23] Ibid., 7.
[24] Documentation of the Riga Sarkankalns hospital, 20 December 1933: Latvijas Valsts Arhīvs (Latvian State Archive – further: LVA), 2917/1/6, 249.
[25] Ibid., 251-255.
[26] Ibid., 248.
[27] Bumke, 1926, 108-109.
[28] Latvian Physicians Society meetings: P. Stradiņa medicīnas vēstures muzeja arhīvs, Laksberga fonds (P. Stradina Museum of History of Medicine, Laksbergs' fund – further: PMVMA-LF), 1/0/47,18;1/0/48,22. The fund preserves only the names and dates of Bumke's lectures in 1933, and the dates of the second course of lectures in 1934.
[29] An advertisement in: *Latvijas Kareivis*, no. 65 (21 March 1931), 3. The content of lecture is not known.
[30] Buduls, 1936, 4.
[31] Ibid.
[32] Ibid.
[33] Kraulis, 1928, 80.
[34] Buduls, 1936.
[35] Ibid.
[36] Buduls, 1954, 210.
[37] Buduls, 1939, 5.
[38] Buduls, 1925; "Ueber die Auswirkung der Paralysebehandlung mit Malaria" *Allgemeine Zeitschrift für Psychiatrie und psychisch-gerichtliche Medicin*, vol. 104 (1936), 168-176.
[39] Buduls, 1939.
[40] Vīksne, 2007, 323-350.
[41] For his biography see: Felder, 2009, 284-287.
[42] Personal file of docent Verners Kraulis, of Latvian University: LVA, 7427/13/862.
[43] "Liepājas psihiatriskas slimnīcas traģēdija": PMVMA-LF, 1/0/39, 40.
[44] Kraulis, 1939a, 161-168.
[45] Latvian Physician Society meetings: PMVMA-LF, 1/0/47, 56.
[46] Kraulis, 1939a, 161.
[47] Ibid., see also for commentary of this important statement in: Felder, 2009, 287-288.
[48] Kraulis, 1939a, 161.
[49] Zelče, 2006, 102.
[50] Felder, 2009, 287-288, 296-297.
[51] Kraulis, 1939a, 162. About Kraulis' evaluation of the Eugenic Legislation 1937 see: Zelče, 2006, 102.
[52] Kraulis, 1939a, 162.

[53] Ibid., 163.
[54] Ibid.
[55] Roelcke, 2002, 1027.
[56] Hansen, 2005, 9-76.
[57] Kraulis, 1939a, 165.
[58] Zelče, 2006, 96-98.
[59] Zelče, 2006, 96; Felder, 2009, 287.
[60] Felder, 2009, 288.
[61] Kraulis, 1939b, 73-79.
[62] Kraulis, 1932, 228.
[63] Kraulis, 1939c, 49-72.
[64] Ibid., 55-56.
[65] Ibid., 65.
[66] Ibid., 62-63.
[67] Felder, 2009, 285-286.
[68] Roll-Hansen, 2005a, 151-94; Roelcke, 2002, 1024-1025.
[69] Eugenics Commission correspondence, 1938-1940: Latvian Valsts Arhivs (Latvian State Archive – further: LVA), 4578/1/202; "Pārskats par Veselības departamenta darbību 1939.g [Overview on the activity of the Healt department]" *Tautas labklājības ministrijas mēnešraksts*, no. 3, 1940, 217-228; also: Zelče, 2006, 104-105; Felder, 2009, 286; Roll-Hansen, 2005b, 195-271.
[70] Kraulis, 1939c, 71.
[71] Kraulis, 1938, 36.

References

Bleuler, E. (1937), *Lehrbuch der Psychiatrie*. Sechste Auflage. Berlin: Julius Springer.

Broberg, G. and Roll-Hansen, N. (eds.) (2005), *Eugenics and the Welfare State. Sterilization Policy in Denmark, Sweden, Norway, and Finland*. East-Lansing, Mich.: Michigan State University Press.

Buduls, H. (1909), *Laulība un cilvēka dzīves mērķis. Bioloģisks un ētisks apcerējums* [Marriage and the Goal of Human Life. Biological and Ethical Essay]. Rīga, Sirģēļa apg.

Buduls, H. (1925), "Par progresīvas paralizes ārstēšanu ar malariju [About general paresis treatment with malaria]" in: *I. Latvijas un ārstu un Zobārstu Kongresa darbi. 11.-13.Sept.1925*. Rīga: Kongresa izpildu komitejas izdevums.

Buduls, H. (1926), "Par pašnāvībām Latvijā [About suicide in Latvia]" *Jaunākās Ziņas*, no. 70 (27 March), 1.

Buduls, H. (1936), "Rases labiedzimtība [Eugenics of Race]" *Jaunākās Ziņas*, no. 125 (6 June), 4.

Buduls, H. (1939), "Vai garīgās slimības iet plašumā [Do mental diseases spread?]" *Jaunākās Ziņas*, no. 58 (11 March), 5.

Buduls, H. (1954), *Cilvēks dzīves spogulī* [Men in the mirror]. Stokholma: Daugava.

Bumke, O. (1926), *Kul'tura i virozdenije* [Culture and Degeneration]. Moskva: Sabashnikovi.

Felder, B. (2005), "'Mazvērtīgo samazināšana' – eigēnika Latvijā [Reducing the Inferior – Eugenics in Latvia]", *Kultūras Diena*, no. 28 (23 April), 16-17.

Felder, B. (2009), *Lettland im Zweiten Weltkrieg. Zwischen sowjetischen und deutschen Besatzern 1940-1946*. Paderborn: F. Schoeningh.

Fērmanis, E. (1935), "Eugenikas tagadējas problēmas [Eugenics and Problems of our Time]" *Universitas*, vol 15, no. 1, 3-8.

Friedlander, H. (1995), *The origins of Nazi Genocide: from Euthanasia to the Final Solution*. Chapel Hill: University of North Carolina Press.

Hansen, S. (2005), "Something Rotten in the State of Denmark: Eugenics and the Ascent of the Welfare State" in: Broberg, G. and Roll-Hansen, N. (eds.), *Eugenics and the Welfare State. Sterilization policy in Denmark, Sweden, Norway, and Finland*. East-Lansing, Mich.: Michigan State University Press, 9-76.

Kraulis, V. (1928), "Kultura un degenerācija [Culture and Degeneration]" *Tautas veselība*, vol. 3, no. 3, 78-83.

Kraulis, V. (1932), *On Hysterical Reaction and it Constitutional Basis. Doctoral thesis*, Rīga.

Kraulis, V. (1938), "Ne kultūra, bet aprēķins mazina tautas dzīvo spēku. Pārrunām par tautas dzīvo spēku [Not culture but calculation diminish a people's life strength. To the discussion about National Vitality]" *Brīvā Zeme*, no.125 (4 June), 36.

Kraulis, V. (1939a), "Eigenikas ceļi Lavijā [The Way of Eugenics in Latvia]" *Tautas labklājības ministrijas mēnešraksts*, no. 3,161-168.

Kraulis, V. (1939b), "Tautība un gara slimības [Nationality and metnal illneses]" in: *Tautas dzīves spēka pētīšanas institūta raksti*, vol.1, Rīga: Veselības veicināšanas biedrības, Tautas dzīves spēka pētīšanas institūta izdevums, 73-79.

Kraulis, V. (1939c), "Par iedzimto plānprātību Latvijā [On hereditary Feeble-mindedness in Latvia]" in: *Tautas dzīves spēka pētīšanas institūta raksti*, vol. 1, Rīga: Veselības veicināšanas biedrības, Tautas dzīves spēka pētīšanas institūta izdevums, 49-72.

Reinhards, G. (1940), *Eugenika. Latviajs universitatā 1935.un 1936. gadā visu fakultāšu studentiem lasītas lekcijas* [Eugenics. Lectures read to students of all faculties at the University of Latvia in the years 1935-36]. Rīga: Valters un Rapa.

Rikshe, V. J. (1978), ""Rassovaja gigyena" na slu'zhbe latishskoy natsionalisticheskoy bur'zhuazii ["Racial hygiene" in the Service of the Latvian Nationalist Bourgeoisie]" in: *Medico-socialnye izsledovanya*, Riga, 139-142.

Roelcke, V. (2002), "Zeitgeist und Erbgesundheitsgesetzgebung im Europa der 1930er Jahre. Eugenik, Genetik und Politik im historischen Kontext" *Nervenarzt*, vol. 73, 1019-1030.

Roll-Hansen, N. (2005a), "Norwegian Eugenics: Sterilization as Social Reform" in: Broberg, G. and Roll-Hansen, N. (eds.), *Eugenics and the Welfare State. Sterilization policy in Denmark, Sweden, Norway, and Finland.* East-Lansing, Mich.: Michigan State University Press, 151-194.

Roll-Hansen, N. (2005b), "Scandinavian Eugenics in the International Context" in: Broberg, G. and Roll-Hansen, N. (eds.), *Eugenics and the Welfare State. Sterilization policy in Denmark, Sweden, Norway, and Finland.* East-Lansing, Mich.: Michigan State University Press, 195-271.

Schmuhl, H. W. (2008), *The Kaiser Wilhelm Institute for Anthropology, Human Heredity, and Eugenics, 1927-1945. Crossing boundaries*. Berlin: Springer.

Stranga, A. (2008), *Ebreji Baltijā. No ienākšanas pirmsākumiem līdz holokaustam. 14. gadsimts – 1945 gads* [Jews in the Baltics. From the Beginning of the Arrival to the Holocaust]. Riga: Latvijas Vēsture.

Upners, T. (1943), *Eugenikas nozīme tautas un valsts dzīvē* [The Meaning of Eugenics for the Life of the Nation and The State]. Rīga: Latvju grāmata.

Veidemanis, M. (1935), "Latvijas 1935.g. 22. marta aborta likums [The Lativan abortion law of 22 march 1935]" *Latvijas ārsts*, no. X-XI, 571-578.

Vīksna, A. (1990), *Pa ārstu takām* [On the Traces of the Physicians]. Rīga: Avots.

Vīksna, A. (ed.) (2002), *Profesors Hermanis Buduls. Biobibliogrāfiskais rādītājs*. Rīga: Medicīnas zinatniska biblioteka.

Vīksne, R. (2007), "Garīgi slimo iznicināšana Latvijā nacistiskās okupācijas laikā [The elimination of the mentally ill in Latvia under Nazi occupation]" in: Caune, A (ed.), *Holokausta izpētes jautājumi Latvijā. The Issues of the Holocaust Research in Latvia*, (Symposium of the Commission of the Historians of Latvia), Rīga: Latvijas vēstures institūts, 323-350.

Zelče, V. (2006), "Vara, zinātne, veselība un cilvēki: Eigēnika, Latvijā 20.gs.30.gados [Power, Science, Health and Men: Eugenics in Latvia]" *Latvijas Arhīvi*, vol. 3, 94-134.

"Over-Latvianisation in Heaven" - Attitudes towards Contraception and Abortion in Latvia 1918-1940

Ineta Lipša

Abstract
This chapter shows the complex process of formation of societal attitudes towards contraception and abortions over 20 years during the independent Republic of Latvia (1918-1940), putting emphasis on the influence of existing public rhetoric on popular opinion, the availability and accessibility of sexual knowledge. Knowledge-based argumentation aimed to change the policy of sexual and reproductive matters and the reversing of the legal achievements towards legalization of abortions caused by changing of the state political system from parliamentary democracy to an authoritarian regime with its emphasis on pro-natalist policy.

There is vast and solid historiography about the abortion issue in Europe. The subject has been studied by historians from the field of the history of sexuality.[1] Problems of abortion have been examined through a cultural approach[2] that seeks "to make visible the web of meanings attached to terminations of unwanted pregnancies in everyday life"[3], as well as by analysing the representation of abortion in feature films, plays and novels. Kate Fisher has used an approach of oral history in her research.[4] This discloses cultures of contraceptive practice that argue against the earlier assumption that women were the driving force behind the radical decline in family size in Britain. Her research suggests that "at least between the 1920s and the 1950s, it was husbands, not wives, who rooted out birth control information, framed contraceptive strategies for the family, and put these into practice".[5] Regardless of class and regional differences, both husband and wife saw contraception as a man's duty.

The policy on abortions in the Republic of Latvia during this period was influenced not only by the ideology of other European countries, but also by neighbouring Soviet Russia, the only European state where abortion was fully legalised from 1920 to 1936. It has to be taken into account that a faction of communists was seated in the Parliament of Latvia from 1928 to 1934. In contemporary Russia, abortion history has been studied as part of Soviet social policy focusing on ideology and everyday life.[6] The abortion politics of the 1920s-1930s has been explored as a mirror of Soviet social care[7], as well as attempts by early Soviet ideology to regulate social policy in the sphere of sexuality.[8]

The situation is different in the Baltic States, where the scientific debate on the history of abortion has not yet taken place. This is because research into social history of the 20th century, inter-war period in Estonia, Lithuania and Latvia started mainly after the end of the Soviet era and the reestablishment of independence in 1991. Moreover, the historiography of Latvia since then has been focused mainly on issues of political history dealing with the political and military history of the Republic of Latvia during the inter-war period, and with the problems caused by World War II and the Nazi and Soviet occupations. So the history of abortion during the 1920s-1930s, as well as the attitude towards birth control, has not yet been presented in any monograph in the historiography of Latvia, nor of Estonia and Lithuania, though there has been a continuous focus on research into these questions. The issue of abortion was analysed in the author's doctoral dissertation about the problems of public morality in Latvia from 1918 to 1940.[9] The debate on the legalisation of abortion at the end of the 1920s and beginning of the 1930s has also been discussed briefly in the historiography of Latvia.[10] Additionally, the problem of abortion from 1938 to 1940 has been highlighted in publications focusing mainly on eugenics. These articles have been written by historians Björn M. Felder[11] and Vita Zelče[12], both of whom focus on the analyses of activities inspired by the Institute for Research of National Vitality (*Tautas dzīvā spēka pētīšanas institūts*) from 1938 to 1940. The publication of Zelče could even be viewed as a reaction to Felder's popular science article in *Diena*, at that time the most influential national newspaper in Latvia. Historian Aivars Stranga valued Felder's conclusions as innovative, though he pointed out that the article exaggerated the significance of biological nationalism in Latvia during the rule of the authoritarian regime (1934-1940).[13]

The situation with the historiography of Estonia and Lithuania is similar. Estonian historian Ken Kalling has analysed the public debate on abortion that took place in Estonia in the 1920s.[14] His research also deals with eugenics.[15] Lithuanian ethnologist Rasa Paukštytė-Šaknienė has focused her studies on the attitude toward abortion in the Lithuanian region of Aukstaitija, as well as aspects of reproduction in the culture of the Lithuanian countryside. She has also examined the historiographical approach to reproduction and its control.[16]

The aim of this article is to analyse the social attitude toward contraception and abortion in Latvia during the 1920s and 1930s. The structure is focused on the following main research issues: (1) the influence of existing public rhetoric on popular opinion; (2) the availability and accessibility of sexual knowledge; (3) knowledge-based argumentation of people in public debate aimed to change the policy of sexual and reproductive matters; and (4) the change of the state political system to an authoritarian regime (1934-1940), which reversed the legal achievements towards liberalisation of the attitude toward abortion during the period of parliamentary democracy (1918-1934).

1. "Angel makers": The Prevailing Rhetoric

This article considers the structuralist idea that language does "not only express but constitute[s] social reality" and "shape[s] the political and social consciousness of the classes that speak it and hear it".[17] Language is not neutral, as it produces meaning. Metaphors used in Latvia during the 1920s and 1930s constitute an important historical context that could help understand the peculiar significance of the issue of sexual matters in Latvian society during this period. It is already acknowledged that "metaphorical clothing of moral codes, taboos, prohibitions and value systems" links "discipline to desires, politeness to policing".[18] Therefore, analysis of metaphors of birth regulation reflects not only facts but also once-existent emotions in the research.

It is acknowledged that the birth rate in Europe has been declining since the 1870s. In the Republic of Latvia, the particular importance of this problem was recognised after World War I, when Latvia declared its independence on 18 November 1918. Various types of intellectuals and professionals[19] stressed that Latvians for the first time in history had created their own state, and therefore the nation of Latvia shall prove that the new state deserves to stay strong among other states, which could be achieved by demonstrating the nation's best qualities. The moral condition of the population of Latvia was considered the criterion for the status of national vitality and was characterised by the publicly expressed negative valuation of unacceptable social behaviour. Because concerns about the quantity and quality of the population were discussed from time to time in periodicals, abortion became a focus of public anxiety.

The gender disproportion caused by World War I, particularly in the 20-29 age group, was perceived as a disturbing problem during the entire inter-war period in Latvia.[20] The birth rate was considered low and the number of divorces alarming. Therefore, until the mid-1930s, the issue of single men, and afterward the issue of single individuals of both genders, were constant subjects of public concern. Simultaneously, this generated debates about the necessity of imposing limits on unmarried people and divorces, emphasising the immorality of separating sex from reproduction. In 1928, physician Eduards Akermanis named the "super-modern spirit of today" as the most essential cause of the decreasing birth rate, which has set apart the human "aspiration for reproduction" from the consequences of the irresistibility of instincts and allowed people to choose their preferences freely.[21] Having analysed the birth rate data of 1926, he found that Lutherans accounted for about 3.4 times more marriages than Catholics, but they had only 20% of children born in 1926, while about 50% of that year's babies were born to Catholic families. Akermanis made the incisive comment that "a huge mountain gave birth to a small mouse" and concluded that sex life had come to the end in itself, facilitated by scientific achievements in the development of contraceptives.[22]

The debate on abortion in Latvia's public sphere in the 1920s-1930s was associated with so-called "sexual love", using the parlance of that time. It reflected the attempts to value the separation of sex from reproduction that already was the reality of everyday life of the people. The formulation "sexual love" identified a sexual relationship in which the aim was mere delight and not the intention to give birth to descendants, which was demanded by public opinion. Experts appealed to people to reject prejudices connected with sexual life[23], in light of the fact that "sexual desire is associated with happiness of individual persons, fate and future of the nation and its strength and endurance"[24] – hence the nation's reproduction.[25] Nevertheless, the clergy and politicians from Christian parties emphasised that the actual aim of marriage was procreation. So-called "pseudo-wives" were considered to be the most detrimental phenomenon, because they avoided the duty of child-bearing and thus brought the danger of extinction to the entire nation.[26] Those who formed public opinion condemned childless parents. Childless marriages were labelled as embodying principles that were dangerous and detrimental for the future, and for the life and vigour of the nation, and partners in such marriages were called co-facilitators of the destruction of the life of the Latvian people.[27] Delight was deemed to be the most detrimental tenet, according to which a childless marriage was qualified as being false, as well as a surrogate marriage or a union of two people for mere amusement[28] – or the most convenient form of prostitution.[29] The so-called illegal partners of a conjugal life were compared to evil rust over the cornfield of people's souls[30], and the propagators of free love were called the enemies of Latvia.[31] However, this rhetoric indicates only what was desired instead of what had actually been accomplished.

At the beginning of the 1930s, Lutheran pastor Jānis Kundziņš declared that "before the First World War mankind had God, but now they have only delight and joy"; and that "earlier they had sought for love and beauty in heart and soul, but now only in sexual delight and passion".[32] In 1932, Berta Pīpiņa, a leader of the women's movement who was elected to the 4th Saeima (1931-1934) as the only woman, acknowledged that the flesh had become omnipotent in its desire for delight, that people had lost the sense of proportion in changing their partners[33], and that "due to lust, even old men lose their minds and get married to young girls".[34] However, in 1934 Pīpiņa suggested that wives should bear in mind that their husbands needed not only a spouse, cook and babysitter, but also a friend and a lover, since this was the possible way of averting the "disintegration of a Latvian family".[35] Thus, the importance of sexuality in a married couple's life, and sexuality as a norm within the family, was emphasised. This was one of the two most important developments of the inter-war period.[36]

The liberal views existing in society[37], although not reflected directly in the press, still influenced the conservative views that were dominant in the press – some women admitted that it was necessary to integrate them into marriage. Life itself urged people to acknowledge the value of sexual activity. However, in periodicals this idea was short-lived, since two weeks after the speech by Pīpiņa on 15 May 1934, Prime Minister Kārlis Ulmanis undertook a *coup d'état* and established an authoritarian regime whose pro-natalist ideology supported reproduction. The press had to grow accustomed to these new requirements, since before the coup it had emphasised the priority of the nation's future interests.

During the inter-war period, condemnation of sexual love in the periodicals was prominent in the name of nation's future interests. In this regard, abortions were judged as a hazardous endangerment of the nation's strength and national independence, and the avoidance of giving birth as a sign of the nation's weakness.[38] In 1925 physician Aleksandrs Baidiņš even asserted that prostitutes and women having abortions should be treated equally as negative, since both commit sins against their own nation and were partly responsible for its extinction.[39]

When sex life-related issues were discussed, language rich in euphemisms and metaphors was used. "To make angels" meant to have an abortion, women who had abortions were called "angel makers",[40] and "over-Latvianisation in heaven" meant an increase in the abortion rate.[41] The most popular symbol of the birth rate problem was the stork. Legalisation of abortion due to social conditions was depicted as a crying stork with his head cut off[42], while repealing it was represented as setting the bird free from a prison cell.[43] Slow growth in the birth rate was represented by a stork with its beak wrapped in a condom wearing striped prisoner slippers[44], or by a message that the bird had become unemployed[45].

However, social attitudes toward birth control gradually relaxed. Although this process was disapprovingly defined as the "contemporary super modern spirit" by physician E. Akermanis in 1928, that spirit was not exactly super-modern in Latvia because it was not acknowledged as valued in the periodicals or the literature.

2. "An Opportunity to avoid the Stork's Visit": Availability of Sexual Knowledge

Attitudes toward contraception and abortion can be approached through studies of the process by which an understanding of sexuality is acquired by individuals. Historian Lesley A. Hall has argued that sex education in Britain was "about controlling and regulating, if not entirely preventing, sexual exploration and activity".[46] Her study provokes the question of whether sexual knowledge was freely available in Latvia or was cast "within the framework of a strategy for damage limitation", as it was, for example, in

Britain. The issue of sex education represents the evolving social control of sexuality and sexual behaviour that was gendered. Therefore, this chapter focuses on the availability of scientific films, lectures and literature on sex-related issues, as well as sex education in schools.

Along with the emphasis on the important role of sexuality in a married couple's life and sexuality as a norm in the family, the other of the two most important developments during the interwar years was a new emphasis on the need to scientifically explain sexuality. But members of the intelligentsia in Latvian periodicals were rather preoccupied with moralisation – a majority of them according to the tradition established by the Neo-Latvians (the nationalist element of the Latvian intelligentsia in the 1850s-1880s) who regarded their individual and public mission as promoting and strengthening the nation and giving preference to the interests of the nation over individual interests. Therefore, the task to educate people on the basis of science was supported by showing scientific films and organising public lectures on issues of conjugal life. It must be stressed that sex education was not focused on sex life but on conjugal life particularly, defining it as conjugal sex life.

There were some peculiarities. In the 1920s, these issues were discussed separately in male and female audiences, who listened to lectures or watched scientific films. In Liepāja in 1923, only women were invited to attend the lecture "Abortion and its consequences", whereas both men and women were invited to attend the lecture "Criminal abortion", the tickets for which were one-third cheaper.[47] Regardless, the attendance of the first lecture was higher.[48] Thus, from the position of the organisers, it was more advantageous to arrange such undertakings separately for the sexes, which indicates to a certain extent the uneasiness of listening to sex-related issues in the presence of people of the opposite sex. These issues included woman and marriage as viewed by public opinion, the attitude of men towards women, sex-related questions in the life of the woman and family, prostitution and the male, reasons for the disintegration of family life, and the kind of love demanded by men.[49] The formulation supposed that the type of love sought by men differed from the kind of love imagined by women. The tradition of organising lectures only for women was also observed during visits to Latvia by German-speaking lecturers. In 1928, Cologne writer Elza Forstmann spoke about the love life of a woman, sexual strength and restoration of life – only to women older than 18.[50] In 1929, Vienna physician Juta Gartner gave a lecture "only to women and adult girls about man's and woman's sexual life, beauty care, sport, pregnancy and terrible consequences of its prevention, about pregnancy hygiene".[51]

This was also practiced when scientific films were shown. The film *Venereal and Other Contagious Diseases* was shown to both sexes in Liepāja in 1923,[52] but it was screened separately in Rīga,[53] in Liepāja in 1925,[54] and in Jelgava in 1927.[55] The film *Abortion* was shown separately in Jelgava in November 1928, but it was shown to women and men together in December.[56] The film *Tragedy of Sex Life* was shown to both sexes in 1929, but the movie theatre was divided into zones – women in the balcony and men in the hall.[57]

The separation of sexes could reflect the view that it was impolite to discuss sex-related issues in public in the presence of the opposite sex. However, this could also be considered an effective publicity measure. Contemporaries themselves acknowledged that the closer the issue lies to intimate life, the wider is the approval – "reading for only one of the sexes is the best kind of publicity".[58] This aroused curiosity. A man dressed as a woman arrived at a lecture about women's "love weapons" in Riga in 1924.[59] He was told to leave the room by a policeman (and fined two lats or one day of imprisonment by the court). It was said that a few more men in disguise listened to the lecture until its conclusion. The separation of sexes also could have been due to the enduring 19th -century perception of a woman as a pure creature who at least publicly had to maintain the impression that she knew nothing about sex-related issues, as they were the prerogative of men. Nevertheless, real life in the post-war years had its own considerations. Therefore, separating cinema and lecture audiences by sexes could be viewed as a midway point for maintaining the image of a woman's purity while simultaneously providing her with necessary knowledge. It also reveals that the existing double moral standard increased the curiosity of men, especially a belief that sexual knowledge explained to women differs from what is presented to man. Such an attitude was indirectly supported by the wing of conservative feminism that was the strongest among Latvian women, emphasising the need to raise the prestige of family life. They accepted traditional sexual roles, but emphasised the necessity of the equal treatment.[60]

Taking into consideration the fact that during the interwar period only one-third of inhabitants of Latvia lived in cities and had access to cinema and public lectures, it is important to find out whether the literature on sexual issues was freely accessible. As it turns out, this accessibility was viewed as a problem because it was linked to concerns about the sexual education of youth. Sexual knowledge was regarded as a privilege of adults, so attempts were made to protect young people from "becoming adults" before their time, thus ensuring certain moral protectionism. The need to secure the future existence of the nation, which considered the new generation as its symbol, was an effective way to put restrictions in place. In the Irish Free State, the Evil Literature Committee was established in February 1926 (the Censorship of Publications Act was passed in 1929). Irish advocates of censorship, however, did not argue for the need to protect youth; rather, their concerns were directed to adults.[61] In Germany, the Law to Protect Youth from Trash and Dirt was passed in 1926.[62] In Latvia the respective regulation was adopted in 1927, with the quite similar title "Regulations to Protect Youth from Trash and Smut Literature" (*Noteikumi par jaunatnes pasargāšanu no sēnalu un neķītrību literatūras*).[63]

As a result, only until summer 1927 could any person freely buy literature about issues pertaining to sex life. Afterwards, some of this literature was included on the List of Trash and Smut Literature. Generally, there were three lists of restricted literature in Latvia: (1) the List of Forbidden Literature published abroad and brought to and distributed in Latvia (the first order was passed on 1 December 1924)[64]; (2) the List of Forbidden Literature published in Latvia[65] (first order passed on 23 October 1925); and (3) the List of Trash and Smut Literature[66] (first published on 20 July 1927). The latter was a list of books inappropriate for youth. This meant that shop owners were not allowed to stock books on this list in a well-accessible and visible place, and he could only sell them to adults (18 and older). These books were also not allowed to be kept in school and youth libraries. In this way, the ruling power attempted to prevent youth under 18 from buying books that adults had acknowledged as being destructive to them, while adults were allowed to buy such books but were not allowed to view them in a public space.

The discussion about sex-related issues in publications (abortion methods, pregnancy prevention, and prophylaxis of sexual diseases) led to these publications being restricted as obscene literature.[67] The primary target of the Censorship of Publications Act of Ireland was printed material related to birth control. Historian Senia Pašeta has concluded that "one of the main objections to birth-control literature was its frank discussion of sexuality".[68]

Advertising for the film *Abortion* in the Latvian press was characterised as an "actual blossom of pornography" in 1928.[69] Censors allowed the film to be shown at a movie theatre, though with the title, *Love Tragedy of Grand Dukes Michailovski*. An official of the Ministry of the Interior asked the Department of Press and Societies to make sure that this did not happen in the future, since "newly invented film titles not only mislead the public but mainly are meant for stimulation of its lower instincts".[70] Thus, interpretations of this kind attracted public attention. For its part, the Committee for Fighting Literature Harmful to Youth (*Jaunatnei kaitīgas literatūras apkarošanas komisija*) in 1928 considered the book *Marriage Technique* by Theodor Hendrick van de Welde "one of the most obscene books known in world literature".[71] The book by sexologist August Forel, *Practical Guide for a Married Couple: How to Avoid Conception (only for Adults)*, was included on the List of Trash and Smut Literature just after it was published in Latvian in 1932.[72] Both of these books were popular among youth, as memoirs suggest. Restrictions and prohibitions with the aim of eradicating so-called pornography maintained the belief in people that it was improper to discuss sex-related issues in public. However, they also attested to the fact that people were so interested in these issues that restricting this interest was only possible by prohibiting availability.

From 1927 to 1937, at least 33 popular science books were included on the list because they contained material linked to sex life issues. This number does not seem very large, as the list contained a total of 255 books, though this figure does not include fiction. The intense attention to this problem at the beginning of the 1930s could be explained by concern expressed by more active contemporaries about the sudden overflow of literature about contraception,[73] which drew the Minister of the Interior's attention[74] to the advertising of books that promised to inform readers about "an opportunity to avoid [the] stork's visit".[75]

Even the liberal newspaper *Jaunākās Ziņas* (The Latest News) expressed concern about the allegedly sudden overflow of literature on contraception on the book market in spring 1933.[76] However, a representative of the Ministry of the Interior pointed out that this was not correct, since the Committee for Fighting Literature Harmful to Youth was doing a competent job. A disabled veteran anonymously drew the Minister of the Interior's attention to an advertisement in the newspaper *Jaunākās Ziņas* about the 1933 book by Dr. E. Bock *What Must a Modern Woman Know*, which promised to inform readers about how to avoid pregnancy.[77] Due to the intimate nature of the book, the publisher did not plan to release it for open sale in bookshops, but instead to deliver in the mail. This seems to be evidence of the shyness of people to request such books in public. The disabled soldier "in the name of a number of former liberators" requested a prohibition on advertising such publications, indicating that they could facilitate the extinction of Latvians.[78] In 1933 another person sent the Minister of the Interior an advertisement published in the newspaper *Jaunākās Ziņas* about a booklet that discussed the contraception method "Antifrukt", which "offered an opportunity to avoid stork's visit and regulate the family growth at one's own will".[79] He asked whether the Department of Health had granted permission for this "means of national extinction" to be advertised. The department had not granted this permission. Regrettably, archival documents do not indicate if any consequences resulted.

However, not all books were only restricted. For example, the book *How to Avoid Pregnancy and Means of Prevention* by Jevgenija Dembskaja was prohibited only in 1938.[80] The implication is that for seven years the book which was published in 1931 was on open sale. The Department of Health stated the need to prohibit the book because of its harm to the state's interest of increasing the birth rate.[81] Police in Riga confiscated 269 copies of the book, one from the bookstore and the others from the publisher's storehouse.[82] Thus, even if the sale of this publication was not restricted or prohibited, they could not always be bought in bookstores in the second half of the 1930s. It seems that bookstore owners avoided accepting for sale literature that could be deemed questionable by authorities. Films were treated similarly. Several previously allowed movies were prohibited after the authoritarian regime took over in 1934 – including *Woman's Joys and Sorrows*, subtitled *Living Together of Both Sexes, Procreation, Abortion and its Consequences*, which had been allowed to be shown in 1930.[83]

In Latvia the debate about the effects of "smutty and trashy literature" on youth stimulated discussions about sex education in a similar way as in Germany.[84] In the second half of the 1920s, there were public discussions about the necessity of sex education in schools. In 1928 the discussions led to the inclusion of sex education in the secondary school curriculum, with the chapter "Hygiene of Woman" in the natural sciences syllabus envisaging materials about sexual issues, venereal diseases, pregnancy and abortion.[85] The attitude towards the advisability of public discussions of these issues changed at the beginning of the 1930s. Around 1923 the Director of the Jelgava First Classic State Grammar School, Jānis Lapiņš, organised among the last year high schools students a public opinion poll about issues of sex life, and he later presented the results at the Congress of Latvian Secondary School Teachers. Some residents of Jelgava complained about the poll and had Lapiņš' dismissed in 1930-1931. The following questions in the poll aroused indignation[86]: "Do thoughts about sex life bother you much? Then set your time in hours! Does training together with the opposite sex heat or chill down your sexual passion? What type of women do you like – passionate or tranquil ones?" The assertion was made in 1931 that such questions "destroy [the] nation's morality" and "tempt students to think about things that otherwise would not occur to them".[87] Critics of Lapiņš also claimed that such questions would lead pupils to make inappropriate conclusions, such as "it is time to start" their sexual life.

Restrictions on sexual knowledge, following the strict state policy, led to a situation in which information that people were able to obtain was often ambiguous. The above mentioned attitudes toward scientific films, public lectures and literature on sexual issues, as well as the rhetoric used by publicists, characterised the discourse that simultaneously shaped and reflected the public moral environment in interwar Latvia and directly influenced the determination of state policy concerning the issues of abortion and contraception. Scientific films, lectures and literature were the part of process that developed individuals' understanding of sexuality. Scientific films and lectures focused on the dangers caused by abortion and venereal diseases thus shaping the view that the sex-life is about difficulties, not about pleasure. The Latvian intelligentsia promoted the opinion that the interests of the nation, not of the individual, should be first when dealing with sexuality. The inclusion of sex education in the curricula of biology in the secondary schools of Latvia, such as the "Hygiene of Woman", reaffirmed "double moral standards, with sex defined for girls as an experience confined to marriage and procreation and with an enduring focus on the impact of female promiscuity on public health and morality".[88] Sexual knowledge obtained at school, along with the content of scientific films and public lectures, promoted gender-based ideology. In its turn, the literature that could provide a more open and liberated discussion of sexuality was restricted.

3. "Exploitation with the consent of the Holy Spirit": The Public Debate

The Republic of Latvia, as a former territory of the Russian Empire, adopted Russia's Penal Laws of 22 March 1903, thus continuing criminal liability for abortion. A woman who had an abortion could be imprisoned for up to three years in a correctional facility, while the person who performed the abortion was also sentenced to a correctional facility. A doctor or midwife found guilty of this offence could be prohibited from continuing their practice for one to five years.[89] An abortion could be performed only if going into labour would endanger a woman's life – which had to be confirmed by at least two doctors – or if the pregnant woman had third-stage tuberculosis, or a kidney or heart disease.[90] At the beginning of 1933, the penalty was mitigated.[91] Additionally, the circumstances for performing a legal abortion were expanded to allow the procedure if it was justified by social conditions (if the birth of baby will financially ruin the expectant mother or her family). The penalty for physicians and midwives was not changed. The public debate over the social and cultural politics of fighting illegal abortions was the impetus for the mitigation of criminal penalties.

The public was involved in the formation of state policy concerning the abortion issue. In 1923 it was concluded that "there are opinions that abortions are not to be punished, since each mother can treat herself according to her own free will".[92] The author believed that this opinion could not be fully agreed upon because "Latvia needs inhabitants", though he acknowledged that the law should be amended to expand the number of cases in which abortion could be allowed.

Statistics attest only to the number of abortions that were registered. From 1921 to 1939, this figure grew by 3.8 times (from 1,241 to 4,760).[93] Most were performed illegally (the lowest number – 81.1% – was in 1933) and for married women. However, the total number of abortions performed was calculated to be from 16,000[94] to 60,000[95] per year.

So-called "anti-fertilisation preparations" were discussed in the press in 1924. The representative of the Department of Health of the Ministry of the Interior (since 1925 the Ministry of Public Welfare) asserted in 1924 that in preceding years the "means endangering the growth of the nation" have dangerously flooded the local drug market. Therefore, the Board of Pharmacy should remove contraceptive methods from the list of preparations allowed to be imported into Latvia.[96] The chief of the Board of Pharmacy explained that the preparations mentioned by the representative of the

Department of Health were medications for treating bile and intestinal diseases, not contraceptive methods.[97]

At the end of 1924 the Department of Health, in another project focused on increasing the birth rate, even planned to prohibit the sale of contraceptive devices and increase the penalty for illegal abortions fivefold, meaning that a woman could be sent to a correctional facility for up to 15 years. However, the charge could be dropped if the person who performed the abortion was named by the woman.[98] These proposals were not implemented, and the availability of contraceptive methods in Latvia was not restricted (as in Britain, but unlike in France, where they were banned from 1920 and 1923,[99] and shortly thereafter in Belgium and Italy).[100] In the Irish Free State, the import and sale of contraceptives was made illegal by the 1935 Criminal Law Amendment Act.[101]

Public lectures on abortions began in 1925.[102] The problem was also discussed at the First Congress of Physicians and Dentists in autumn 1925.[103] It was shown at the congress that the rise in abortions was influenced not only by material conditions but also by "selfishness, striving for life conveniences, the nudity cult, desire for delight, perversion, revolutionary propaganda, turning away from the church, shame and fear".[104] The Congress acknowledged that abortion could not be fought solely by imposing punishment, but that more attention needed to be paid to preclusion of the causes of terminating the pregnancy.[105] During debates, controversial theses were brought forward that were not included in the resolution, but which illustrated the diverse opinions of professionals.[106] One delegate advised punishing men who were fathers of expectant babies with the same sentence as women who have abortions. Another delegate pointed out the importance of increasing the birth rate in Latvia and suggested popularising the detrimental effects of contraceptive methods.

Various sex and marriage manuals published during the interwar period recommended mechanical contrivances such as condoms and pessaries as well as chemical methods.[107] Nevertheless, women could not afford to buy these items regularly even if they were well aware of them. The most popular methods of contraception during the 1920s and 1930s were traditional – withdrawal by the man and abortion by the woman, as well as preventive measures such as folk healing to cause spontaneous miscarriage counselled by so-called witch-midwives and quack-doctors.[108] This proves that contraception was considered a male prerogative and duty. Therefore, the conclusion of historian Kate Fisher that in Britain "male methods were largely preferred precisely because they put the onus on the husband" can also apply to the social reality of interwar Latvia.[109] The viewpoint that contraception was regarded as a male responsibility was also reflected in proposals to bring men to justice as well as women. Such a proposal was put forward during the 1st Women's Conference of Latvia, which was held a few weeks after the congress of physicians.[110] Beginning in 1926 the problem of abortion was also frequently discussed by members of the Women's Centre of the Latvian Social Democrat Workers' Party. This women's group pointed out that the number of abortions could be reduced by the use of contraception. Therefore, they advised opening a consultation station where workers' wives could be offered similar forms of contraception as women from more prosperous classes.[111] The motivation was ideologically grounded: "In order that the woman would have enough strength for battles for improving of her own existence she must not be caught in the flood of labours and abortions."[112] An appeal was made for society to reject prejudices popularised by religion that "force the woman to procreate without end", since they make the woman "an object of the most shameless exploitation with the consent of the Holy Spirit". Obviously, the interest to legalise abortion was expressed not in terms of sexual freedom but with reference to social problems of poverty and women's health.

The women's Social Democrat group opened the first Counselling Station on Issues of Marriage and Conjugal Life in Riga on 10 October 1929.[113] The three physicians employed there had three reception hours per week. In 1930 the press asserted that about 60% of all visitors of such consultation stations abroad wished to receive contraceptive methods, but in Latvia only 34.5% of visitors did.[114] Another station was established in 1934, which was not connected with the Social Democrat Workers' Party, where one physician had one reception hour per week.[115] Regrettably, no data has been found thus far to enable the effectiveness of these consultations to be judged. The activities of Social Democrats were prohibited after the *coup* of 1934.

The women's Social Democrats group proposed making abortion legal, indicating in 1927 that the Social Democrat fraction in German Reichstag had already submitted such a draft law.[116] (In Germany a decree of the Supreme Court permitted therapeutic abortion in 1927. Historian Cornelie Usborne has argued that it made the German regulation of abortion one of the most liberal in the world.[117]) It contained a provision stating that abortion was not punishable if performed by a physician authorised by the government, in a hospital, and during the first three months of pregnancy. The Latvian Social Democrat Workers' Party also held that abortions should be allowed if permission had been given by physicians, taking into consideration the social conditions (if the birth of the child would endanger the mother's economic situation or her ability to raise children she already had). The women's Social Democrats group pointed out that they did not support the full legalisation of abortion, but they concluded that abortions could not be extinguished through punishment, and that the punishment for women was often unjustly harsh while men were released from responsibility.[118]

The discussion of the abortion issue was not limited to merely debates. In summer 1930 the Department of Health of the Ministry of Public Welfare had finished preparing the draft of a law that would repeal the penalty for having an abortion in certain situations (if there was a possibility of inheriting diseases from the parents or if the pregnancy was the result of a rape), also taking social conditions into consideration ("when the woman due to poverty cannot support her child and when her organism is weakened after several labours").[119] An abortion would be allowed only in clinics and with permission from the appropriate commission.

At the beginning of the 1930s, the proposal to allow women to treat their bodies according to their own will was being actively pursued by the Council of the Latvian Women's Organisations, which representatives from the bourgeoisie had joined.[120] Discussions of the issue became emotional. When a demand was voiced to punish the father 10 times tougher than the mother, the entire conference hall burst in applause. Women used whistles and foot-stamping to force speakers who seemed to be disagreeable to leave the platform, while supporters of the men shouted from their seats, "You are killing our children!"[121]

The Department of Health finished preparing the draft law on abortion in summer 1930. Representatives of women's organisations, doctors and scientists took part in the discussion of the draft law at the Saeima Commission on the Penal Laws. Consequently, doubts were raised about the idea floated at the outset of the discussions about the general legalisation of abortions. At the meeting of the Board of the Council of Latvian Women's organisations on 14 November 1932, Berta Pīpiņa even declared that she did not undertake to support the decision that the council supported in the Parliament two years earlier about the full legalisation of abortion, since even the Latvian Social Democrat Workers' Party had changed its opinion and agreed that full legalisation would endanger women's health.[122] It was also acknowledged that women living in the countryside were too conservative, and that they even considered abortion as shameful.

On 25 October 1932, the Saeima submitted the bill on Amendments to the Penal Laws for consideration to the Commission on the Penal Laws. The Latvian press labelled the amendments the Abortion Law to describe their substance more precisely. It was suggested that abortion should be legalised if it was performed by a physician during the first three months of pregnancy in cases where the child could be born with severe mental or physical defects, or if the child's birth could cause a major disaster to the pregnant woman or her family. The Amendments to the Penal Laws were characterised as "not extreme" because they were the outcome of a compromise between the political parties represented in Parliament.[123] During the meetings of the Commission on the Penal Laws, stronger motions were put forward based on the modern emphasis on woman's rights, but they were not supported by the majority. Such a motion was proposed by the Faction of Workers and Farmers (*Strādnieku un zemnieku frakcija*), whose members were communists, that sought to use the resources of the state or local governments to cover abortion expenses, on the grounds that needy women and female workers in particular suffered from the prohibition on abortion. They also demanded full legalisation of abortion and amnesty for people sentenced for this crime.[124] After this motion was rejected, the Social Democrats proposed a motion to limit the term in prison for this crime at six months,[125] which was also rejected by a majority of Parliament.

The debate in the Parliament was extensive and heated during all three readings. Two contradicting views dominated. Factions that backed legalising abortion were the Latvian Social Democrat Workers' Party; the Democratic Centre (*Demokrātiskais Centrs*), which included the only female member of the Fourth Saeima (1931-1934) Berta Pīpiņa; the New Farmers and Small Landowners Party (*Jaunsaimnieku un sīkgruntnieku partija*); the Progressive Union of Latgalian Farmers (*Latgales zemnieku progresīvā apvienība*); and the faction of Workers and Peasants. Opposing legalisation were the Christian parties – the Party of Christian Peasants and Catholics (*Kristīgo zemnieku un katoļu partija*) and the Christian Labour Block (*Kristīgo darba bloks*) – and small factions of national minorities. Though the debates were open, the vote on the bill was closed. The results suggested that the Latvian Farmers' Union – the second-largest party during that period of Latvian parliamentary democracy, ranking behind the Latvian Social Democrat Workers' Party[126] – had not voted at all.

Supporters of the amendments characterised the existing punishment as extremely severe and inappropriate for the contemporary situation and values. Social Democrat Fēlikss Cielēns argued that one reason for a woman to have an abortion was the scornful attitude of contemporaries toward women who were pregnant or had already given birth to children out of wedlock. Cielēns stressed that society in particular should be blamed for suicides committed by pregnant women who "are killed by the existing laws".[127] Pēteris Zālīte from the Karlis Balodis Labour Party (*Kārļa Baloža darba partija*) emphasised the need for the government to take care of the "fallen woman and her children" because then there would be no need for abortion.[128] Communist Emīls Sudmalis pointed out that the negative view of sex outside of marriage "put high pressure on the new, modern woman of our times to perform an abortion to escape a role of a laughing-stock for the outside world".[129] His faction's member Fricis Bergs asserted that this attitude mainly came from the clergy. On one hand the clergy argues against abortion, but on the other hand it curses a woman who bears a child out of wedlock, thus pushing the woman to commit crime.[130]

The opposing view was presented by the bishop of the Latvian Roman Catholic Church, and the deputy of the Party of Christian Peasants and Catholics, Jezups Rancāns. In formulating his faction's viewpoint, Rancāns argued that abortion was an "unnatural and illegal deed and therefore it is crime and great sin", and that the "state, that allows by law to wipe out the foetus, digs the grave for itself".[131] (Social Democrat Cielēns spoke ironically about the bishop's motion and even suggested as joking amending the Penal Law to prohibit the sale and use contraceptives.[132]) Lutheran pastor Kārlis Beldavs of the faction of the Christian Labour Block described the proposed amendments as an endangerment to the virtue of a woman's chastity. He declared that if a potential pregnancy had been one of the reasons for sexual abstinence, then a woman now would be in moral danger – because the man would promise to cover abortion expenses and the woman will fall into sexual debauchery as a result.[133]

The Saeima legalised abortions performed because of social conditions.[134] This was done by a secret vote of 50 ayes, 33 nays and 3 abstentions (The Saeima was chamber of 100 members). The Saeima also tasked the government with submitting to it a draft law on amnesty in abortion matters.

The law on amnesty in abortion matters entered into force on 20 February 1933.[135] It stipulated that penalties for pregnant women and physicians must be repealed if the abortion was done before 13 January 1933, and if the abortion was requested by the pregnant woman during the first three months of pregnancy. Penalties for doctors who performed abortions also had to be repealed in cases when the pregnant woman's life or health was in danger. For abortions after the third month of pregnancy, the penalty had to be reduced for doctors and pregnant women. In all cases for midwives, imprisonment had to be reduced to two-thirds of the term, and the term in a correctional facility and consignment to forced labour had to be reduced by half. Additionally, consignment to forced labour was replaced by imprisonment in a correctional facility but the imprisonment in the correctional facility was replaced by imprisonment in prison.

In Estonia, the active public debate on abortion took place during the 1920s, earlier than in Latvia.[136] But it is worth emphasising that there were no references to debate in the national press in Latvia. The Estonian debate, which involved politicians, doctors and socially active women, led to a new Criminal Law being adopted in 1929. It legalised abortions performed by qualified specialists during the first three months of pregnancy. However, it never entered into force because it was determined that the provisions should be implemented by a special implementation law. This law was drawn up in 1934, by which time Estonia had an authoritarian government that, accordingly, deleted the articles concerning the legalisation of abortion from the new Criminal Law. Historian Indrė Karčiauskaitė has argued that Lithuanian women rejected abortions because they said the practice facilitated the irresponsible behaviour of men rather than improving women's opportunities.[137] This attitude was significantly influenced by the Catholic church, which considered abortion to be a grave sin. Moreover, Catholic women were requested to confess the use of contraceptives.[138] During the 1920s and the 1930s, abortion in Lithuania was a crime, and women who had abortions could be punished by up to three years in prison.

In the Baltic context, the amendments to the Latvian Penal Laws that dealt with abortion can be deemed as proof of the successful implementation of democratic traditions in Latvia.[139] In some respects, these amendments were passed amid a feeling of injustice strengthened by socially active women who fought to abolish the state's guardianship over women's bodies and to give this decision-making right to women. The public discourse on punishment for abortion at the end of the 1920s and beginning of the 1930s triggered an extended debate not only over the social versus the individual, but also over female versus male responsibility. The demand to reduce the penalties was expressed in feminised political rhetoric of personal liberty for all classes of women, equal responsibility for both genders and social responsibility for crime. This rhetoric could be viewed as an attempt to make criminal justice more socially relevant by identifying the root causes of crime in the structures of everyday life – namely, the number of children already living in poor conditions. The idea of socially sensitive law envisaged that the courts should take these social factors into account when handling criminal cases.

4. "It is necessary to re-educate the Society": The Period of the Authoritarian Regime

The authoritarian regime in Latvia again declared abortion performed for social reasons illegal, and the relevant articles of the Penal Law[140] were repealed on 22 March 1935. This was done due to changes in social policy. It is worth mentioning that Kārlis Ulmanis, head of the authoritarian government before the *coup*, was also the leader of the Latvian Farmers' Union, which had a reserved attitude towards legalisation of abortion. However, the penalty was not changed in 1935. According to its pro-natalistic ideology, the authoritarian government declared that with the destruction of the family, the very basis of the entire nation and the state would be destroyed. Therefore, "it is necessary to re-educate the society honouring the ancient virtues and sanctity of the family", since "while the public moral does not improve, the number of abortions will not decrease".[141] Restriction of the avoidance of becoming a mother: this was how the press interpreted the amendments.[142] It could be viewed as using criminal law as an instrument of social governance and an effort to re-masculinise criminal justice. Additionally, to aid single mothers, the authorities used criminal sanctions to intrude on the civil rights of fathers. Through amendments to the Law on Social Custody, which came into force on 7 August 1937, the state was entitled to interfere in the civil relations of its citizens to make them pay alimony by threatening criminal sanctions.[143] Under the amendments, people – mainly fathers – who avoided paying alimony could be consigned to forced community service or imprisoned in a workhouse for up to two years.

The number of prosecutions brought for performing illegal abortions rose from 49 cases in 1924 to 195 cases in 1937.[144] Beginning in 1938, when the Law on Medical Treatment came into effect, the decision to allow an abortion to be performed could only be made by the Eugenics Commission.[145] In 1938, 285 women received permission from this commission to terminate their pregnancies for reasons related to eugenics (65% of those who made a request). This increased to 325 women in 1939 (65% of applicants).[146] All abortions performed by the decision of the Eugenics Commission were substantiated for reasons related to eugenics; nevertheless, they were voluntary. The number of rejected applicants (35%) also suggests that it was the rhetoric that changed, absorbing the term of eugenics while the intention of women was the same as before 1938. The female concern was interrupting of the unwanted pregnancy. After 1938 women only tried to adapt their causes of abortion to the law in order to receive the permission, thus avoiding the illegal abortion. The law thus limited the possibilities of legal abortion. From 1939 there was a ban on information about contraceptives in advertisements. “Malicious advertisements”, namely those “offering advice and means” to unmarried or childless couples, were prohibited by an order by the Minister of Public Affairs in summer 1939.[147] Thus, restriction of information about contraceptives was also extended to adults.

The motion to place in a workhouse midwives who had been sentenced for repeatedly performing abortions was drawn up by lawyers. The intention was to reduce the number of persons qualified to perform abortions. The first midwife was imprisoned in the Labour House on 4 October 1938 – in fact, her fifth sentence.[148] On 5 March 1939, the Institute for Research on the National Vitality, the Latvian eugenic research institute, opened a marriage counselling station on issues related to eugenics, but not to contraception.[149]

5. Conclusions

The discourse that shaped and reflected the public moral environment in inter-war Latvia, and which simultaneously had a direct influence on state policy concerning the issues of abortions and contraception, reflects the attitude towards acquiring sexual knowledge at that time. The separation of women and men while watching scientific films and attending public lectures, and restrictions on literature about sexual issues, reveal the conventional beliefs that sex-related issues are too sensitive to be discussed publicly and in the presence of members of the opposite sex. This opinion was also promoted through pro-natalist and moralising rhetoric used by publicists in response to the evolving social process of separating sexual activity from reproduction. Nevertheless, sexuality as a conjugal norm, and the need to scientifically explain sexuality, were important social developments that characterised the interwar period in Latvia. The public discourse on punishing abortion at the end of the 1920s and the beginning of the 1930s is proof of this, because views were finally aired that prior had not been openly discussed. Though the issue of legalising abortion was not a part of the public discussion on sexual freedom, it was linked to social problems such as poverty and women's health. In the broader context, this issue involved an extended debate over not only the social versus the individual, but also female versus male responsibility. It was an important issue for organisations that united female representatives of the bourgeoisie who categorically demanded that the principle of gender equality be observed. The use of gender categories proved to be an effective argument in discussions about the legalisation of abortion. The growing public demand to lower penalties could be viewed as an attempt to make criminal justice more socially sensitive, identifying the root causes of crime in the structures of everyday life.

In conclusion, the legalisation of abortions performed because of social reasons during the period 1933-1935 must be deemed as a successful implementation of democratic traditions in Latvia. In its turn, the amendments of 1935 must be viewed as the use of the criminal law as an instrument of social governance and an effort to re-masculinise criminal justice.

Notes

[1] Szreter, 1999, 159-194; Fisher, 1999, 213-232; Cook, 2006, 19-40; Weeks, 1981.
[2] Usborne, 2007; Usborne, 2005, 321-338; Blecourt, 1999, 195-212.
[3] Usborne, 2007, 3.
[4] Fisher, 2006.
[5] Ibid., 236.
[6] Romanov and Yarskaya-Smirnova, 2007, 432.
[7] Lebina, 2007, 228-241.
[8] Pushkarev and Pushkareva, 2007, 199-227.
[9] Lipša, 2009a, 89-100. See also: Lipša, 2009b, 53.
[10] Lipša, 2006, 16-33.
[11] Felders, 2005, 16-17.
[12] Zelče, 2006, 94-137.

[13] Stranga, 2005, 2.
[14] Kalling, 2005, 359-365.
[15] Kalling, 2007, 253-262; Kalling, 2000, 133-140. See also: Kalling, 1999, 31-42.
[16] Paukštytė-Šaknienė, 2008, 31-39; Paukštytė-Šaknienė, 2004a, 14-23; Paukštytė-Šaknienė, 2004b, 89-110.
[17] Iggers, 1997, 129, 131.
[18] Porter, 2001, 252.
[19] See for example: Buduls, 1924, 1-2.
[20] Lipša, 2009c, 53-80.
[21] Akermanis, 1928, 831.
[22] Ibid., 827.
[23] "Vai ir vajadzīga seksuāla informācija skolās? [Is there a Need of Sexual Information in Schools?]" *Jaunākās Ziņas*, no. 232 (15 October), 1926, 4
[24] "Dr. M. Hiršfelds par „seksuālo mazākumu" tiesībām [Dr. M. Hirschfeld about the Rights of „Sexual Minority"]" *Jaunākās Ziņas*, no. 94 (27 April), 1929, 9.
[25] "Par seksuālo audzināšanu [About Sexual Education]" *Jaunākās Ziņas*, no. 265 (25 November), 1925, 2; "Sabiedrisko tiesu par pornogrāfiju [The Public Court on Pornography]" *Jaunākās Ziņas*, no. 240 (24 October), 1925, 6.
[26] "Sabiedriskās tiesas spriedums par latvju sievieti [The Verdict of the Public Court about Latvian Woman]" *Jaunākās Ziņas*, no. 281 (14 December), 1925, 4.
[27] Breikšs, 1935, 243, 247.
[28] Ibid., 219.
[29] Lazdiņa, 1931, 36.
[30] Pīpiņa, 1936, 32.
[31] Krastiņš, 1925, 37.
[32] Kundziņš, 1932,1-2.; "Kad dzīve irst... [When Life crumbles...]" *Zemgales Balss*, no. 36 (13 February), 1930, 1-2.
[33] Pīpiņa, 1932, 2.
[34] "500 sievietes līdz pusnaktij disputē par mīlestību un laulības dzīvi [500 Women discuss about the Love and the Married Life till the Midnight]" *Jaunākās Ziņas*, no. 88 (21 April), 1934, 12.
[35] "Mīlestība, draudzība un laulība [Love, Friendship and Marriage]" *Aizkulises*, no. 17, 1934, 2.
[36] Weeks, 1981, 200.
[37] The fact that the "little man" – mainly the youth – did not take into consideration the opinions dominating in the periodicals is attested by the anxiety of the "older ones" about the popularity of modern dances. Those who condemned such dances declared that modern dances imply impolite behaviour in public space. They were acknowledged to be "flagrant

allurements of the lowest instincts of man" which serve "only to erotic, partly even pornography, as well as generally sexual amusement". They asserted that modern dances implied "modern debauchery" in a dancing hall since in these dances the main role was played by the opposite sex, and thus, passion was aroused. This was an urge to demand the combating of pornography in social life in the light of the fact that it was put down in literature and art. Even some doctors asserted that by restricting "modern erotic dances" it would become possible to reduce the number of abortions. However, attempts to restrict "modern dance" were unsuccessful. The "little man" used to dance passionately and freely and refused the demand to conceal the expressions of sexuality in contacts with the opposite sex in public space. See: Lipša, 2007a, 51-62.

[38] "Kas traucē rases uzlabošanos? [What does trouble the Improvement of Race?"]" *Jaunā Balss*, no. 3, 1924, 4.; "Brīvību sievietei lemt par abortu! [Freedom to Woman to decide on Abortion!]" *Jaunākās Ziņas*, no. 65 (21 March), 1931, 12.

[39] Baidiņš, 1925, 271.

[40] "'Eņģelīšu' klīnika Rīgā [The Clinic of "Angels" in Riga]" *Jaunākās Ziņas*, no. 261 (16 November), 1928, 7.

[41] *Svari*, no. 41, 1924, 324.

[42] *Sikspārnis*, no. 336, 1931, 1.

[43] *Vecais Sikspārnis*, no. 505, 1934, 5.

[44] *Vecais Sikspārnis*, no. 540, 1935, 1.

[45] *Sikspārnis*, no. 426, 1932, 1.

[46] Hall, 2009, 20.

[47] "Tautas namā [In the Society House]" *Kurzemes Vārds*, no. 29 (7 February), 1923, 3.

[48] "Divās lekcijās [In two lectures]" *Kurzemes Vārds*, no. 35 (14 February), 1923, 3.

[49] "Priekšlasījums "Kas jāzin katrai sievietei" [Lecture "What must every Woman know"]" *Zemgales Balss*, no. 48 (28 February), 1925, 3; *Kurzemes Vārds*, no. 79 (5 April), 1930, 2; "Noslēpumi, kas tiek atklāti tikai sievietēm [Secrets that are unraveled only to Women]" *Aizkulises*, no. 8, 1931, 3.

[50] "Populāru lekciju sievietēm [Popular Lecture to Women]" *Kurzemes Vārds*, no. 256 (10 November), 1928, 3.

[51] Junona. "Vīnietes padomi Rīgas dāmām [Advices of the Viennese to Ladies of Riga]" *Jaunākās Ziņas*, no. 9 (11 January), 1929, 6.

[52] "Zinātnisku bilžu „Veneriskās un citas lipīgās slimības" demonstrēšana [The Displaying of Scientific Slides „Venereal and Other Contagious Diseases]" *Kurzemes Vārds*, no. 107 (17 May), 1923, 3.

[53] *Jaunākās Ziņas*, no. 29 (6 February), 1925, 12.

[54] *Kurzemes Vārds*, no. 79 (7 April), 1925, 4.

[55] *Zemgales Balss*, no. 97 (3 May), 1927, 2.
[56] *Zemgales Balss*, no. 266 (22 November), 1928, 1; *Zemgales Balss*, no. 271 (28 November), 1928, 1.
[57] *Zemgales Balss*, no. 83 (14 April), 1929, 1.
[58] Rāmaviete [Anna Rūmane-Ķeniņa]. "Kādēļ sievietēm jāvēl sievietes [Why do Women have to vote for Women?]" *Jaunākās Ziņas*, no. 24 (31 January), 1925, 4.
[59] "Ziņkārīgs vīrietis [The Curious Man]" *Jaunākās Ziņas*, no. 25 (31 January), 1924, 5; "Par M. Ferliņas lekciju [About the Lecture of M. Ferliņa]" *Jaunākās Ziņas*, no. 30 (6 February), 1924, 5; "Sieviešu mīlas ieroči tiesā [Love Weapons of Women in the Court]" *Jaunākās Ziņas*, no. 66 (20 March), 1924, 3.
[60] Lipša, 2007b, 91-92.
[61] Pašeta, 2003, 198, 200.
[62] Sauerteig, 1999, 28.
[63] "Noteikumi par jaunatnes pasargāšanu no sēnalu un neķītrību literatūras [Regulations to Protect Youth from Trash and Smut Literature]" in: *Likumu un Ministru Kabineta Noteikumu Krājums. 1927. gads* [Collection of Laws and Regulations of the Cabinet of Ministers. 1927]. Rīga: Kodifikācijas nodaļas izdevums, 1927, 317-318.; "Pārgrozījumi noteikumos par jaunatnes pasargāšanu no sēnalu un neķītrību literatūras [Amendments to Regulations to Protect Youth from Trash and Smut Literature]" in: *Likumu un Ministru Kabineta Noteikumu Krājums. 1930. gads* [Collection of Laws and Regulations of the Cabinet of Ministers. 1930]. Rīga: Kodifikācijas nodaļas izdevums, 1930, 15.; "Likums par jaunatnes pasargāšanu no sēnalu un neķītrību literatūras [Law to Protect Youth from Trash and Smut Literature]" in: *Likumu un Ministru Kabineta Noteikumu Krājums. 1932. gads*[Collection of Laws and Regulations of the Cabinet of Ministers. 1932]. Rīga: Kodifikācijas nodaļas izdevums, 1932, 107-108.
[64] *Valdības Vēstnesis*, no. 273 (1 December), 1924, 1-2.
[65] Lapsiņš, 1939a, 73-176. [There were 810 items on the list.]
[66] Lapsiņš, 1939b, 61-102. [There were 255 items on the list.]
[67] "Neķītrības literatūras plūdi Rīgā [The Flow of Smut Literature in Riga]" *Jaunākās Ziņas*, no. 74 (31 March), 1933, 11. ; "Aizpleivurōta pornogrāfija [The Veiled Pornography]" *Latgolas Vōrds*, no. 17 (29 April), 1936, 2.
[68] Pašeta, 2003, 203-204, 206.
[69] The letter from a resident of Ventspils to the Section of Press and Societies at the Ministry of Interior on February 28, 1928: Latvijas Valsts vēstures arhīvs (Latvian State Historical Archives – further: LVVA), 3724/1/606, 1.
[70] The letter of the Board of General Matters (Vispārējo lietu pārvalde) at the Ministry of Interior to the Section of Press and Societies at the Ministry of Interior on February 2, 1928: LVVA, 3724/1/606, 2.

[71] The reference written by head of the Committee for Fighting Literature Harmful to Youth to Minister of Education on March 22, 1928: LVVA, 1632/2/1628, 49.
[72] Lapsiņš, 1939b, 89.
[73] "Pilsoņiem jābūt palīdzīgiem neķītrību literatūras apkarošanā [Population has to be Obliging in the Fighting of Smut Literature]" *Jaunākās Ziņas*, no. 75 (1 April), 1933, 10.
[74] The letter of an anonymous war veteran to the Minister of Interior, 16 August 1933: LVVA, 3724/1/10865, 371.
[75] The advertisement from newspaper *Jaunākās Ziņas*, 15 August 1933: LVVA, 3724/1/10865, 372; The advertisement from newspaper *Jaunākās Ziņas*, 27 September 1933: LVVA, 3724/1/132, 101.
[76] "Pilsoņiem jābūt palīdzīgiem neķītrību literatūras apkarošanā [Population has to be Obliging in the Fighting of Smut Literature]" *Jaunākās Ziņas*, no. 75 (1 April), 1933, 10.
[77] The advertisement from the newspaper *Jaunākās Ziņas*: 15 August 1933: LVVA, 3724/1/10865, 372. The exact title of the book is: *Kas jāzin modernai sievietei* [What the modern women has to know] (1933). Pēc dr. E. Bock'a, prof. Dennerta u. c. pārstrādājis Brantoms. Rīga: Zinātne un Daile.
[78] The letter of an anonymous war veteran to the Minister of Interior, 16 August 1933: LVVA, 3724/1/10865, 371.
[79] The advertisment from newspaper *Jaunākās Ziņas*, 27 September 1933: LVVA, 3724/1/132, 101.
[80] Lapsiņš, 1939a, 73-176.
[81] Paeglis, 1996, 59-60.
[82] The report of Prefecture of Riga to the Department of Press and Societies at the Ministry of Public Affairs in 1938: LVVA, 1376/2/525, 14.
[83] "Rīkojums kinematogrāfiem [Instruction to Cinematographs]" *Valdības Vēstnesis*, no. 247 (1 November), 1934, 1.
[84] Sauerteig, 1999, 12.
[85] *Vispārīgās izglītības vidusskolu programmas* [Curriculas of Universal Secondary Education Schools]. Rīga: A. Gulbis, 1928, 115.
[86] "Labā roka nezin, ko kreisā dara [The Right Hand does not know what the Left does]" *Zemgales Balss*, no. 259 (17 November), 1931, 2.
[87] "J. Lapiņa kungs, patiesību pirmā vietā! [Mr. J. Lapiņš, Truth in the First Place!]" *Zemgales Balss*, no. 287 (20 December), 1931, 2.
[88] Sauerteig and Davidson, 2009, 5.
[89] *1903. gada 22. marta Sodu likumi. Tulk. ar paskaidr. un ar motīviem par Latvijas valdības laikā izdotiem grozījumiem* [Penal Laws of March 22, 1903. Translation with Explanations and Causes on Amendments Passed during Latvian Government]. Rīga: Tieslietu ministrijas atsevišķas komisijas sagatavojumā, 1930, 116.

[90] "Laulības dzimumdzīve un aborti [Conjugal Sexual Life and Abortions]" *Jaunākās Ziņas*, no. 49 (1 March), 1929, 3.
[91] "Amnestijas akts aborta lietās [Amnesty Act on Abortion Cases]" in: *Likumu un Ministru Kabineta Noteikumu Krājums. 1933. gads* [Collection of Laws and Regulations of the Cabinet of Ministers. 1933]. Rīga: Kodifikācijas nodaļas izdevums, 1933, 77.
[92] "Aborts un likums [Abortion and Law]" *Jaunākās Ziņas*, no. 23 (30 January), 1923, 2.
[93] Lipša, 2009a, 89.
[94] "Latvijas Republikas IV Saeimas IV sesijas 16. sēde 1932. gada 2. decembrī [The 16th Meeting of the 4th Session of the 4th Saeima of the Republic of Latvia on December 2, 1932]" in: *Latvijas Republikas IV Saeimas stenogrammas, 4. sesija* [Shorthand Records of the 4th Saeima of the Republic of Latvia, the 4th Session]. Rīga: Latvijas Republikas Saeimas izdevums, 1933, 659, 663.
[95] "60 000 aborti vienā gadā [60 000 Abortions in a Year]" *Jaunākās Ziņas*, no. 147 (5 July), 1930, 8.
[96] "Apaugļošanas novēršana [The Prevention of Conception]" *Jaunākās Ziņas*, no. 225 (3 October), 1924, 3.
[97] „Apaugļošanas novēršanas jautājumā [On the Issue of the Prevention of Conception]" *Jaunākās Ziņas*, no. 238 (18 October), 1924, 4.
[98] "Kā pacelt dzimstību Latvijā. Vecpuišu nodoklis vecmeitām par labu [How to increase the Birth-Rate in Latvia. The Bachelor's Tax for the Grant of Spinsters]" *Rīgas Ziņas*, no. 45 (10 December), 1924, 1.
[99] Sohn, 1994, 111.
[100] Pašeta, 2003, 217.
[101] Ibid., 212.
[102] "Par abortiem un mātes mīlestību [About Abortions and Mother's Love]" *Jaunākās Ziņas*, no. 53 (7 March), 1925, 6.
[103] "Abortu jautājums Latvijā [The Issue of Abortions in Latvia]" *Jaunākās Ziņas*, no. 204 (12 September), 1925, 7.
[104] Putniņš, 1926, 298-300.
[105] *Latvijas ārstu un zobārstu 1. kongresa darbi* [Works of the First Congress of Latvian Physicians and Dentists], 558.
[106] Ibid., 329-330.
[107] Zelče and Sprugaine, 2005, 70.
[108] Zuments, 1926, 28, 29, 30; Bētiņa, 1928, 13; ""Mums jābūt modrām par savu eksistenci" ["We have to be on the Alert of our Existence"]" *Vecmāšu Vēstnesis*, no. 59, 1935, 3.
[109] Fisher, 2006, 12.
[110] Protocols of the first conference of Latvian women, 26 to 27 September 1925: LVVA, 2412/2/11, 9.

[111] Hibšmane, 1928, 4.
[112] Kļava, 1929, 7.
[113] "Rīgā atver padomu punktu laulības un dzimumdzīves jautājumos [Counselling Station on Issues of Marriage and Conjugal Life is opened in Riga]" *Jaunākās Ziņas*, no. 47 (27 February), 1929, 3; "Padomu punktu laulības un dzimumdzīves jautājumos [Counselling Station on Issues of Marriage and Conjugal Life]" *Jaunākās Ziņas*, no. 230 (10 October), 1929, 5; "Kur sniegs padomus laulību un dzimumdzīves jautājumos [Where will be offered Advice on Marriage and Conjugal Life]" *Jaunākās Ziņas*, no. 231 (11 October), 1929, 3; "Kas meklē padomus laulības un dzimumdzīves jautājumos [Who seeks for Advices on Marriage and Conjugal Life]" *Jaunākās Ziņas*, no. 239 (22 October), 1929, 2.
[114] "Ko veicis padompunkts laulības un dzimumdzīves jautājumos [What is done by Counselling Station on Issues of Marriage and Conjugal Life]" *Jaunākās Ziņas*, no. 276 (5 December), 1930, 5.
[115] "Padomu punkts [The Counselling Station]" *Jaunākās Ziņas*, no. 35 (13 February), 1934, 5.
[116] Kalniņa, 1927, 5.
[117] Usborne, 2007, 5.
[118] Kalniņa, 1927, 6.
[119] "60 000 aborti vienā gadā [60 000 Abortions in a Year]" *Jaunākās Ziņas*, no. 147 (5 July), 1930, 8.
[120] "Sieviete ir savas tautas sirds [Woman is the Heart of her Nation]" *Jaunākās Ziņas*, no. 265 (22 November), 1930, 1-2; "Pret likumu par abortu protestē 15 000 latvju sievietes [15 000 Latvian Women demonstrate against the Law on Abortion]" *Jaunākās Ziņas*, no. 29 (6 February), 1931, 4; Pīpiņa, 1931, 2; "Brīvību sievietei lemt par abortu! [Freedom to Woman to decide on Abortion!]" *Jaunākās Ziņas*, no. 65 (21 March), 1931, 12; "Sievietes un ārsti prasa abortu legalizēšanu [Women and Physicians demand the Legalization of Abortion]" *Jaunākās Ziņas*, no. 106 (15 May), 1931, 6.
[121] "Brīvību sievietei lemt par abortu! [Freedom to Woman to decide on Abortion!]" *Jaunākās Ziņas*, no. 65 (21 March), 1931, 12.
[122] Protocols of the Council of Latvian Women Organizations, 14 November 1932: LVVA, 2497/1/69, 18.
[123] "Latvijas Republikas IV Saeimas IV sesijas 7. sēde 1932. gada 25. oktobrī [The 7th Meeting of the 4th Session of the 4th Saeima of the Republic of Latvia on October 25, 1932]" in: *Latvijas Republikas IV Saeimas stenogrammas, 4. sesija* [Shorthand Records of the 4th Saeima of the Republic of Latvia, the 4th Session], 1933, 277.; "Latvijas Republikas IV Saeimas IV sesijas 16. sēde 1932. gada 2. decembrī [The 16th Meeting of the 4th Session of the 4th Saeima of the Republic of Latvia on December 2, 1932]" in: Ibid., 659.

[124] "Latvijas Republikas IV Saeimas IV sesijas 17. sēde 1932. gada 9. decembrī [The 17th Meeting of the 4th Session of the 4th Saeima of the Republic of Latvia on December 9, 1932]" in: Ibid., 701, 708.
[125] Ibid., 700.; "Latvijas Republikas IV Saeimas IV sesijas 22. sēde 1932. gada 20. decembrī [The 22nd Meeting of the 4th Session of the 4th Saeima of the Republic of Latvia on December 20, 1932]" in: Ibid., 928.
[126] On their complicated relations see: Bleiere; Butulis; Feldmanis; Stranga and Zunda, 2006, 159-160.
[127] Latvijas Republikas IV Saeimas IV sesijas 16. sēde 1932. gada 2. decembrī [The 16th Meeting of the 4th Session of the 4th Saeima of the Republic of Latvia on December 2, 1932]" in: *Latvijas Republikas IV Saeimas stenogrammas, 4. sesija* [Shorthand Records of the 4th Saeima of the Republic of Latvia, the 4th Session], 1933, 673.
[128] "Latvijas Republikas IV Saeimas IV sesijas 17. sēde 1932. gada 9. decembrī [The 17th Meeting of the 4th Session of the 4th Saeima of the Republic of Latvia on December 9, 1932]" in: Ibid., 684.
[129] Ibid., 687.
[130] "Latvijas Republikas IV Saeimas IV sesijas 16. sēde 1932. gada 2. decembrī [The 16th Meeting of the 4th Session of the 4th Saeima of the Republic of Latvia on December 2, 1932]" in: Ibid., 662.
[131] Ibid., 668., 669.
[132] Ibid., 672.
[133] "Latvijas Republikas IV Saeimas IV sesijas 22. sēde 1932. gada 20. decembrī [The 22nd Meeting of the 4th Session of the 4th Saeima of the Republic of Latvia on December 20, 1932]" in: Ibid., 928-929.
[134] "Pārgrozījumi un papildinājumi Sodu likumos [Amendments to Penal Laws]" in: *Likumu un Ministru Kabineta Noteikumu Krājums. 1932. gads* [Collection of Laws and Regulations of the Cabinet of Ministers. 1932]. Rīga: Kodifikācijas nodaļas izdevums, 1932, 311.
[135] "Amnestijas akts aborta lietās [Amnesty Act on Abortion Cases]" in: *Likumu un Ministru Kabineta Noteikumu Krājums. 1933. gads* [Collection of Laws and Regulations of the Cabinet of Ministers. 1933]. Rīga: Kodifikācijas nodaļas izdevums, 1933, 77.
[136] Kalling, 2005, 359-365.
[137] Karčiauskaitė, 2007, 15.
[138] Gineitienė, 1998, 82.
[139] Zelče, 2006, 104.
[140] "Pārgrozījumi un papildinājumi Sodu likumā [Amendments to Penal Law]" in: *Likumu un Ministru Kabineta Noteikumu Krājums. 1935. gads* [Collection of Laws and Regulations of the Cabinet of Ministers. 1935]. Rīga: Kodifikācijas nodaļas izdevums, 1935, 71; "Papildinājumi Ārstniecības

nolikumā [Amendments to Regulation of Medical Treatment]" in: Ibid., 172-173.; Mincs and Lauva, 1938, 219-221.; Veidemanis, 1935, 573.
141 "Kas apdraud ģimenes pamatus [What does endanger the Foundations of Family?]" *Jaunākās Ziņas*, no. 21 (25 January), 1935, 8.
142 "Ierobežo izvairīšanos kļūt par māti [The Avoidance of Becoming a Mother is restricted]" *Rīts*, no. 119 (1 May), 1935, 11.
143 "Pārgrozījumi un papildinājumi likumā par sociālo apgādību [Amendments to Law on Social Custody]" in: *Likumu un Ministru Kabineta Noteikumu Krājums. 1937. gads* [Collection of Laws and Regulations of the Cabinet of Ministers. 1937]. Rīga: Kodifikācijas nodaļas izdevums), 1937, 1018-1022; "Valsts vara gādā par katra pilsoņa pienākumu izpildīšanu pret apgādāmajamiem. Pārgrozījumi un papildinājumi likumā par sociālo apgādību [State Power takes Care so that Every Dweller would do his Duty to Dependants. Amendments to Law on Social Custody]" in: *Ceturtais gads. 1937. 15. V – 1938. 15. V* [The Fourth Year, 1937. 15. V – 1938. 15. V]. Rīga: Leta, 1938, 46.
144 Lipša, 2006, 91.
145 "Ārstniecības likums [Law on Medical Treatment]" in: *Likumu un Ministru Kabineta Noteikumu Krājums. 1937. gads* [Collection of Laws and Regulations of the Cabinet of Ministers. 1937], 1937, 1443-1445.
146 Zelče, 2006, 104-105; "Sekmīgi kārtojas mūsu izglītības un veselības jautājumi (Norādījumi iekšlietu resora darbiniekiem) [Issues of Our Education and Health proceed successfully (Instructions to Employees of Internal Affairs Department]" *Jaunākās Ziņas*, no. 93 (25 April), 1940, 2.
147 "Aizliegti ļaunprātīgi sludinājumi [Malicious Advertisments are forbidden]" *Kurzemes Vārds*, no.161 (21 July), 1939, 3.
148 "Pirmā vecmāte – darba namā [The First Midwife – in a Work House]" *Jaunākās Ziņas*, no. 225 (4 October), 1938, 10.
149 "Ievadīts plašs darbs tautas dzīvā spēka pētīšanai (Rīgā atvērs padomu došanas punktu iedzimtības jautājumos) [Extensive Work is started on Research on the National Vitality (Counselling Station on Issues of Heredity will be opened in Riga)]" *Jaunākās Ziņas*, no. 219 (27 September), 1938, 4; "Jāprecas jauniem, jādzīvo saskanīgi [The Demand is to marry while Young and to live in Harmony]" *Jaunākās Ziņas*, no. 54 (6 March), 1940, 6.

References

Akermanis, E. (1928), "Apdraudēta Latvju tauta [The Endangered Latvian Nation]" *Burtnieks*, no. 9,819-831.

Cook, H. (2006), "Demography" in: Cocks, H. G. and Houlbrook, M. (eds.), *Palgrave Advances in the Modern History of Sexuality*. Hampshire, New York: Pagrave Macmillan, 19-40.

Baidiņš, A. (1925), "Aborts un tautas labklājība [Abortion and the Well-Being of Nation]" *Sieviete*, no. 12,267-272.

Bētiņa, G. (1928), "Sieviete un aborts [Woman and Abortion]" *Darba Sieviete*, February,11-14.

Blecourt, W. de (1999), "Cultures of abortion in the Hague in the early twentieth century" in: Eder, F. X.; Hall L. A. and Hekma, G. (eds.), *Sexual cultures in Europe. Themes in sexuality.* Manchester, New York: Manchester University Press, 195-212.

Bleiere, D.; Butulis, I.; Feldmanis, I.; Stranga, A. and Zunda, A. (2006), *History of Latvia. The 20th Century*. Riga: Jumava.

Breikšs, L. (1935), "Mūsu ģimene [Our Family]" in: Bērsons, I. and Bisters, M. (eds.) (2002), *Leonīds Breikšs. Dzīve un darbi. III sēj.* [*Leonids Breikss. The Life and Works. III Volume*]. Rīga: Valters un Rapa, 209-375.

Buduls, H. (1924), "Kokainisms (Beigas) [Cocainism (End)]" *Jaunākās Ziņas*, no. 251 (4 November), 1-2.

Felders, B. (2005), "'Mazvērtīgo samazināšana'- eigēnika Latvijā ['Reducing the Inferiors' – Eugenics in Latvia]" *Diena*, no. 95 (23 April), pielikums *Kultūras Diena*, 16-17.

Fisher, K. (1999), "'Didn't stop to think, I just didn't want another one': the culture of abortion in interwar South Wales" in: Eder, F. X.; Hall, L. A. and Hekma, G. (eds.), *Sexual cultures in Europe. Themes in sexuality.* Manchester, New York: Manchester University Press, 213-232.

Fisher, K. (2006), *Birth Control, Sex, and Marriage in Britain, 1918-1960.* New York: Oxford University Press.

Gineitienė, D. (1998), "Church, Nationalism, and the Reproductive Rights of Women" in: LaFont, S. (ed.), *Women in Transition. Voices from Lithuania.* Albany: State University of New York Press, 81-89.

Hall, L. A. (2009), "In Ignorance and in Knowledge. Reflections on the History of Sex Education in Britain" in: Sauertaig, L. D. H. and Davidson, R. (eds.), *Shaping Sexual Knowledge. A Cultural History of Sex Education in Twentieth Century Europe*. New York: Routledge, 19-36.

Hibšmane, K. (1928), "Sievietēm jāizsargājas no aborta [Women have to avoid Abortion]" *Jaunākās Ziņas*, no. 35 (13 February), 4.

Iggers, G. G. (1997), *Historiography in the Twentieth Century. From Scientific Objectivity to the Postmodern Challenge*. Hanover and London: Wesleyan University Press.

Kalling, K. (1999), "Introduction to the History of Estonian Eugenics" in: *Annual Report 1998. Tartu University History Museum*. Tartu: Tartu University Press, 31-42.

Kalling, K. (2000), "The Story of Estonian Eugenics" in: Arons, K. Ē. and Salaks, J. (eds.), *Acta Medico-Historica Rigensia*, vol. V no. XXIV, 133-140.

Kalling, K. (2005), "Abort sõjaeelses Eestis – legaliseerimisest keelustmiseni [Abortions in pre-war Estonia – from Legalisation to Prohibition]" *Eesti Arst*, vol. 84, no. 5, 359-365.

Kalling, K. (2007), "The Self-Perception of a Small Nation: The Reception of Eugenics in Interwar Estonia" in: Turda, M. and Weindling, P. J. (eds.), *„Blood and Homeland". Eugenics and Racial Nationalism in Central and Southeast Europe, 1900-1940*. Budapest, New York: Central European University Press, 253-262.

Kalniņa, K. (1927), "Apdraudētā dzimstība [The Endangered Birth-Rate]" *Darba Sieviete*, March,3-7.

Karčiauskaitė, I. (2007), *The Catholic Women's Movement in Lithuania (1907-1940). Summary of Doctoral Dissertation*. Kaunas: Vytautas Magnus University, The Institute of Lithuanian History.

Kļava, L. (1929), "Uzspiestas dzemdības [The Imposed Deliveries]" *Darba Sieviete*, November, 7-9.

Krastiņš, K. (1925), "Par nākotnes cilvēka morāli [About the Morals of the Man of Future]" *Nākotnes Sieviete*, no. 3, 37.

Kundziņš, J. (1932), "Pašnāvību sērga. II [The Plague of Suicides. II]" *Zemgales Balss*, no. 83 (15 April),1-2.

Lapsiņš, R. (1939), *Aizliegto grāmatu, brošuru un citu Latvijā iespiesto poligrafisko ražojumu saraksts. Ziņas par laiku līdz 1939. g. 31. janvārim*

[List of Forbidden Books, Booklets and Other Printing Matters Published in Latvia. Information till January 31, 1939]. Rīga: J. Rozes izdevums.

Lapsiņš, R. (1939b), *Sēnalu un neķītrību literatūras saraksts. Komisijas lēmumi līdz 1937. gada 31. decembrim* [List of Trash and Smut Literature. Resolutions of Commission Passed to December 31, 1937]. Rīga: J. Rozes izdevums.

Lazdiņa, E. (1931), "Vecpuišu jautājums [The Issue of Bachelors]" *Sieviete*, no. 3, 36.

Lebina, N. (2007), ""Navstrechu mnogochislennym zayavleniyam trudyashchichsya zhenshchin..." abortnaya politika kak zerkalo sovetskoy sotsal'noy zaboty ["Towards Many Appeals of Working Women..." Abortion Politics as a Mirror of Soviet Social Care]" in: Romanov, P. B. and Yaskaya-Smirnova, E. R. (eds.), *Sovetskaja sotsial'nya politika 1920-1930-ch godov: ideologiya I povsednevnost'. Sbornik statey* [Soviet Social Policy in 1920-1930s: Ideology and Everyday Life]. Moskva: OOO „Variant", TSSPGI, 228-241.

Lipša, I. (2006), "Sabiedriskā tikumība Latvijas Republikā (1918-1934): priekšstatu maiņa [Public Morality in the Republic of Latvia (1918-1934): the Shift of Notions]" *Latvijas Vēsture*, vol. 62, no. 2, 16-33.

Lipša, I. (2007a), "Arvien tuvāk. Priekšstati par moderno starpkaru Latvijā [All the Closer. Notions of the Modern in inter-war Latvia]" *Rīgas Laiks*, January, 51-62.

Lipša, I. (2007b), "Sievietes tēla veidošana Kārļa Ulmaņa autoritārajā režīmā (1934-1940): ideoloģija un īstenība [The Formation of Woman's Image in the Authoritarian Regime of Mr. Kārlis Ulmanis (1934-1940)]" in: Zellis, K. (ed.), *Sieviete Latvijas vēsturē* [Woman in the History of Latvia]. Rīga: LU Akadēmiskais apgāds,80-99.

Lipša, I. (2009a), *Sabiedriskā tikumība Latvijā, 1918.-1940. gads: promocijas darbs* [Public Morality in Latvia, 1918-1940: Doctoral Thesis]. Rīga: Latvijas Universitāte.

Lipša, I. (2009b), *Sabiedriskā tikumība Latvijā, 1918.-1940. gads. Promocijas darba kopsavilkums. Public Morality in Latvia, 1918-1940. Summary of the Promotion Paper*. Rīga: Latvijas Universitāte.

Lipša, I. (2009c), "Sabiedriskā tikumība Latvijā demogrāfisko problēmu aspektā, 1918-1940 [Public Morality in Latvia in the Facet of Problems of Demography, 1918-1940]" *Latvijas Vēstures Institūta Žurnāls*, vol. 70, no. 1, 53-80.

Mincs, P. and Lauva, J. (1938), *Sodu likums ar komentāriem – izvilkumiem no Senāta Kriminālā kasācijas departamenta spriedumiem un ar alfabētisko un salīdzināmiem rādītājiem. 2. izd.* [Penal Law with Commentaries-Excerpts from the Sentences of the Department of Criminal Appeal of Senate and with the Alphabetical and Comparative Index. The Second issue]. Rīga: Valsts tipogrāfijas izdevums.

Paeglis, J. (1996), *Kas bija liegts Pirmās republikas lasītājiem. Preses likumi un izplatīšanai aizliegtā Latvijā izdotā literatūra 1918. – 1940. gadā* [What was Forbidden for the Reader of the First Republic. Press Rules and the Publications Forbidden for Distribution in Latvia, 1918-1940]. Rīga: Zinātne.

Pašeta, S. (2003), "Censorship and its Critics in the Irish Free State, 1922-1932" *Past and Present*, vol. 181, no. 1, 183-218.

Paukštytė-Šaknienė, R. (2004a), "Požiūris į abortus Aukštaitijoje [An Attitude towards Abortions in Aukštaitija]" *Liaudies kultūra*, vol. 98, no. 5, 14-23.

Paukštytė-Šaknienė, R. (2004b), "Kultūrinių reprodukcijos aspektų raiška kaimo kultūroje [Expression of Cultural Aspects of Reproduction in Village Culture]" *Lietuvos etnologija: socialinės antropologijos ir etnologijos studijos*, vol. 13, no. 4, 89-110.

Paukštytė-Šaknienė, R. (2008), "Gimimų reguliavimo būdai ir jų raiška istoriografiniame kontekste [The Reproduction and its Control: Historiographical Approach]" *Liaudies kultūra*, vol. 122, no. 5, 31-39.

Pīpiņa, B. (1931), "Sievietes likteni liek svaru kausos [The Destiny of Woman is weighed on Scales]" *Jaunākās Ziņas*, no. 63 (19 March), 2.

Pīpiņa, B. (1932), "Liekat dvēselei valdīt pār miesu [Bid the Soul to rule over the Flesh]" *Jaunākās Ziņas*, no. 170 (2 August), 2.

Pīpiņa, B. (1936), "Brīvība un tikumi [Liberty and Virtues]" *Latviete*, no. 6-11, 31-32.

Porter, R. (2001), “History of the Body Reconsidered” in: Burke, P. (ed.), *New Perspectives on Historical Writing. Second edition*. University Park, Pennsylvania: The Pennsylvania State University Press, 233-260

Pushkarev, A. and Pushkareva, N. (2007), “Ranyaya sovestkaya ideologiya 1918-1928 godov i “polovoy vopros” (o popytkch regulirovaniya sotsal’noy politiki v oblasti sekual’nosti) [Early Soviet Ideology of 1918-1928 and the „sexual question” (about the Attempts to Regulate Social Policy in Sphere of Sexuality)]” in: Romanov, P. B. and Yaskaya-Smirnova, E. R. (eds.), *Sovetskaja sotsial’nya politika 1920-1930-ch godov: ideologiya I povsednevnost’. Sbornik statey* [Soviet Social Policy in 1920-1930s: Ideology and Everyday Life]. Moskva: OOO „Variant”, TSSPGI, 199-227.

Putniņš, E. (1926), “Aborta jautājums Latvijā [The Issue of Abortion in Latvia]” in: Skuja, J.; Brants, J.; Hiršbergs, M. and Storoženko, V. (eds.), *Latvijas ārstu un zobārstu 1. kongresa darbi* [Works of the First Congress of Latvian Physicians and Dentists]. Rīga: Kongresa izpildu komitejas izdevums,285-302.

Romanov, P. V. and Yarskaya-Smirnova, E. R. (eds.) (2007), *Sovetskaja sotsial’nya politika 1920-1930-ch godov: ideologiya I povsednevnost’. Sbornik statey* [Soviet Social Policy in 1920-1930s: Ideology and Everyday Life]. Moskva: OOO „Variant”, TSSPGI.

Sauerteig, L. D. H. (1999), “Sex education in Germany from the eighteenth to the twentieth century” in: Eder, F. X.; Hall L. A. and Hekma, G. (eds.), *Sexual cultures in Europe. Themes in sexuality*. Manchester, New York: Manchester University Press,9-33.

Sauerteig, L. D. H. and Davidson, R. (2009). “Shaping the Sexual Knowledge of the Young. Introduction” in: Sauerteig, L. D. H., Davidson, R. (eds.), *Shaping Sexual Knowledge. A Cultural History of Sex Education in Twentieth Century Europe*. London, New York: Routledge, 1-18.

Sohn, A. (1994), “Between the Wars in France and England” in: Thébaud, F. (ed.), *A History of Women in the West. V. Toward a Cultural Identity in the Twentieth Century*. Cambridge, Massachusetts, London, England: The Belknap Press of Harvard University Press, 92-119.

Stranga, A. (2005), “Replika [The Remark]” *Diena*, no. 101 (30 April), pielikums *Kultūras Diena*, 2.

Szreter, S. (1999), "Falling fertilities and changing sexualities in Europe since c. 1850: a comparative survey of national demographic patterns" in: Eder, F. X.; Hall, L. A. and Hekma, G. (eds.), *Sexual cultures in Europe. Themes in sexuality*. Manchester, New York: Manchester University Press, 159-194.

Usborne, C. (2005), "Rebellious Girls and Pitiable Women: Abortion Narratives in Weimar Popular Culture" *German History*, vol. 23, no 3, 321-338.

Usborne, C. (2007), *Cultures of Abortion in Weimar Germany*. New York, Oxford: Berghahn Books.

Veidemanis, M. (1935), "Latvijas 1935. g. 22. marta aborta likums [Abortion Law of March 22, 1935]" *Latvijas Ārstu Žurnāls*, no.10,571-578.

Weeks, J. (1981), *Sex, politics and Society. The regulation of sexuality since 1800*. London and New York: Longman.

Zelče, V. and Sprugaine, V. (2005), *Marginālās jeb 1376. fonds* [The Marginal Ones, Or Fund No. 1376]. Rīga: Latvijas Valsts arhīvu ģenerāldirekcija, Latvijas Valsts vēstures arhīvs.

Zelče, V. (2006), "Vara, zinātne, veselība un cilvēki: Eigēnika Latvijā 20. gs. 30. gados [Power, Science, Health, and People: Eugenics in Latvia during the 1930ies]" *Latvijas Arhīvi*, no. 3, 94-137.

Zuments, K. (1926), *Vīrietis un sieviete savā kopējā dzimumu dzīvē* [Man and Woman in Their Common Sex Life]. Rīga: P. Liepa.

Eugenics against State and Church: Juozas Blažys (1890-1939), Eugenics, Abortion and Psychiatry in Interwar Lithuania 1918-1940

Björn M. Felder and Arūnas Germanavičius

Abstract

The Lithuanian Psychiatrist Juozas Blažys was influential both for psychiatry and the eugenic debate in interwar Lithuania, asking for sterilisations since 1926. He was director of the main Lithuania psychiatric clinic and became first chair of psychiatry at Kaunas University. He stood exemplarily for the Lithuanian elite which was eugenic-orientated, and open to "modernity". But the regime of Antans Semetona had to rely on the Catholic Church, and therefore applied eugenics was introduced very slowly. However, the reform of Lithuanian abortion law of 1935 shows the inability of Lithuanian regime to stay back from the international tendency to accept eugenic selection by abortion.

1. Introduction

With the sudden death of the psychiatrist Juozas Blažys in April 1939, Lithuania lost not only its leading psychiatrist and university professor, but also its most influential eugenicist. The figure of Blažys is interesting not merely in terms of the development of the eugenics debate, but also regarding the transfer of eugenics ideas and methods between countries, and the perception of eugenics-based ideas in Lithuania. Born in 1890, Blažys represented a group of academics trained before the First World War who became main figures in science during the 1930s and 1940s, and who maintained a positivistic worldview, at least regarding science. The example of Blažys also illustrates the different but not contradictory influences of the Russian and German schools of psychiatry – in turn, the former being strongly influenced by the latter. As a student of Vladimir Bechterev, a key figure in Russian psychology and psychiatry,[1] Blažys was exposed to the "problem of degeneration" and eugenics at the Military Medical Academy in St. Petersburg prior to 1917. He was a neo-Darwinist in terms of genetics: his first article in the scientific journal *Medicina* focused on genetics and "Mendelism", and he frequently referred to the Russian eugenicist and geneticist Juri Filipchenko.[2] In Blažys' 1935 textbook on psychiatry, the first of its kind written in Lithuania, he quoted Bechterev and his successor Osipov. But with the same frequency he mentioned German psychiatrists such as Emil Kraepelin and Ernst Rüdin, who both influenced his orientation toward eugenics, especially his focus on genetic pathology and sterilisation.[3]

Further, he was following the latest developments in psychiatry and adopted a programme of mental hygiene that became embedded into his eugenics agenda after attending the World Congress of Mental Hygiene in Washington, D.C. in 1930.[4]

Blažys is a good example of the dominance of the psychiatrist in the field of eugenics. Unlike Estonia and Latvia, the eugenics debate in Lithuania had no prominent racial biologist or racial anthropologist. Blažys also demonstrates the antagonism between biomedicine and the Catholic church, similar to the situation in Poland.[5] Blažys had a strong scientific consciousness oriented toward natural sciences. Scientifically, he opposed Catholic views such as the 1930 *Casti connubii* by Pope Pius XI, which becomes obvious by his intensified eugenics writings in the 1930s.
Blažys and other Lithuanian eugenicists not only acted in opposition to the church, but also against government policy. Dictator Antanas Smetona introduced agro-nationalism as part of his para-fascist state cult that glorified the small farmer as the model citizen.[6] Bio-politics in the form of welfare and pro-natalism came into practice in the late 1930s, and applied eugenics was officially rejected until the Soviet occupation began in 1940. Smetona had to rely on the Catholic Church during his rule, so he did not want to question Catholic bio-power by introducing applied eugenics. Blažys and other eugenicists therefore acted against the church and against the state by propagating eugenics. In the European context, eugenics in Catholic Lithuania may have been peculiar, not only compared to its Baltic neighbours Estonia and Latvia but also compared to other "Catholic" nations such as France, Italy and Poland.

In Lithuania the history of eugenics was closely connected with the process of modernisation. Historians' recent narratives on eugenics and bio-politics in the 20th century have emphasised strong ties with modernity.[7] Eugenics was linked to the growing influence of natural science, and especially biology, in the 19th century. In general, Charles Darwin and his theory of evolution had strongly impacted on biology. But eugenics was designed as an applied science and as a direct reaction to the effects of industrialisation, known as the "social question". Eugenics was touted as an idea to solve social problems through biological means. Early German eugenicists including Alfred Ploetz and Wilhelm Schallmayer were initially connected (mostly informally) to the social democratic party. A utopian approach was a fundamental element of eugenics. The main actors in promoting eugenics-based modernity were academic elites and the middle class which sought to preserve their social status, wealth and political power.[8] In the comparative European-North American perspective, the classical formation that led from debates over eugenics to applied eugenics passed through three steps. During the late 1980s, Paul Weindling demonstrated this scheme of the development of implementing applied eugenics – from the debate stage to the actual practice – citing the German example. The first step was the eugenic debate and promoting eugenics, not least through the founding of the German Society of Racial Hygiene in 1905, which was the first eugenic society worldwide. Its prominent members included Emil Kraepelin, Ernst Rüdin and Max von Gruber, who further lobbied for a national eugenics institute. Institutionalisation of eugenics - the second step - was finally realised in 1926 with the founding of the Kaiser Wilhelm Institute for Anthropology, Human Heredity and Eugenics in Berlin. Lobbying by eugenicists achieved its goal in a range of positive welfare oriented public health measures, and also with the draft of a sterilisation law by the Prussian interior minister in 1932. The "Law of Preventing Hereditarily Sick Offspring" (*Gesetz zur Verhütung erbkranken Nachwuchses*) was enacted only after the Nazi takeover in 1933. This meant the political implementation of eugenics as a state agenda, which was the final and third step in Weindling's scheme.[9]

This three-step theory was challenged not only by "Catholic" eugenics in countries such as France, Italy and Poland,[10] where the Catholic church had a powerful position in society, and its own values and claim to bio-power and bio-politics. Lithuania and to some extent Latvia were also exceptions. In general, the Baltic states seemed to be more peculiar within the classical narrative of European eugenics. Estonia may fit within the scheme, considering its eugenics society from 1924 and its sterilisation law of 1936. But Latvia was different, as its eugenics institute was established in 1938, the same year its eugenic-based sterilisation law came into force. Further, Latvia had no special eugenics society, beside the Society for Promoting Health that added a eugenic agenda to its program in the late 1930s. Lithuania had no eugenics society, eugenics research institute, or eugenic-based sterilisations until the Second World War, though the country had a vibrant eugenics debate and a strong eugenics movement. Eugenics in the Baltics developed in agrarian societies within a small academic elite and a middle class of small farmers, instead of being cultivated by *Bildungsbürger (as in Germany)*. In Latvia and Lithuania, where no eugenics societies were established, eugenics though became an issue of broader social discourse, as demonstrated by the eugenics agenda of the feminist movement.[11] Another Baltic peculiarity was the topic of a "small nation" that should be biologically secured through eugenics. This type of Baltic eugenics with a focus on sustaining the nation was evident in debates prior to World War I.[12] Lithuania was also a special case, as historically it was not part of the "Baltikum" nor was it influenced by German aristocratic culture. In Lithuania, eugenics in state and society was mainly promoted by the bio-medical elite. Still, there was a strong eugenics community, as most of the bio-medical elite were educated in pre-war Russia, where the eugenics debate was strongly influenced by German racial hygienists.[13] Under German and

Russian influence, eugenics was discussed in the fields of psychiatry, criminal biology and "social diseases" (tuberculosis, alcoholism and sexually transmitted diseases).

The literature on Lithuanian psychiatry, not to mention eugenics, is very limited. The 1959 book on racism and eugenics by the Soviet scholar Irmja Zaksas was a large but problematic compendium about the eugenics debate. Zaksas' aim was to condemn "bourgeois" science as "fascism". The book has methodological problems in terms of differentiating racial biology from eugenics, and in contextualising eugenics. [14] The first article on euthanasia in Nazi-occupied Lithuania, from 2004, ignored interwar eugenics in Lithuania, as did all of the major books on Lithuania under Nazi rule.[15] But eugenics and related patterns of thinking were broadly received by Lithuanian bio-medical experts, as shown by a recent article about Lithuanian psychiatry under Nazi occupation. [16] Beyond the unpublished 1960 dissertation by J. Šurkus there are only smaller works, for example, on Blažys, which have the character of Soviet hagiography – or demonology.[17] Blažys only recently has been the subject of more balanced research.[18] This chapter provides an overview of the eugenics debate in Lithuania through the example of Blažys in the context of psychiatry and government bio-politics. The study is meant as a comparative contribution to the volume's chapters on Latvia and Estonia, as well as a motivation for further research on Lithuanian eugenics and nation-building in the context of bio-politics during the interwar period. To contextualise the figure of Blažys, first we summarise the situation of psychiatry in Lithuania between the wars.

2. The Beginning of Psychiatry in Lithuania

Until 1918 Lithuania was ruled by tsarist Russia, where psychiatry developed into a modern form in academic terms. In 1857 the St. Petersburg Medical Surgical Academy launched a special branch for psychiatry, though the number of graduates in such a huge empire was inadequate. Laws that regulated the treatment of mentally ill people and limited some of their civil rights dated to the rule of Katherina II, and the tsars who followed enacted provisions for the entire country. The first Lithuanian in tsarist Russia to become a psychiatry professor before World War I was Jonas Balinskis (Balinskij) (1827-1902), who headed a small department for mental illness at St. Jacob Hospital in Vilnius. In 1903 a modern, regional mental hospital in Naujoji Vilnia (Vileika) was established, with a capacity of 1,000 patients. Nikolaj Krainskyj (1869-19?) served as its director. The hospital was built in accordance with a master plan for mental hospitals along with six other projects in different locations in the Russian empire. Ten doctors were assigned to each hospital, and each featured modern treatment and nursing, as well as daily activities. Work therapy was also introduced. In 1915, staff members and documents from the Naujoji Vilnia mental hospital were evacuated to Slaviansk in the Ukraine due to Germany's advance during World War I. Most of the patients were released and settled in neighbouring villages.[19]

In 1916, while under German occupation, Tauragė (Tauroggen) established a 100-bed *Landesirrenanstalt für Litauen* (National Psychiatric Hospital for Lithuania). After the war, it served as Lithuania's main psychiatric hospital and was known the State Psychiatric Hospital. Its first director was Juozas Blažys, who worked there from 1918-24.[20]

The Lithuanian independence movement accomplished its goal during final months of World War I by re-establishing Lithuanian statehood – the Republic of Lithuania – on 16 February 1918. Many prominent physicians such as Vincas Kudirka (1858-1899), physical anthropologist Jonas Basanavičius (1851-1927) and Jonas Šliūpas (1861-1944) participated in the political movement. Medical professionals trained in Russia, Germany, Poland, Scandinavia and the US were asked to take positions in state and local governments, in science and in the health system. King Stefan Batory University was established in Vilnius as a national university in 1918, and was the country's only university until 1922. It had no special department or chair for psychiatry. In 1918-19 the Lithuanian army fought Bolshevik troops, pushing them back to Russia and consolidating Lithuanian independence. On 9 October 1920, Vilnius and south-eastern Lithuania were occupied by the Polish army, which remained until 1939. Due to political events and the loss of Vilnius, the academic landscape including psychiatry was separated. When Kaunas became the new capital, a new national university was founded in 1922: Vytautas Magnus University.[21]

3. Psychiatry at Vilnius University

In 1922 a chair of psychiatry was established at the University of Vilnius, which was a Polish University at the time. Professor A. Mikulski served as chair, followed by Professors R. Radziwilowicz and S. Vladiczko. A prominent figure in both local and European psychiatry, Dr. Maximilian (Maksymilian) Rose (1883-1937), was appointed as chair in 1931. He was said to be an able public speaker and was beloved by students and public. He gave public lectures on issues including nervous and mental disorders, and soon became an influential figure in the scientific world. He had a solid academic standing, having studied and researched with T. Ziehen and H. Oppenheim in Berlin, and later under K. Brodmann in Tübingen. In 1925 Rose was appointed prospective director of the neurology department at the Kaiser Wilhelm Institute in Berlin for Anthropology, Human Heredity and Eugenics, where he remained until 1928. Later he moved to the University of Warsaw, where he founded the first institute for brain research.

In 1923 the Manson house on Vasaros Street in Vilnius was donated to city by an aristocratic family to become a hospital for children and teenagers who suffered from tuberculosis. After a patient committed suicide, the clinic was converted to psychiatric hospital in 1927. It remains open, as the longest continuously operating psychiatric institution in Lithuania. It also constitutes the Clinic of Psychiatry of the Faculty of Medicine at Vilnius University. In 1930 a modern building was erected at Vasaros Street, and the Polish Brain Research Institute was transferred there from Warsaw. Only three such research institutes existed in Europe at the time. Rose was appointed director, and in addition to clinical work he implemented new neuropathological staining techniques for diagnostics. The method of a Nobel Prize winner, Spanish histologist Ramon y Cajal (1852-1934), was introduced to the Brain Research Institute in Vilnius and further developed by local assistants.[22]

In 1933 the neurology department was also transferred to the Vasaros hospital, and the mental clinic was renamed the Clinic for Nervous and Mental Disorders. The latest treatments were used, such as the malaria therapy of Viennese psychiatrist and Nobel Prize winner Julius Wagner-Jauregg (1857-1940), to "cure" progressive paralysis, a form of syphilis, and possibly also mental diseases. In particular, this type of therapy was demonstrated and taught to medical students: Rose asked that soldiers infected with malaria be transported from Pinsk to Vilnius by military aircraft in order to carry out blood transfusions with syphilis patients. After some weeks of febrile episodes due to malaria, the cases of syphilis paralysis and mental disorders in psychiatric patients were reduced remarkably, following Rose's assumptions.[23]

In May 1935, the heart and brain of Marshal Josef Pilsudskyj was transferred to Vilnius. According to his testament, Vilnius University was entitled to keep Pilsudskyj's brain, which Rose studied with a research group under his supervision at the Brain Research Institute. Among them was prominent Polish-Canadian neuropathologist Jerzy (George) Olszewski, who later described the Steele-Richardson-Olszewski disease in neurological literature. Rose found no exceptional post-mortem brain morphology indicating that Pilsudskyj was a genius, and therefore Rose was condemned by the public.[24] This disappointment caused him great anguish, and he died from myocardial infarction in 1937. In 1938 his report on the study of Pilsudskyj's brain was published in French.

Psychiatry at Vilnius University was based on the latest concepts of clinical psychiatry, which benefited greatly from its confluence with neurology. This developed into neuropsychiatry, which searched for causes and possible treatments. Possible social causes for mental illnesses were not sought after in this research. There is no evidence that eugenics or social psychiatry were discussed or focused upon in Vilnius University - though this possibility seems reasonable.

4. Juozas Blažys: The "Godfather" of Lithuanian Eugenics

In 1920 university-level courses were established in Kaunas to compensate for the loss of Vilnius University. In 1924 the courses were incorporated into Vytautas Magnus University. In Kaunas the development of psychiatry was closely linked to Professor Blažys, who became important and influential not only for the development of Lithuanian psychiatry, but also for the launching a national eugenics debate, which began in Lithuania after World War I. Blažys remained a leading figure in the debate into the 1930s.

Blažys was the first chief of the department of nerve and psychiatric illnesses and the vice rector at Lithuania's Vytautas Magnus University. He was a highly prominent figure and productive scientist of enormous erudition who knew eight languages and wrote comprehensive reviews. He wrote textbooks including "Introduction to Psychiatry", the monograph "Tolerance as a Basis of Culture", and many articles in *Medicina* and various journals.[25] To this day his ideas are recognised by Lithuanian psychiatrists. His interests included the causes of psychosis, alcoholism, somatogenic psychosis, forensic psychiatry, military expertise, heredity and evaluations in the field of eugenics. Because one article called him the unofficial "godfather" of Lithuanian eugenics, he was later condemned during Soviet times and his texts forbidden.[26]

Blažys was born in Šiauliai. In 1914 he graduated from the Military Medical Academy in St. Petersburg and began working at various Russian military hospitals. In 1918 he returned to Lithuania and to work at the psychiatric hospital in Tauragė, first as a department head and later as director, while also working as a physician in Tauragė. The Tauragė hospital was established in the former castle of Duke Radvila (Radziwillo), which had been used as a customs office. Blažys was committed to improving conditions for patients during this difficult post-war period, and tried to persuade local and state governments to fund psychiatric services. He wrote annual reports on hospital performance that were published in the Kaunas-based medical journal *Medicina*. In addition to detailed statistics that illustrated various problems and clinical case histories, Blažys emphasised the lack of financial support and the need for a larger and up-to-date psychiatric hospital that could serve the entire country. He published principles for such a hospital, emphasising modern care, nursing and diagnostics. This became an effective lobbying tool. In 1923 the Lithuanian government increased the number of beds for psychiatric patients by moving the State Psychiatric Hospital to Kalvarijos, where the facility was renovated. In 1920-21 Blažys taught psychiatry at the Higher Courses of Study in Kaunas. In 1924, after the University of Kaunas was established, he was elected chief of the department of nerve and psychiatric illnesses. In 1935 he was granted the degree of professor, and in 1938 he was appointed vice rector of Vytautas Magnus University. He published articles in various fields including psychiatry, psychology,[27] mental hygiene, criminality and criminal anthropology, and "degeneration". In particular, he wrote about the negative impacts on heredity caused by alcoholism, smoking and prostitution,[28] and heredity and genetics in general.[29] He also wrote about "improving" the Lithuanian genotype with methods of eugenics, such as sterilisation.[30] Blažys died of a heart attack in 1939.

Blažys strove to understand and follow the latest ideas, and he carefully examined their scientific foundations. In his 1935 textbook "Introduction to Psychiatry", he wrote with enthusiasm about the latest discoveries in genetics while highlighting their limitations. He discussed some ideas as being not based on science, revealing their weaknesses. He strongly linked hereditary mental conditions with social and biological "degeneration", and he actively promoted sterilisation.[31]

He wrote two articles and a book about sterilisation and put forth a comprehensive review of these ideas, from antiquity through the beginning of the 20th century. He described examples of eugenic sterilisation all over the world, emphasising that the US was the first country where sterilisation became widespread. He wrote that Switzerland and Denmark, beginning in 1929, were the first European countries where eugenic sterilisation was carried out. Blažys argued that Lithuania should work to pass eugenics laws and pursue applied eugenics. He presumed that about 15,000 "mentally defectives" lived in Lithuania, of whom about 300 should be sterilised each year. He also suggested three categories for indications of sterilisation:

> "First group: mental patients discharged from a psychiatric hospital, when it is established that their disease is inherited in its origins; there are about 200 such patients annually.
> Second group: recruits who are not suitable for military service due to oligophrenia, epilepsy or mental disorder; there are about 70 such individuals annually.
> Third group: criminal offenders who show signs of mental degeneration (defect) and are recognised as legally irresponsible; there are about 30 annually."[32]

Though Blažys began promoting eugenic sterilisations in 1926, the idea never came to fruition in Lithuanian. It is interesting that he targeted similar groups as his colleagues in Latvia and Estonia, including individuals suffering from mental diseases that easily could be connected to the imaged group of so called "feeble-minded" (*silpnapročiai*), who become victims of eugenic sterilisation in Latvia, Estonia and had been also the main target of Nazi compulsory sterilisation programme.[33]

It should be mentioned that he also followed what would even be called today a modern approach: educating people about tolerance as a measure against psychiatric morbidity and the stigma of mental illness. This is far-reaching idea that was not generally present among the leading psychiatric thinking of the 1930s. Blažys wrote extensively about examples of tolerance, the crisis of tolerance at that time, and the limits of tolerance. He harshly criticised the "*Massenmensch*" (popular mass) fanaticism and totalitarian regimes such as Nazi Germany, fascist Italy and Stalinist Russia. However, he did not specify his approach or mention any methods for implementing tolerance, except by educating society and maintaining high moral values in families and in public life. In the end, the mentally ill and disabled were not included in his agenda of tolerance.[34]

5. The Eugenics Debate in Interwar Lithuania

The dominance of Blažys' thought becomes obvious when reviewing the eugenics debate in Lithuania during the interwar period. It was Blažys himself who launched the debate in 1921 with an article on genetics. He published at least 17 articles and monographs on eugenics or related topics such as criminal biology, alcoholism and mental hygiene, which contributes to his eugenics viewpoints. The articles were published not only in academic journals such as *Medicina*, but also in journals for the intellectual elite, including as *Gamta*.[35] The sheer quantity of Blažys' writings on eugenics served to elaborate and promote his ideas on the topic. But his most influential role was that as a university professor. He lectured eugenics and wrote a psychiatry textbook in 1935 that included a chapter that promoted eugenics, and which also provided a scientific foundation and legitimising element for eugenics, focusing on genetics and hereditary mental pathology.[36]

The eugenics debate in Lithuania developed slowly during the mid-1920s and became more intense toward the end of the decade. Blažys published his first major work in 1926, which coincided with the political takeover by Smetona.[37] There seemed to be a reason behind it: Blažys became the most active lobbyist for eugenics during the Smetona regime. It seems that in his perspective an authoritarian nationalist ruler as Smetona might be much interested in applying eugenics. His lobbying and the eugenics debate peaked in 1933, when Blažys published two works and gave public lectures on the issue. This was clearly due to the enactment of Nazi Germany's eugenic law that same year.[38] At the time, the interest in and discussion on eugenics grew as a response to the Nazi German law. Blažys' programmatic paper of 1933 embraced the law and proposed similar measures for Lithuania.[39] Some physicians such as O. Pūdymaitis welcomed

Germany's eugenics agenda,[40] but other Lithuanian eugenicists distanced themselves from Nazi policy. Even Blažys emphasised that eugenic sterilisation should only be conducted following the presence of a strict "medical indication". The well-known physician L. Epšteinas also underlined that the eugenics measure he proposed for Lithuania should be conducted as a "medical operation", not a "political" task. He emphasised that "eugenics is not racism" and that it did not "aim to breed a master race".[41] Also in 1933, critical views from publicists and politicians condemned eugenic sterilisations. Valentinas Gustainis, who served as Lithuania's ambassador to the US, wrote critically about the Nazi Regime. His close relationship with the Smetona regime and the Catholic church became obvious through his comments about the anti-Catholic agenda of Nazi racial hygiene. And, his articles in the journal *Vairas*, the ideological organ of the Smetona regime, demonstrate Gustainis' connection to it. Despite his critical works, Gustainis supported eugenics in general.[42]

Most authors who wrote about eugenics came from medical professions or at least were academics. As a result, major contributions appeared in the bio-medical journal *Medicina*. Because Blažys came from the psychiatric field, he focused on supposed hereditary mental diseases, mental pathology and "feeble-mindedness" as the main factors of "degeneration". Given his approach to psychiatry and genetics, and because of the influence of German racial hygienists, it is not surprising that he favoured a quite radical form – with 300 people to be sterilised each year. He had a number of followers, including physicians O. Pūdymaitis and A. Babianskis, as well as K. Vainius. Also veterinarian K. Aleksa was in favour of sterilisation, his contribution even appeared in *Vairas*.[43] It is interesting to note that some of them wrote their pro-eugenics and pro-sterilisation works anonymously.[44] One of them can be identified as Professor Jonas Vabalas-Gudaitis, an academic pedagogue and psychologist, who extensively cited the works of Blažys. Also interesting is the case of V. Bendoravičius, a physician and military officer who directly followed Blažys by focusing on genetics and mental diseases. He wrote the book "Degeneration and Eugenics" in a series edited by the "Kaunas City Health Insurance" (*Kauno miesto ligonių kasa*). He also contributed to the regime's organ, *Vairas*.[45]

Affirmation of eugenic sterilisation came not only from bio-medical experts but also from lawyers and generated public discussion. In 1933, when Lithuanian eugenicists (spearheaded by Blažys) lobbied strongly for eugenic sterilisations in response to the Nazi Germany's eugenics law, a lawyer named Vazbys quoted Blažys' article in which Blažys said that eugenic sterilisation would be illegal and a crime under contemporary Lithuanian law. Vazbys stated that due to tsarist lawyers (part of Lithuania's penal law stemmed from tsarist times), sterilisation due to any indication would be legal with the person's agreement, in contrast to the Nazi Germany's law, which allowed for coercion. We have no information on governmental reactions on voluntary sterilisation or any practical application of sterilisation in Lithuania until 1940.[46]

A special case worth noting is Juozas Šliupas, a well-known member of Lithuanian society, a founding father of the pre-WWI national movement, and an important politician in the early republic. He was also a physician who studied in the US and was a eugenics enthusiast. At the Second Lithuanian Congress of Culture in 1926, he lectured on "racial hygiene" and continued this activity into the 1930s.[47] He supported German-style racial hygiene and sterilisation. In 1943, Šliupas became the first Lithuanian author to say that mentally ill and other chronically sick people should be killed.[48]

Other voices questioned eugenic sterilisation, to varying degrees, due of the low level of research and incomplete knowledge about heredity and genetics. Ethics also played a role as an argument against eugenic sterilisation. Physicians such as K. Epšteinas and M. Marcinkevičius mostly promoted measures of "positive eugenics", including eugenics counselling, genealogical research, genetic prenuptial certificates but also "negative eugenics," such as banning marriage for certain groups of people.[49] It should be mentioned that Balžys' critics frequently referred to him and his research, particularly the estimated numbers of mentally ill and "feeble minded" people in Lithuania that he calculated. These calculations appear in nearly every work related to the eugenics debate in Lithuania.

In addition to psychiatry, another field of medicine and hygiene in which eugenics developed a strong presence was the category of so-called "social diseases". The widespread forms of alcoholism, tuberculosis and sexually transmitted diseases were a major public health problem in Lithuania, and it had been known since the late 19th century that they could be genetically harmful to offspring. The "fight" against these diseases shaped the rhetoric behind the ideology of eugenics; during the interwar period, alcoholism and its consequences were seen as major health and social problems in Lithuania.[50] As a psychiatrist, Blažys played a role in this debate within the context of the Europe-wide consensus among members of the bio-medical elite that "social diseases" were producing "feeble-minded" people, "idiots" and "criminals" who psychiatrists would have to treat. Therefore, Blažys also wrote about the "dysgenic" effects of alcoholism and sexually transmitted diseases.[51] Other authors who took part in this debate included the mentioned Epšteinas and psychiatrist Viktoras Vaičiunas.

While some eugenicists questioned sterilising mentally and hereditarily ill people, surprisingly they demanded sterilisation of alcoholics, syphilitics and tuberculosis patients. Dr. A. Babianskis, who headed a tuberculosis dispensary in Panevėžys, asked for tuberculosis patients to be sterilised and for their children to be isolated from their parents. His idea was not only intended to limit the spread of the diseases but also to reduce the number of "a-social", quoting Blažys. Obviously, for some authors, shrinking the underclass was also a prophylactic method of fighting tuberculosis.[52] In the European context there was a strong eugenics element within the arguments and means of "fighting" social diseases, which was also the case in pre-revolutionary Russia. Babianskis, therefore, was not an isolated case. Physician L. Gutmanas also favoured sterilising patients with tuberculosis, rather than the mentally ill. Also politician V. Gustainis asked to enlarge of the group of "inferiors" to include syphilitics and alcoholics – though within the frame of eugenics.[53]

Lithuanian eugenicists who focused more on the hereditary and alleged biological-social effects of "social diseases" than on mental illnesses targeted another group that they defined biologically: criminals. Lithuania as pre-revolutionary Russia had a strong group of doctors who followed Cesare Lombroso's theory of "atavism" and its implications for hereditary criminality.[54] Lombroso was frequently mentioned as an expert on "degeneration" in the Lithuanian eugenics debate. For many Lithuanian eugenicists, Lombroso's thesis was the main argument of extraordinary eugenics means: sterilising criminals. This was the eugenics goal of Lombroso: to end deviance and criminality by sterilisation and "root" it out forever, for the sake of future generations. At the time, some US states were still castrating criminals; this was well known in Lithuania, as Blažys mentioned on several occasions.[55] Again it was Blažys who led the debate, which partially was staged in a specialised media outlet, the *Criminological News (Kriminalistikos žinynas)*, published from 1935 to 1940. It was the only journal on criminology published in the Baltics before the Second World War. There, heredity of criminality and sterilising criminals and the mentally ill was discussed mostly in an affirmative manner.[56] Leading eugenicists Gustainis and Gutmanas both recommended sterilising criminals, but condemning sterilising the mentally ill.[57] This interesting aspect of Lithuanian eugenics was discussed in particular in Estonia but not Latvia, nor was it the focus of eugenics debates in Western Europe at that time.

A common feature of European eugenics was the basic idea of biological "degeneration". Some eugenicists also considered possible social causes of degeneration, but in general, genetic-based explanations were favoured. We have seen the predominance of Lombroso in the debate about criminality. Concerning heredity and genetics, there was a domination of neo-Darwinism or "Mendelism", the classical genetics of Gregor Mendel and August Weismann. Hardly anyone followed neo-Lamarckian directions and the theory of the possible heritability of acquired characteristics, besides some Communist enthusiasts.[58] But the mainstream bio-medical elite, represented in the journal *Medicina*, was also convinced of determined genetics. The natural science journal *Kosmos* published a special issue on the anniversary of Mendel's death in 1935.[59]

There was a strong association between ideas about degeneration and genetics in debates on inherited "feeble-mindedness", criminality, and the genetically harmful effects of alcohol and sexually transmitted diseases. All of these discourses had a common feature that formed the nucleus of eugenics: the idea of "inferior" human beings who represented a dead-end in evolution: accidents of nature who did not deserve the treatment of normal, "superior" individuals, and who should not procreate or mingle with "normal" people. Within academic debates, we find no open racism against ethnic groups or "races" of the usual categories of "white", "black" or "yellow". Instead, there was a categorisation that implied a hierarchy of values. Blažys frequently used the term "inferior" to label mentally ill and mentally disabled people, which was a psychiatric tradition of the 19th century. This is evidenced by a quote by Emil Kraepelin concerning the "inferior", which Blažys used in his 1935 psychiatry textbook. Further, there were the "a-social", criminals, prostitutes and so on, who were perceived as being quite near the "inferior", as these forms of deviance were also seen as hereditary or "pathological genotypes".[60] Eugenicists who were more left-wing or social democratic, such as the physician Epšteinas, spoke about "social pathology". So, there was a strong agreement regarding undesirable people who were biologically beneath the imagined norm – those who had hereditary or possibly hereditary defects. Actually, in the interwar debate in Lithuania, no one spoke in favour of killing these "inferiors", while the discussion on euthanasia was also noted in Lithuania.[61] Interestingly, many eugenicists rejected Nazi racism and also officially labelled that phenomenon as "racism". "Reducing of the inferior", such as the mentally ill, the mentally disabled, invalids and the socially weak, by eugenics was seen as something totally different to racism.

In general, Lithuanian eugenicists took a collective and national approach. Even when Blažys called eugenics a programme of "prophylactic psychiatry", it was clear that the eugenic programme he proposed was dedicated to the Lithuanian nation. His aim was to reduce the "degenerates" and lower the costs incurred by the state. Epšteinas spoke about the importance of "health" for the "national economy", and the economic "deficit" caused by the mentally and physically unhealthy, which Blažys also emphasised.[62] Strengthening the nation and the state biologically and economically was an approach, which Marius Turdas described as being typical for Central Europe. Similar to Estonia and Latvia, the question of national health was linked to the issue of the "small nation", which in turn was connected to social-Darwinism. In the eyes of Baltic eugenicists, politicians had to take extra care of the nations because they were under permanent threat due to their small size. Bendoravičius wrote in the context of eugenic sterilisation: "We are a small nation, surrounded by nations with vigorous inhabitants. We cannot let feeble-minded and other degenerates procreate: we have to strengthen our nation."[63]

The Lithuanian eugenics movement also shared with its Baltic neighbours the Nordic-Protestant approach. There was the clear concept of neo-Darwinist genetics and a broad support for sterilisation. There was a pronounced openness to German-style racial hygiene, even though the Nazi regime itself was opposed. Meanwhile, the Russian approach contained a strong element of hygiene by "fighting" against "social diseases", as well as a eugenic agenda. Interestingly, the Lithuanian debate differed from "Catholic" eugenics in Italy, for example, where neo-Lamarckism was very influential, methods of "positive" eugenics were predominate, and sterilisation was rejected.[64] The main eugenicists in Italy were themselves influenced by Catholic values; in fact, the church was involved in the eugenics debate and promoted genetic research. In Lithuania, eugenicists were critical toward the church, and we could hardly speak of Catholic eugenics here. It seems obvious that the Lithuanian bio-medical elite, who were mostly educated in Tsarist Russia, shared an anti-church agenda.[65] There was a similar situation in Poland, where eugenicists also faced resistance from the Catholic church.[66] Unlike the other Baltic States, Italy and Poland, "race" in terms of racial anthropology was absent from the eugenics discourse in Lithuania. While Blažys referred to "racial pathology", only a few works emphasised the role of race, in particular anthropological categories. And unlike their Baltic and Italian counterparts, Lithuanian racial anthropologists did not write extensively about eugenics. The main figures in Lithuania's race debate, ethnographer Jonas Balys and racial anthropologist Jurgis Žilinskas, emphasised the "Nordic" origins of Lithuanians and their membership of the "Nordic race". Both strongly opposed Nazi German viewpoints toward the "low" racial "value" of Lithuanians.[67] Balys and Žilinskas did not include eugenics-based arguments in their writing.

Probably the biggest difference from their Baltic neighbours of Latvia and Estonia was the fact that Lithuanian eugenicists could not convince the Lithuanian government, in particular the Smetona regime, to implement a programme of applied eugenics until 1940.

6. The Smetona Regime and Eugenics

To date, little historical research has been conducted on the authoritarian regime of Antanas Smetona, who took over Lithuania though a military coup d'état in 1926.[68] Smetona developed a para-fascist regime that became more authoritarian during the 1930s and provided a blueprint for regimes in Estonia and Latvia followed in 1934. Like his Baltic dictator colleagues, Smetona was a veteran of the war of independence, and he played an important role when the independent Lithuanian state was formed in 1918.

He aggressively held down right-wing radicals, such as Augustinas Voldemaras and his fascist *Iron wolf* organisation. He initiated a political cult of personality and adopted fascist aesthetics and rituals, including large rallies by his party organisation, *Tautininkai* (Nationalists). Like Päts in Estonia and Ulmanis in Latvia, he emulated political structures from Italian fascism.

While Smetona was eager to establish a good relationship with minorities, at least with the Jewish minority, the regime was tolerant toward the minority question. There was a problematic relationship with Lithuanian Germans and Germany itself because the Memel region was occupied by Lithuania as compensation for the loss of Vilnius in 1920. Smetona developed the rhetoric of "blood and soil" and the practice of "Lithuanisation" in certain aspects of society, and his policies implied a homogeneous nation-state dominated by Lithuanians. Nationalism under Smetona included biological concepts of nation and sometimes anti-Semitism that was not wholeheartedly encountered.[69] Like in Latvia and Estonia, the Smetona regime built up not only a cult around its leader, but also one around the small farmer, as Lithuania more so than its neighbours was an agrarian country.[70]

In contrast to the regimes in Latvia and Estonia, Smetona was slow to enact social policies in the field of bio-politics, such as methods of pro-natalism and "positive" eugenics. Even the social insurance system was not fully developed until 1940. Only on the eve of war were special programmes to foster births and secure the national health proposed.[71]

Estonia and Latvia frequently are described in this volume as forms of racial states. In the case of Lithuania, at least a concealed biological concept of nation, the tendency toward an ethnic state and the late and cautious introduction of bio-political legislation may support this categorisation. Based on Michael Burleigh's definition of the bio-political state, the applied eugenics agenda is missing. Without a doubt, the main reason for this was the influence of the Catholic church. The church in Lithuania had strong social and political positions, and Smetona obviously needed the church to support his regime. At the same time, the church did not resist Smetona. In terms of eugenics, Smetona had to respect the church's position, and its claims regarding bio-power and bio-politics.

During the eugenics debate of the 1930s, Catholic authors strictly followed the doctrine of Pope Pius XI and his 1930 encyclical on eugenics, the *Casti connubii.*[72] The pope condemned eugenics in nearly every form, although the encyclical allowed eugenics counselling. Catholic authors promoted the encyclical and opposed the programmes of Blažys and other Lithuanian eugenicists.[73] Authors close to the church also discussed issues such as euthanasia, the debate that followed the book of Binding and Hoche, and the political and bio-medical changes that took place in Germany after

1933.[74] In the fields of eugenics and bio-politics, the Catholic church in Lithuania tried to influence the government, as demonstrated by the Lithuanian abortion law of 1935.

7. Eugenic Abortion in Lithuania

It is clear that Smetona officially rejected the implementation of applied eugenics in Lithuania. No laws were enacted that provided for programmes such as eugenic sterilisation, eugenics counselling or eugenic-based bans on marriages. On first examination, it seems that the regime went along with the Catholic church's criticism of eugenics. State agencies directly intervened in the debate on applied eugenics by opposing eugenic sterilisations.

There are signs that support the view that the Smetona regime thought favourably of eugenics, but it could not implement them because it needed the church's support. At first, the regime allowed public discussions on eugenics in media outlets that reached not only specialists and the academic elite, but also the broader society. Blažys and others, for example, could speak openly in favour of eugenics sterilisations. Smetona and his dictator counterparts in Estonia and Latvia had introduced censorship, giving them the power to block eugenics debates. In 1934 private efforts to implement practical eugenics peaked with the founding of a eugenic counselling point, which was initiated by the Kaunas health insurance and physicians' society "Friends of Health". There are no signs that the state opposed the project.[75]

Another means of eugenics was directly permitted by the Lithuanian state: eugenic abortion. In general, eugenic abortion is underestimated as a method of eugenics, and there is no further research conducted on that field in European context yet. For example, a book has been written on sterilisation but not on eugenic abortion in Nazi Germany.[76] This may be due to the fact that eugenic abortion hides behind the term "medical indication", but is part of daily life today and is probably conducted in every European country – while certainly avoiding the use of the term "eugenic". This was also the case in interwar Lithuania. The issue of abortion is an obvious case, when the Smetona regime inclined to the scientific and public debate demanding applied eugenics.

The public debate over abortion began in Lithuania in the mid-1920s, both in the public media and in scientific journals. The debate was associated with the question of demography, as well as the social and economic situation of women in general.[77] The rate of illegal abortion was perceived as being far too high. This was due to the dual role of women – as mothers and workers – as well as low levels of social and governmental support for mothers, the lack of knowledge about contraception, and moral norms. Stemming from tsarist law, abortion was illegal – punishable by three years' imprisonment for the women who had the abortion. Additionally, physicians, medical personnel and non-medical personnel could be sued. The only exception was if the mother's life was in danger.[78]

As a result, some authors came out in favour of liberalising abortion and introducing "social" indications, as in Germany and the Soviet Union before 1936. This was particularly the case after Latvia permitted "free" abortions in 1933.[79] While the Catholic church and Christian democrats rejected any form of abortion except to save the life of the mother, most authors favoured the liberalisation of the ban of abortions, reducing punishment for illegal abortion and carrying out reforms concerning the social and economic situation of women in Lithuania.[80] The Smetona government had no interest in changing the general ban on abortion because it followed a pro-natalist course, due to its "Catholic" agenda of bio-politics. In 1933 Smetona finally reacted to the public debate on abortions and to the changes in Latvia's abortion law. That year, the medical journal *Medicina* reported that the cabinet was discussing liberalising abortion.[81] In 1934, the projected new medical law was published in newspapers.[82] This provoked a harsh reaction by the archbishop of Kaunas, who wrote Smetona that "following the laws of the Catholic church, abortion is seen as a crime, and the culprit and anyone else involved will face heavy punishment from the church".[83] The medical law was finally enacted in 1935. The new regulations permitted an abortion following a rape and medical indication. The law did not use the term "medical indication", but required women to "provide two attests on the necessity of the termination of pregnancy by two physicians".[84] The law allowed physicians to define medical indication. On the issue of abortion, the Smetona regime reacted to the bio-medical debate and contrary to the requests of the Catholic church. But the government's approach, particularly due to its open definitions of medical indication, went further. Through the 1935 medical law, Smetona introduced eugenic abortion in Lithuania – or at least legalised an on-going practise.

There was a general consensus within the Lithuanian debate that abortions could be conducted if the mother had a disease that could damage the child's genes and cause conditions such as heart problems, tuberculosis, syphilis and mental diseases.[85] Some writings openly demanded eugenic abortion as a prophylaxis for hereditary and alleged hereditary diseases, but also as means to improve national health. In 1929 physician T. Goldbergas, for example, called for "prophylactic abortion" for cases such as tuberculosis, epilepsies, vitum cordis and chorea minor.[86] For this reason, it is not surprising – and it can be concluded – that abortions were conducted routinely in Lithuanian hospitals. The operation registers of the private Women's Hospital (*Mažylio Moterų Ligoninė*) in the capital of Kaunas indicate that abortions were done frequently, in most cases following diagnoses of heart problems and tuberculosis.[87]

Interestingly, eugenic abortion was not called an "achievement" of the lobbying activities by Lithuanian eugenicists, nor was it a main issue in the general eugenics debate before enacting the new medical law implementing eugenic abortion.

Eugenic counselling and the permitting of eugenic abortions show the hidden eugenic agenda of the Smetona regime and the influence of bio-medical ideas from the Lithuanian academic elite. Despite its alliance with the Catholic church, the Smetona government could not totally reject the "modern" demands from medicine and bio-politics, nor could it avoid being influenced by the policies of neighbouring Latvia, Estonia and the Soviet Union.

8. Conclusions

Based on the very few historical reviews of Lithuanian psychiatry that appeared after 1949, Professor Blažys was either condemned by Soviet psychiatrists and historians as a racist, nationalistic supporter of eugenics,[88] or, more recently, idealised as a preventive and quite modern psychiatrist whose ideas were not recognised – the typical "fate" of a genius.[89] Regardless, his influence on the development of Lithuanian eugenics and clinical psychiatry cannot be denied or underestimated. Viewed comparatively, Blažys was an outstanding figure in context of eugenic movement: neither Latvia nor Estonia had a physician who at the same time was a founding father of a psychiatric discipline, the director of clinical psychiatry and a eugenics enthusiast. Similar figures such as Latvian Hermanis Buduls and Estonian Maximilian Brezovsky (a Baltic German) were early and important psychiatrists, but both were reluctant to promote radical forms of eugenics as extensively as Blažys did. A possible comparison would be Estonian Hans Madissoon, who began preaching eugenics in the early 1920s.[90] He was much younger and only became a professor in 1939, however, and in terms of science and medicine his influence was limited because he was not a physician. He obtained a Ph.D. in natural science at the end of the 1920s and changed academic fields several times, from physical anthropology to criminal biology and finally to eugenics.

The debate on eugenics in Lithuania was similar to that in Latvia and Estonia. The significant influence from German racial hygiene fostered what is called "Nordic-Protestant" eugenics, similar to eugenics in Germany and Scandinavia with its neo-Darwinist approach in genetics, and its focus psychiatric, hereditary pathology, propagating sterilisations as adequate solution to the problem of degeneration. Linked to the German influence was the Russian school, which was somehow closely connected, as the eugenics debate in Russia was strongly influenced by German racial hygiene, but with a strong focus on hygiene; the "fight" against the "social diseases", alcohol, tuberculosis and sexually transmitted diseases; and the strong reference to Lombroso's "atavism". Following Marius Turda, the palingenetic approach was largely a contemporary central-eastern European issue. The focus on a "small nation" that should be saved and secured by a programme of national eugenics is clearly a Baltic peculiarity linked to social-Darwinism.

Further, Lithuania can be counted to "Catholic eugenics" – but is still special. In contrast to Italy, the Catholic church in Lithuania did not actively participate in the debate on eugenics, nor were Lithuanian eugenicists on that scale influenced by Catholicism as in Italy. Lithuanian eugenics is more comparable to Poland, following the secular religion of "nation" and "nature", and with its "strict" scientific elite who opposed the church, which was seen as a kind of "backward" institution.[91] That was typical for the generation of scientists educated in Russia. At the same time, Catholic ethics influenced the Lithuanian eugenics debate, as a number of voices rejected sterilisation and favoured "positive eugenics" such as eugenics counselling, prenuptial certificates or, at least, marriage bans. To some extent, this might be less "Catholic" and more a direct reaction to Nazi Germany's legislation and practice of sterilisation under the German eugenics law of 1933.

Particular to Lithuania were figures in the eugenics debate who rejected sterilising mentally ill and feeble-minded people, but who demanded it for tuberculosis patients, syphilitics and alcoholics. This clearly originated from the proto-eugenics and eugenics debate occurring in Russia before the revolution, where "social diseases" played a major role. Also peculiar in the comparison to the Baltic neighbours and the European debate at that time was the support for sterilising or castrating criminals that was going on among Lithuanian criminal biologists and among eugenicists – they often were part of both groups as Blažys. Interestingly, some critics of sterilising mentally ill people and some Catholic-minded figures such as Gutmanas favoured sterilising criminals. In general, Lombroso and his thesis of hereditary criminality were known and strongly embraced in Lithuania – probably also a heritage from Russia. In Soviet Russia, until 1930 criminal biologists were likely to follow his ideas.[92]

Catholic doctrine had a major influence on the state's eugenics policy, as evidenced by the authoritarian Smetona regime's official rejection of eugenics. This was caused by the Catholic ethics espoused by members of the regime. Smetona was acting pragmatically because he obviously needed the church to help stabilise and legitimate his regime. Like Mussolini in Italy and Hitler in Germany, Smetona also had an alliance with the Catholic church. Introducing applied eugenics at a time when the church was following the Pope's anti-eugenics *Casti Connubii* clearly would have jeopardised support from the church. Still, the regime's hidden eugenics agenda endorsed the promotion and popularisation of eugenics; the establishment of private or semi-private eugenics counselling points in Kaunas; and the introduction of eugenic abortion under the label of "medical indications" in 1935. We could presume that Lithuania may not have had a national eugenics project as in Estonia and Latvia, but still the situation was one of applied eugenics with governmental support. The allowing of eugenic abortions demonstrates the eugenic-mindset of the Smetona regime, and how the regime followed the demands of the medical elite. Lithuania in fact fits within the framework of Baltic authoritarianism and the racial state: the regime promoted the *voelkisch* ideology of "blood and soil", implying a racial-biological definition of the Lithuanian nation. Indeed, there was a strong tendency towards an ethnic state. There was no national eugenics programme, and eugenics may not have been high on the agenda. But there was a hidden eugenics agenda, and various means of bio-politics were introduced or at least suggested at the end of the 1930s.

Practically speaking, promoting and lobbying for eugenics was more difficult for Lithuanian eugenicists such as Blažys and here the situation was markedly different to the situation in the other Baltic states. In Lithuania, the struggle must have been perceived by the active eugenicists as a fight against both state and church. Though this chapter has brought many key findings to light, the complex history of Lithuanian eugenics and psychiatry is in need of further examination.

Notes

---

[1] On Bechterev see: Pfrepper, 2007.

[2] Blažys, 1921, 65-73; Blažys, 1936a.

[3] Ibid.

[4] Blažys, 1938a, 115-26; Bendoravičius, 425.

[5] For Polish eugenics see: Gawin, 2007, 167-184.

[6] On the Smetona regime see: Lopata, 2001, 95-142; Richter, 2010, 111-136; in general: Rauch, 1977; Misiunas and Taagepera, 1983.

[7] Turda, 2010.

[8] Weindling, 1989.

[9] Ibid., 522.

[10] For Catholic eugenics see: Löscher, 2009; Cassata, 2011; Schneider, 1990, 69-109; Gawin, 2007.

[11] The strong eugenics approach among the Latvian social-democratic feminists movement demonstrate several contributions to the journal *Darba Sieviete*.

[12] See the chapter of Ken Kalling on eugenics in this volume.

[13] See: Felder, 2011, 335-365.

[14] Zaksas, 1959.

[15] Andriušis and Dembinskas, 2006, 80-89; see for example the recently released voluminous book by Dieckmann: Dieckmann, 2011.
[16] Felder, 2013.
[17] Šurkus, 1960; Dembinskas and Matonienė, 2005, 442-445.
[18] Germanavičius, 2012, 12-14.
[19] Šurkus, 1960, 140-318.
[20] Blažys, 1923, 577-582.
[21] For the events see for example: Rauch, 1977.
[22] On Rose see: Kulczycki, 2010.
[23] For malaria therapy on syphilis and mental diseases in Vilnius see: Dembinskas, 2003; Spatz and Hossmann, 2008, 170-171.
[24] Rose, 1938.
[25] Blažys, 1936a; Blažys, 1936b.
[26] Dembinskas and Matonienė, 2005.
[27] See i.e.: Blažys, 1937a, 604.
[28] Blažys, 1926, 653-677; Blažys, 1932.
[29] Blažys, 1921.
[30] Blažys, 1933a; Blažys, 1933b, 582-596; Blažys, 1938b, 8-17; Blažys, 1939a.
[31] Blažys, 1936a.
[32] Blažys, 1933b, 596.
[33] For Latvia see the chapter by Björn Felder in this volume, for Estonia the chapter by Ken Kalling.
[34] Blažys, 1936b.
[35] See: notes 18, 19, 20 above.
[36] Blažys, 1936a.
[37] Blažys, 1926, 653-677.
[38] Blažys, 1933b; Blažys, 1933a.
[39] Blažys, 1933a.
[40] Pūdymaitis, 1933, 138-142.
[41] Epšteinas, 1934, 74-75.
[42] Gustainis, 1933a, 109; Gustainis, 1933b, 53; on his eugenics affirmation see: Gustainis, 1930, 2; Gustainis, 1933c.
[43] Contributions that favoured eugenics sterilisations: Vainius, 1926, 130; Pūdymaitis, 1933; Babianskis, 1933, 635-638; Aleksa, 1928, 105-122; Aleksa, 1938, 151-158.
[44] G. (1929), "Eugenika arba žmonių pagerinimas [Eugenics or the Improvement of Humans]" *Kultura*, vol. VIII, no. 3, 113-120; Dr. F. (1935), "Kokybine tautos politika [The Quality of the National Policy]" *Naujoji Romuvas*, no. 40, 705-709.
[45] Bendoravičius, 1932; Bendoravičius, 1935, 344-350.
[46] Vazbys, 1933, 673.

[47] On his 1926 speech see: G. (1929), "Eugenika arba žmonių pagerinimas [Eugenics or the Improvement of Humans]", *Kultura*, no. 3, vol. 8, 113-120. For his eugenics publication see: Šliūpas, 1934, 2-4; Šliūpas, 1935, 300.
[48] Šliūpas, 1943, 363-365. About Nazi Euthanasia in Nazi occupied Lithuania see: Felder, 2013.
[49] Epšteinas, 1934; Marcinkevičius, 1933, 577-582.
[50] Grinius, 1930.
[51] Blažys, 1929, 14-25; Blažys, 1936a; Blažys, 1938c, 261-266; Blažys, 1939b, 426-437.
[52] Babianskis, 1933, 637-636.
[53] Gustainis 1930, Gustainis 1933c, Gutmanas 1926, Gutmanas 1934.
[54] For the perception of Lombroso in Tsarist Russia see: Beer, 2008. In connection with early Russian debates on eugenics see: Felder, 2011.
[55] Blažys ,1933b, 589-593.
[56] Blažys, 1935, 57-68; Blažys, 1937b, 281-287. Other contributors to the debate have been: Oželis, 1935, 120-125.
[57] Gustainis, 1930; Gustainis, 1933c; Gutmanas, 1926, Gutmanas, 1934.
[58] Examples for communist activists: Zakas, 1959, 178-186.
[59] *Kosmos*, vol. 14, no.4-6 (1935).
[60] Blažys, 1936a, 35.
[61] See for example the review on the book of Bindung and Hoche by Blažys: *Medicina*, vol. II, no. 3 (1921), 425; Blažys, 1926, 659.
[62] Epšteinas, 1934, 74.
[63] Bendoravičius, 1932, 68.
[64] Cassata, 2011.
[65] Examples for anti-Catholic rhetoric provides: Blažys, 1937b, 286-287.
[66] For Catholic eugenics see: Gawin, 2007.
[67] Balys, 1927, 186; Balys, 1933a, 51-52; Balys, 1933b, 321; Balys, 1934, 40-42; Žilinskas, 1927; Žilinskas, 1930, 137-146, 215-225; Žilinskas, 1935; Žilinska and Jurgutis, 1939; Žilinskas and Masalskis, 1937.
[68] See: note 6.
[69] Vareikis, 2004, 119-172.
[70] Richter, 2010, 124.
[71] "Aus der Tätigkeit des Mutter- und Kinderschutzkomitees der baltischen Staaten im Jahre 1937" *Pro Juventute Baltica*, vol. 2, no. 1 (1938), 19.
[72] See: www.stjosef.at/dokumente/casti_connubii.htm (7 April 2012).
[73] See for example: Barauskas, 1934, 145-149; Gobis, 1938, 565-568; Gobis, 1939, 492.
[74] Gruodis, 1936.
[75] Blažys, 1936a, 264.
[76] Bock, 2010.
[77] Jurėnienė, 2008, 34-45.

[78] Kavolis and Bieliackinas, 1934; See: paragraph 465, 466.
[79] For Latvian abortion law see the chapter by Ineta Lipša in this volume; Jurėnienė, 2008.
[80] Ibid.
[81] *Medicina*, vol. XIV, no. 2 (1933), 123.
[82] "Priimtas medicinos gydytojų praktikos į statymas [Adopted Law on Medical Practice]" *Lietuvos aidas*, no. 281 (10 December 1934), 7.
[83] 11 December 19: LVCA, 923/1/811, 1.
[84] *Vyriausybės Žinios*, no. 498 (13 September 1935), 3.
[85] Jurėnienė, 2008.
[86] Goldbergas, 1929, 617.
[87] See the operation books of the fund of *Mažylio Moterų Ligoninė*: Kauno Apstrities Archyvui, fund 122/1.
[88] For the Soviet literature see: Zakas, 1959.
[89] See: Dembinskas and Matonienė, 2005.
[90] For Madissoon see the contribution by Ken Kalling on eugenics in Estonia in this volume.
[91] See also the contribution by Björn Felder on Latvia in this volume.
[92] See: Beer, 2008.

References

Aleksa, K. (1928), "Eugenika" *Kosmos*, no. 9, 105-122.

Aleksa, K. (1938), "Eugenika ir Lietuvos ateitis [Eugenics and the Future of Lithuania]" *Vairas*, no. 3, 151-158.

Andriušis, A. and Dembinskas, A. (2006), "Psychiatric Euthanasia in Lithuania During Nazi Occupation" *International Journal of Mental Health*, vol. 35, no. 3, 80-89.

Babianskis, A. (1933), "Tuberkuliozė ir kova su ja rasės hygijenos (eugenikos) šviesoje [Tuberculosis and the Fight against it in the Light of Racial Hygiene (Eugenics)]" *Medicina*, vol. XIV, no. 11, 635-38.

Balys, J. (1927), "Rasė ir siela [Race and Soul]" *Kultura*, no. 4, 186.

Balys, J. (1933a), "Rasė ir dvasia [Race and Spirit]" *Vairas*, no. 9, 51-52.

Balys, J. (1933b), "Rasė ir dvasia [Race and Spirit]" *Vairas*, no. 8, 321.

Balys, J. (1934), "Anthropologinė ir sociologinė žydijos problema [Anthropological and Sociological Problem of the Jews]" *Akademikas*, vol. 19, no. 2, 40-42.

Barauskas, K. (1934), "Eugenikos klausimas enciklikoje "Casti Conubii" [About Eugenics in the Encyclica "Casti Connubii"]" *Naujoji vaidilutė*, vol. 110, no. 4, 145-149.

Beer, D. (2008), *Renovating Russia. The Humans Sciences and the Fate of Liberal Modernity 1880-1930.* Ithaca, London: Cornell University Press.

Bendoravičius, V. (1932), *Išsigimimas ir Eugenika* [Degeneration and Eugenics]. Kaunas: Kauno Miesto Ligonių Kasa.

Bendoravičius, V. (1935), "Genijus ir išsigimimas [Genius and Degeneration]" *Vairas*, no. 7-8, 344-350.

Bendoravičius, V. (1939), "A.a. Prof. Juozas Blažys kaip psichiatras [Juozas Blažys as Psychiatrist]" *Medicina*, vol. XX, no. 6, 417-26.

Blažys, J. (1921), "Paveldėjimo klausimas ir mendelizmas [The Question of Heredity and Mendelism]" *Medicina*, vol. II, no. 3, 65-73.

Blažys, J. (1923), "Valst. psichiatrinės ligoninės 1922 m. darbuotės apyskaita 1923 [Overview of the work of national psychiatric hospital in 1922]" *Medicina*, vol. IV, no. 10, 577-582.

Blažys, J. (1926), "Išsigimimo ir jo įspėjimo klausimas bendrai ir ypatingai Lietuvoje [Questions of degeneracy and its prevention in general and especially in Lithuania]" *Medicina*, vol. VII, no. 12, 653-677.

Blažys, J. (1929), "Neurosifilio Lietuvoje klausimu [On Neurosyphilis in Lithuania]" *Medicina*, vol. X, no. 1, 14-25.

Blažys, J. (1932), *Alkolizmas ir psichinės ligos* [Alcoholism and mental disorders]. Kaunas: Blaivybės Draugijos Centras.

Blažys, J. (1933a), *Psychiškai defektingųjų sterilizacija* [Sterilization of mental defectives]. Kaunas: Spindulio.

Blažys, J. (1933b), "Sterilizacijos klausymu [The question of sterilization]" *Medicina*, vol. XIV, no. 10, 582-596.

Blažys, J. (1935), "Įgimto nusikaltėlio problema [The Problem of the Inherited (Congenital) Offender]" *Kriminalistikos žinynas*, vol. 1, no. 2, 57-68.

Blažys, J. (1936a), *Įvadas į Psichatrija* [Introducion to Psychiatry]. Kaunas: Kaides.

Blažys, J. (1936b), *Tolerancija kaip kultūros principas* [Tolerance as a principle of the culture]. Kaunas: Spaudos Fondas.

Blažys, J. (1937a), "Circular psychosis" *Encyclopaedia Lituanica*. Kaunas: Varpas, vol. 5, 604.

Blažys, J. (1937b), "Sterilizacija ir kastracija kaip nusikaltimų prevencija [Sterilisation and Kastration as Prevention of Criminality]" *Kriminalistikos žinynas*, vol. 3, no. 18, 281-287.

Blažys, J. (1938a), "Psichinė Higiena Aukštojoje Mokykloje [Mental Hygiene at the High School]" *V. D. Universiteto Žinios*, no. 5-6, 115-126.

Blažys, J. (1938b), "Biologiniai eugenikos pagrindai [Biological foundations of eugenics]" *Gamta*, vol. 3, no.1, 8-17.

Blažys, J. (1938c), "Kova su alkoholizma [The Fight against Alcoholism]" *Kriminalistikos žinynas*, vol. IV, no. 24, 261-266.

Blažys, J. (1939a), "Eugenika [Eugenics]" *Encyclopaedia Lituanica*. Kaunas: Varpas, vol. 7, 1250-1252.

Blažys, J. (1939b), "Lietuvių sveikatingumo apsaugojimo ir pakėlimo klausimas [On the Protection and Improvement of Health of Lithuanians]" *Medicina*, vol. XX, no. 6, 426-437.

Bock, G. (2010), *Zwangssterilisation im Nationalsozialismus. Studien zur Rassenpolitik und Geschlechterpolitik*. Münster: MV-Wissenschaft.

Cassata, F. (2011), *Building the New Man. Eugenics, Racial Science and Genetics in Twentieth-Century Italy*. Budapest: CEU Press.

Dembinskas, A. (2003), *Psichiatrija*. Vilnius: Vaistų žinios.

Dembinskas, A. and Matonienė, V. (2005), "Profesorius Juozas Blažys - psichiatras, mokslininkas, visuomenės veikėjas [Professor Juozas Blažys - Psychiatrist, scientist, public figure]" *Medicina*, vol. XXXXI, no. 5, 442-445.

Dieckmann, C. (2011), *Deutsche Besatzungspolitik in Litauen 1941-1944*, 2 Bände. Göttingen: Wallstein.

Epšteinas, L. (1934), "Rasės higiena teorijoj ir praktikoj [Theory and Practice of racial hygiene]" *Kultura*, no. 2, 67-74.

Felder, B. (2011), "Rassenhygiene in Russland. Der Hygieniker und Bakteriologe Evgenij A. Šepilevskij (1857-1920) und die Anfänge der Eugenik im Russischen Zarenreich. The Hygienist and Bacteriologist Evgenii A. Shepilevsky and the Beginning of Eugenics in Russia" in: Riha, O. and Fischer, M. (eds.), *Naturwissenschaft als Kommunikationsraum zwischen Deutschland und Russland im 19. Jahrhundert. Internationale Tagung. Leipzig, 29.2. - 1.10.2010*. Aachen: Shaker, 335-365.

Felder, B. (2013), "National-Socialist Euthanasia, Human Experiments and Psychiatry in Nazi-Occupied Lithuania 1941-1944" *Holocaust and Genocide Studies*. (Forthcoming).

Gawin, M. (2007), "Progressivism and Eugenic Thinking in Poland, 1905-1939" in: Turda, M. and Weindling, P. (eds.), *"Blood and Homeland": eugenics and racial nationalism in Central and Southeast Europe, 1900 - 1940*. Budapest: CEU Press, 167-184.

Germanavičius, A. (2012), "Lietuvos akademinės psichiatrijos raida 1918-1940 metais [Development of academic psychiatry in Lithuania during 1918-1940]" *Nervų ir Psichikos Ligos*, vol. 56, no. 1, 12-14.

Gobis, J. (1938), "Eugeniškoji sterilizacija ir moralė [Eugenic Sterilisation and Moral]" *Draugija*, no. 17-18, 565-568.

Gobis J. (1939), "Eugeniškoij sterilizacija pedagoginiu atžvilgiu. Is ciklo Eugeniškoji sterilizacija [Eugenic Sterilisation from a Paedgocical Point. Series on eugenic Sterilisations]" *Draugija*, no. 8, 492.

Goldbergas, T. (1929), "Mūsų įstatymai ir mediciniškais abortas [Our Laws and Medical Abortion]" *Medicina*, vol. X, no. 9, 609-621.

Grinius, K. (1930), *Kovai su tuberkulioze Lietuvos Respublikoje planas* [Plan of Fight against Tuberculosis in Republic of Lithuania]. Kaunas.

Gruodis, S. (1936), *Eutanazija* [Euthanasia]. Kaunas: Viltis.

Gustainis, V. (1930), "Susipažinkime su eugenika [Get aquainted with eugenics]" *Lietuvos aidas*, no. 229, 2.

Gustainis, V. (1933a), "Naujieji mitai [New Myth]" *Akademikas*, no. 6, 109.

Gustainis, V. (1933b), "Rasizmo filosofija ir etika [Philosophy and Ethics of Racism]" *Vairas*, no. 8, 53.

Gustainis, V. (1933c), "Kriminaliniai nusikaltimai ir eugenika [Criminal offence and Eugenics]" *Kovokime su nusikaltimai*, 86-92.

Gutmanas, L. (1926), "Susirgimų ir nusikaltimų perspėjimo reikalu. (Eugenetinė problema) [On the Prevention of Illnesses and Crime (Eugenic Problem)]" *Medicina*, vol. VII, no. 5-7, 245-257.

Gutmanas, L. (1934): "Sterilizacijos įstatymo kritika [Critics on the Sterilisation law]" *Medicina*, vol. XV, no.2, 63-73.

Jurėnienė, V. (2008), "Abortų problema pirmojoje Lietuvas Respublikoje [The Problem of Abortion in the First Lithuanian Republic]" in: Pavilionienė, M. A. (ed.), *Abortas Lietuvoje. 2008 m. Birželio. 11 d. Konferencija*. Vilnius: Valstybės žinios, 34-45.

Kavolis, M. and Bieliackinas, S. (eds.) (1934), *Baudžiamasis statutas su papildomaisiai baudžiamisiais įstatymas ir komentarais* [Penal Code with Addition Penal Laws and Comments]. Kaunas: D. Gutmano Knygynas.

Kulczycki, J. (2010), "Historia Polskiej Neurologii i Neurochirugii. Maksymilian Rose, neurolog, psychiatra i wybitny znawca cytoarchitektoniki kory mózgu [The History of Polish Neurology and Neurosurgery. M.Rose Neurologist, Psychiatrist and Specialist of the cyto architecture of the cerebralcortex]"in: http://www.termedia.pl/Czasopismo/Neurologia_i_nbsp_Neurochirurgia_Polska-15/Streszczenie-16060, (accessed 15 August 2011).

Löscher, M. (2009), *"... der gesunden Vernunft nicht zuwider ...?" Katholische Eugenik in Österreich vor 1938*. Innsbruck [u.a.]: StudienVerl.

Lopata, R. (2001), "Die Entstehung des autoritären Regimes in Litauen 1926 - Umstände, Legitimation, Konzeption" in: Oberländer, Erwin (ed.), *Autoritäre Regime in Ostmittel- und Südosteuropa 1919-1944*. Paderborn: Schöningh, 95-142.

Marcinkevičius, M. (1933), "Rasės hygijenos klausymu [On Racial Hygiene]" *Medicina*, vol. XIV, no. 10, 577-582.

Misiunas, R. and Taagepera, R. (1983), *The Baltic States. The Years of Dependence 1940-1980*. London: C. Hurst.

Oželis, K. (1935), "Kovo su išsigimimais-nusikaltimų įspėjimas [The Fight against degeneration as an prevention of crime]" *Kriminalistikos žinynas*, vol. 1, no. 3, 120-125.

Pfrepper, R. (ed.) (2007), *Vladimir Michajlovič Bechterev (1857 - 1927): neue Materialien zu Leben und Werk*. Aschen: Shaker.

Pūdymaitis, O. (1933), "Apie eugeniką [About Eugenics]" *Gamtos draugas*, no. 9, 138-142.

Rauch, G. von (1977), *Geschichte der baltischen Staaten*. München: dtv.

Richter, K. (2010), "Der Kult um Antanas Smetona in Litauen (1926-1940)" in: Ennker, B. and Hein-Kirchner, H. (eds.), *Der Führer im Europa des 20. Jahrhunderts*. Marburg: Herder-Institut, 111-136.

Rose, M. (1938), *Par le Cerveau de Joseph Piłsudski*. Wilno.

Schneider, W. H. (1990), "The Eugenics Movement in France 1890-1940" in: Adams, M. B. (ed.), *The Wellborn Science. Eugenics in Germany, France, Brazil and Russia*. New York, Oxford: Oxford University Press, 69-109.

Šliūpas, J. (1934), "Šis tas dėl mūsų rasės sveikatingumo [Some Thoughts on the Health of our Race]" *Laisvoji mintis*, no. 8, 2-4.

Šliūpas, J. (1935), "Kur pažanga, o kur regresas? [Where is the Progress, where the Regress?]" *Kultura*, no. 5, 300.

Šliūpas, J. (1943), "Ko mums reikia? [What do we need?]" *Lietuviškoji Medicina*, vol. XXIV, no. 5-6, 363-365.

Spatz, M. and Hossmann, K. A. (2008), "In Memoriam of Igor Klatzo 1916-2007" *Journal of Neuropathology and Experimental Neurology*, vol. 67, no. 2, 170-171.

Šurkus, J. (1960), *Psichatrijos vystymasis Lietuvoje* [Development of Psychiatry in Lithuania]. Dissertation, Kaunas: University of Kaunas.

Turda, M. (2010), *Modernism and Eugenics*. Houndsmill: Palgrave Macmillan.

Vainius, K. (1926), "Eugenika" *Kultūra*, no. 3, 130.

Vareikis, V. (2004), "Anti-Semitism in Lithuania (Second half of 19th - First half of 20th century)" in: Truska, L. and Vareikis, V. (eds.), *The Preconditions for the Holocaust. Anti-Semitism in Lithuania (Second half of the 19th Century - June 1941)*. Vilnius: Margi raštai, 119-172.

Vazbys, T. (1933), "Ar galima sterilizuoti ir kastruoti eugeniniais tiksais? [Is it allowed to sterilise and castrate for Eugenic Reasons?]" *Medicina*, vol. XIV, no. 11, 673.

Weindling, P. (1989), *Health, Race and German Politics between National Unification and Nazism, 1870 - 1945*. Cambridge: Cambridge Univ. Press.

Zaksas, I. (1959), *Rasizmas ir Eugenika buržuazinėje Lietuvoje* [Racism and Eugenics in Bourgeois Lithuania]. Vilnius:Valstybinė Politinės ir Mokslinės Literatūros Leidykla.

Žilinskas, J. (1927), *Lietuvių kaukolės dėžė* [Collection of Lithuanian Bones]. Kaunas: Varpos.

Žilinskas, J. (1930), "Lietuvių rasinė sudėtis pagal kraujo biologines savybes ir jų koreliaciją su svarbiausiais morfologiniais požymiais [Racial Composition of Lithuanians with Regard to Biological Characteristics abd their Correlation with main Morphological Traits]" *Medicina*, vol. XI, no. 3-4, 137-146, 215-225.

Žilinskas, J. (1935), *Serologinis Mažosis Lietuvos Gyventojų (ir Kitų Baltų Rasės Tautų) Giminiškumas* [Serological Relatedness of Inhabitants of Small Lithuania (and other Balitc Tribes)]. Kaunas: Spindulio.

Žilinskas, J. and Masalskis, R. (1937), *Senojo geležies periodo Lietuvos gyventojų kaukolių studija: kaukolės dėžės evoliucija ir rasinė priklausomybė* [Study of Ancient Iron Age Bones of Inhabitants of Lithuania: Evolution of Bones with regard to Race]. Kaunas: Viltis.

Žilinskas, J. and Jurgutis, A. (1939), *Crania lithuanica (XX amžiaus ir iškastinės lietuvių kaukolės). su 12 paveikslų atskirose atskirose lentose* [Lithuanian Bones (Lithuanian palaeontological and contemporary 20th-century Bones) with 12 pictures in separate boards]. Kaunas: Spindulio.

Part 2

Eugenics in the Baltic Sea Region

World War One and National Characterology in East-Central Europe

Maciej Górny

Abstract
In the long 19th century the social sciences were rapidly developing. One of their popular research fields was national character. It was believed to manifest itself in high and rural culture, art and architecture, literature and history, forms of religiosity and philosophy, and – last but not least – in psychology. The present paper analyses scientific discourses referring to national, cultural and racial character of the "others" - identified mostly with the war enemy - in the years from 1914 to the mid-1920s. Its geographical range covers Eastern and Central Europe as well as (to a lesser extent) France, Great Britain and the USA. The author refers to the methods of historische Stereotypenforschung along with the discourse analysis and comparative history of science to describe not only the intellectual tools and motifs of the genre but also the interdependences between various sciences within and beyond every one of the countries under scrutiny.

The topic of this paper is part of a broader analysis of the discourses in scientific literature during World War I and afterward, with a special eye on national characterology of "the Other". The territorial range covers East-Central Europe in a broad sense (including Germany, and Russia). As a context, some American and Western European publications will be analysed. In the first stage of research, several main discourses have been identified, including "autochthonist" argumentation of territorial claims, symbolic "feminisation" of war enemies, or cultural and racial theses of national superiority. The "carrier" of these discourses predominantly has been the social and medical sciences.

National characterology belongs to the most popular intellectual endeavours. Even now, general statements concerning the psychological profile of a nation often accompany both scientific and popular descriptions of "the Other". To search for the hidden traces of one's own and other human groups seems to be a response to the vital needs of any social group, as shown by sociological and psychological research from the 1960s onwards.[1] This phenomenon was also examined in the last century through at least several research schools, to mention only historical research on stereotypes (*historische Stereotypenforschung* in the German scientific tradition)[2] or literature-oriented comparative "imagology".[3] None of them, however, examines the characterological statements as objective sources for the ethnic, national or racial characteristics, but solely the representations of national characterology as a specific genre, as a discourse inseparably connected to the psychological and sociological phenomena of stereotype and prejudice. Nevertheless, attempts to place national characterology in the frame of scientific methodology appeared as early as the 18th century.[4] In the mid-19th century, a group of German scientists attempted to regulate the methodology of *Völkerpsychologie* – as the science had been baptised – announcing it to be the new, more "humanistic" division of psychology. It absorbed findings of human sciences and searched for general answers to the question of the psychological profile of entire ethnic groups.[5] Further, notwithstanding the optimistic visions of Moritz Lazarus and Hajim Steinthal, the scientific discourse devoted to the national character went through evolution analogous to other human sciences, paying tribute to the developing nation-state. This was especially evident in the enthusiastic atmosphere of the 1870s in Germany and afterwards. But the mobilisation of "national science" in service of the state and the nation was a general phenomenon, with powerful representations in East-Central Europe.[6]

World War I was certainly not an exception from this trend. It mobilised entire scientific societies . Intellectual production was lower than peacetime levels; however, the total number of publications in circulation does not reflect this. In fact, swarms of texts and public initiatives devoted to any possible aspect of conflict appeared, mostly in the form of short brochures.[7] In East-Central Europe this genre survived the Versailles Treaty and developed further in the inter-war period, concentrating on ethnic and territorial questions of the region. One of its species was the revisionist literature of the 1920s and 1930s. Quite often not only special issues of established scientific and intellectual journals, but also wholly new titles referring to the war and post-war territorial changes were published. The variety of topics connected to the war situation is impressive, including national fertility, national fashion (meant to replace the destructive influence of French fashion in Germany), literary representations of war and heroism, trench humour (at least in the first years of the war), religious literature, and short historical essays on patriotic topics. A considerable number of these publications was of scientific character or aspired to it, using a formal scientific apparatus and referring to the authority of various branches of expert knowledge. Among these publications and other representations (such as ethnographic exhibitions), the studies of national character of the author's own nation and of enemy nations dominate. The variety of registers and genres is huge: national character was researched from the point of view of philosophy, philology, history, linguistics, psychology, anthropology, theology, economics, etc.

This variety mirrored the mobilisation of the professoriate, that was especially visible in Germany, where this group played a prominent role.

Nationalist mobilisation of German professors in the first months of the war and even prior to the war, in the first decade of the 20th century, has been a topic of many publications. In October 1914 thousands of scientists signed memoranda condemning France and "England", and declaring "German militarism" to be right in the struggle for universal values.[8] The national character of Germans, the "spirit of the nation" and its culture, was prominent among the problems mentioned in the scientific memoranda. As the title of one of the most influential publications of this kind indicates, German authors identified Germanness with heroism, and the most "perverse" enemy – Great Britain – with mercantilism.[9] On the other side of the front, the philosopher Henri Bergson announced the war to be a struggle against the barbarians and used the academic authority to label Germany a nation of wilds.[10] In response to this and other megalomaniac declarations of the French Academy, it was stated by the most prominent German intellectuals (among others Friedrich Meinecke, Max Scheler, Ferdinand Toennies, Otto Hoetzsch, Hermann Oncken and Werner Sombart) that Germans represented idealism (versus English materialism and utilitarianism, French frivolity and Russian barbarianism), freedom, higher culture and, finally, individualism (versus mechanised and rotten Western civilisation).[11]

Among various motifs which accompanied the phenomenon of the *Krieg der Geister*, an important role was played by the *völkisch*-racial complex of ideas. The historian Rüdiger vom Bruch points at the aggressive representations of racial thinking in the pre-war congresses of literary scientists and sociologists. Notions of *Volkstum*, race, nation and *Kultur* enriched the vocabulary of social scientists far beyond the circles of the organised movement for racial hygiene.[12]

The same vocabulary and similar concepts influenced not only the scientific community but can also be observed in the wartime mass media, literary works and works of art. In fact, the research on national stereotypes during the war often combined various genres of sources interpreting them along the lines of main motifs.[13] Here, however, I will concentrate on the scientific reflection on national character of the adversary. I am of the opinion that the scientific contribution to the "spiritual war" deserves special treatment – not because it necessarily formulates distinct opinions from the popular press or fiction. On the contrary, the main topics and clues of the scientific wartime literature are close enough to what non-scientific authors formulated. Quite often, authors simply used war propaganda as objective material for their psychological studies. Nevertheless, these publications represented a separate interdisciplinary contribution to the dominant discourses. Their special character manifested itself in the modes of argumentation and formulation of theses as well as in the affiliations of their authors. Moreover, they communicated over the frontlines to launch polemics with the scientific formulations of the war enemy, or to use their own expert authority or that of foreign researchers to support their argumentation.

The idea of racial or national superiority can be found on almost every corner, including such original constructs as Darwinian inspiration in theology.[14] I will concentrate on works in psychiatry and anthropology in the context of war and post-war conflicts. In the second part of the chapter I will try to characterise the interconnections between these scientific genres and other types of characterological thought during World War I and the immediate post-war period.

1. *Psychopathia gallica* – Pathology in the Enemy's Rows

Starting from the first weeks of the war, the contrast between the national character of the French and the Germans was a popular topic in scientific literature. Both were analysed within the categories of individual psychology. Max von Gruber, the Austrian professor of hygiene in Munich and prominent adherent of the racial hygiene movement, explained to the university-educated public Germany's inability to communicate with the enemy, stating, "What we believe by the French […] to be illogical predilection to pathetic idle talk, lack of the love for truth, as hysterical excitability is perceived by them as beautiful power of the inspired souls that overcomes all earthly burdens. And all that we, Germans perceive as most valuable in our contemporary culture: our effort to keep the head high, our strict love of truth and objectivity, the well balanced and modest judgement […], the strict systematics in the selection of individual facts, the compactness of our knowledge, the irrefutability of our critical method, shortly our growing scientific nature […], finally the stubbornness, impassiveness with which we complete what we have recognised as true, all that appears to the Roman and Slavic people as coldness, lack of energy, hardheartedness, shortly as inferiority".[15] On this psychological ground, a special suggestibility of the French and Slavs was diagnosed by Leopold Loewenfeld, a renowned specialist in mental and sexual diseases. It is commonly known, to quote Loewenfeld: "Children are generally more susceptible to suggestion than the adults, women than men. Narrow-mindedness and lack of knowledge support the development of suggestibility to the highest degree. […] Among the nations suggestibility is in general much more developed in the Latin race than in the Germanic race".[16] The suggestibility of the hostile powers made them an easy object of mental manipulations and hypnosis. The latter was used by the governments and ruling circles to manipulate the masses with the fantasies of sea power (in Britain), the siege of Constantinople (in Russia), or irredentism (in Italy).[17] In another brochure, Loewenfeld combined the suggestibility of the French with their aggressiveness, formulating the notion of *psychopathia gallica*. According to the author, they were highly emotional, always circulating between neurasthenia, hysteria and depression, as the Dreyfus case showed.[18] The hysterical blindness of Western countries manifested itself in their acceptance of the unnatural alliance with Asiatic Russia, as well as in the theories of French psychiatrists (notably Édouard Toulouse), according to which the Germans as a nation suffered from mental diseases.[19]

The same psychopathic syndromes diagnosed in France by Loewenfeld could be transferred to the East, whenever the author felt such a need. The Austrian psychiatrist Erwin Stransky began his study on mental diseases during the war from an assessment of the weak psyche of the French and compared it to the "healthy nerves of the *Mittelvölker*".[20] Going further he connected the psychopathological features of the enemy nations to their antipathy towards Germans. This phenomenon was – according to Stransky – even more understandable if one would study the attitude of Slavs, especially "broad masses of people living to the east from Vistula and their distant western peninsula being in constant contact with German *Kulturgebiet*".[21] The Poles and – especially Czechs, to follow the enigmatic geographical description of Stransky – subconsciously hated their German cultural benefactor, seeing in him the prevalent and superior intellectual power.

Not only in Stransky's theories does the impact of psychoanalysis become evident. But it was not solely Freud (who stated that in this war, all of his libido belonged to Austria-Hungary) who provided inspiration for the authors quoted above. The psychiatric war theories border on more research fields than I am able to enumerate and analyse here. However, some of them must be mentioned, if possible with reference to respective theories from a period of publication similar to the chronological range of this study. The moral superiority of Germans over their war enemies, which represented one of the pillars of Loewenfeld's and Gruber's thinking, could be examined in the German debate on sexual diseases and social health, where traditionally France was condemned for her infertility, state of mental health and *morbus gallicus*.[22] The war enemy in the occupied territories was not only a potential guerrilla fighter but also – equally dangerous – a female prostitute in Poland, Belgium and France. As the venereologist Alfred Blaschko put it, "sexual diseases [...] make often almost equally many soldiers unable to fulfil their duty, as enemy's bullets".[23] Even the folklorist research of Friedrich S. Krauss, the editor of the *Anthropophyteia*, provided material for the general statements of another active member of eugenic circles, Iwan Bloch: the sexual perversions typical for the French were held to be equally common among the Slavs.[24] The intellectual superiority of Germans, especially over Eastern Europeans, manifested itself in suicide rates: the highest among the Germanic people, the lowest among the Slavs and extremely low among Poles. The main factor that increased the number of suicides – as Gerhard Füllkrug put it in 1919 – was the stage of cultural development. Cultural poverty reduced the suicide rate.[25]

Using scientific arguments of highly differentiated origins, the authors tended to create a surprisingly unified picture of the enemy nations. The war enemies in the quoted texts were compared to children and women. In fact, the characteristic of the latter in contemporary psychological publications of the period wholly corresponded to the allegedly French tendency to heave between extremes of neurasthenia, depression and hysteria. It was also often stated that women were especially susceptible to hypnosis.[26] All of these quasi-scientific theories would allegedly handicap any attempts to introduce military service for women. As Fritz Giese put it, had this been introduced, "many examples of lack of subordination, quarrels and tetchiness leading to conflicts with the superiors would occur having their source in menstruation".[27] The "feminisation" of the war enemy suggested by Loewenfeld and Gruber was a topic of H. N. Reichl's study, *Die weibliche Kampfesart der Gegner Deutschlands im Weltkrieg* (The Feminine Way of Fighting the World War by Germany's Foes).[28] Leaving aside his Freudian vocabulary, the author stated that the entire war with Germany bore a feminine character. This was caused by the national character of the Southern Europeans, full of fantasy, illogical, unwise, theatrical, and always ready for bloody revenge. Their femininity manifested itself in the need to be ruled, and they would enthusiastically follow any strong personality that flattered their instincts. By way of contrast, the Germans bore the most masculine features; first of all, they knew what they were fighting for. The Southern Europeans needed the strong masculine Germans to be controlled and conquered by them but, at the same time, they feared the inevitable future. On the other hand the liaison between France and Russia proved that Russia did not possess power as manly as Germany, and it resembles France and Italy in its (feminine) tendency for cruelty.[29] In the last sections of his work, Reichl summarises his observation in one general formula: man and woman are not equal. By analogy, there cannot be any equality among the nations.[30]

In looking for explanations of psychological differences between the sides in the conflict, many authors used anthropological arguments. However, they were often combined with different scientific approaches. Hans F. K. Günther, who would soon become an established ideologist, commented in his *Rassenkunde des deutschen Volkes* (Racial Science of the German Nation) (1922) on the false accusations on both sides of the front. As emotions subsided, he admitted, one should say openly that it was untrue that the French did not possess any drop of Germanic blood, as was stated in Germany during the war. Equally false were the accusations of the non-Germanic racial character of England and even Germany itself. All of these theories, according to Günther, ignored fundamental anthropological knowledge.[31] In fact, the German wartime publications on racial characteristics of the fighting sides, even if published by professional craniologists, rarely shared the privilege of Aryan origin with other nationalities. The war edition of Ludwig Wilser's *Die Überlegenheit der germanischen Rasse* (The Superiority of the Germanic Race) interpreted the conflict as an offensive of racially inferior forces against the German Reich.[32] Especially the idea that all of human culture could be derived from the Aryans (or the Nordic race – in works of authors who identified the term "Aryan" as unscientific) influenced German war publications. Author and owner of Anthropologischer Verlag in Breslau, Ferdinand G. Faßhauer, diagnosed the rising mixing of blood both in France and England leading to decadence of the nations. According to Faßhauer, the French actually were already facing the final collapse of their culture; the English, though starting from a quite Germanic position, degenerated due to alcohol consumption, city life and industrialisation, and could succumb to neurasthenia.[33]

As of 1907, as illustrated by the substantial iconographic material collected by Ludwig Woltmann, the most important personalities in the history of mankind were said to have been of Nordic racial origin, even if they lived and worked in a non-Nordic environment.[34] The common identity of Nordic and Germanic was an axiom for most authors who were active during the war, up to the eccentric "Ostara" library, with its quasi-religious cult of blonde, north European Aryans.[35] Even if some anthropologists such as Karl Classen stressed the racially mixed character of the nation and environmental influences on its evolution, the outcome was similar: the Germans were still less "mixed" than the Southern Europeans or Slavs, which automatically made them more talented.[36]

The very description of the Germanic race's physical and psychological profile in contemporary anthropological literature made it clear why so many authors felt the need to identify their nation with the Nordic race. According to one of the most renowned German anthropologists, Eugen Fischer, "The spiritual characteristics of the Nordic race are great tendency towards creative activities, rich fantasy and great intelligence. In accordance to those features are [...] organising talent and artistic abilities. [...] That [the Nordics] can become the most gifted leaders, inventors, artists, thinkers, [...] is clear. It would be a bit biased to claim that this race only is the creator of the humanity, but we may be sure, it belongs to the most gifted".[37] Their active role in the history was acknowledged alike by non-German authors. As Lothrop Stoddard stated: "Ever since the dawn of history the restless, energetic Nordics have been coming down upon the Alpines, overwhelming their territories and setting themselves up as masters. [...] The Nordics are assuredly the most masterful breed that the world has ever seen".[38]

Whatever the theory of the racial origin of war enemies was, it remained certain for the German authors that the enemies were inferior to the Germans.[39] Interestingly enough, the final outcome of the conflict did not lead to abandoning the thesis, but rather let the German authors launch the search for Nordic-Aryan-Germanic political and military leaders in the victorious countries. So, after 1918, it was not only Hindenburg and Ludendorff but also Joffre and Foch, Lloyd George and Wilson who were said to represent the true Aryan blood and be dolichocephalic.[40] Otto Hauser sketched the basic difference between the superior blonde race and the inferior brunettes: the former were dying in the battlefield whereas the latter were hiding behind the frontlines. This racial version of the *Dolchstoßlegende*, which accused Jews and other racially or socially inferior elements of the German nation of treason, proved to be extremely influential during the next decades.[41] Generally speaking, the superiority of Germany was proving to be a result of their racial profile, as it was before through their mental dispositions, strong nerves and masculine character. The fact that the war had been lost did not change the general attitude.

Not surprisingly, American and British authors expressed certain doubts of the purity of German-Nordic composition. As Stoddard claimed, "Germany's fundamental mistake during the generation before the Great War lay in misreading history and perverting biology – the science of race. On this basic error the Germans built a gigantic delusion which in some sections of German public opinion came to amount almost to what insanity experts call a 'mania of grandeur' ".[42] The error should have been the German assumption of their racial purity. Madison Grant attributed the almost complete annihilation of Nordic elite to the military calamities of the 17th century. "To-day" – as Grant ironically commented – "the ghastly rarity in the German armies of chivalry and generosity toward women, and of knightly protection and courtesy toward the prisoners or wounded, can be largely attributed to this annihilation of the gentle classes. The Germans of to-day, whether they live on the farms or in the cities, are for the most part descendants of the peasants who survived, not of the brilliant knights and sturdy foot soldiers who felt in that mighty conflict".[43] The status of a truly Nordic nation was anything but uncontroversial.

German authors of anthropological works tended to bind together Germany's factual and spiritual enemies, similarly to the authors of texts on psychiatry analysed above. In the second post-war issue of *Zeitschrift für Morphologie und Anthropologie*, Max Hilzheimer compared the characterological features of French people to Jews, characterised by the same „nervousness". It is difficult, he claimed, to distinguish between an inhabitant of southern France and a Jewish person, which is the reason why anti-Semitism was almost non-existent in France (sic!).[44] Despite victory, France was doomed for death: it had no intellectual power or no new ideas that could replace the shallow anti-German ideology of revenge.[45] The link between the war enemies and Jews was paralleled by the reference to femininity, this time based on the racial observation. Ludwig Wilser and other authors referred explicitly to the thesis about the feminine character of

Slavs and the French, quite contrary to the masculine character of Germans.[46] For Firmin Coar, it was certain that the French were just as unable to produce culture as women. The French were said to remain always on the surface of problems (whereas Germans always study them to their fundamentals), and they were easy to manipulate by anyone who flattered their undeveloped minds.[47] Hauser claimed that "all the mixed peoples strangely reformulate the role of woman", meaning women's active participation in political and cultural life. "Whereas the purely Nordic woman accepts the superiority of man as the creative spirit; respects him and truly admires, the racially mixed woman very easily starts to feel superior to the men in her milieu. [...] Especially in the case of the Jews the bigger spirituality of woman may be observed. And that is precisely the racial foundation of the contemporary "womens' movement' ".[48]

A Tyrolean nationalist journalist and writer, Karl Felix Wolff, summed up the characterological production of the war period in such a way that a longer passage from his anthropological work should be quoted:
"The mass of notions derived from anthropology and prehistory was growing bigger and the wish to use those findings for our national struggle and aims more urgent. We used to sum up everything under the description *Rassenlehre* without noticing that this new science covers quite a few independent sciences, namely: anthropology, prehistory, political and cultural history, hygiene, ethics etc. This feature of the *Rassenlehre* as a true *universitas litterarum* contributed vastly in Germany and in other countries to the development of the new ideology that was said to have grown organically, from blood and soil, from the national idea and racial feeling which is paramount to everybody".[49]

In psychiatry and anthropology from the period of World War I and immediately afterward, some general tendencies were evident. Having in mind that most of the above-quoted authors were physicians and anthropologists, a question arises: what was their place in the professions they represented and in the German (and Austrian) public discourse? The crucial role of the anti-modernist backlash in German modern history, combined with the specific reading of social Darwinism, has been topic of numerous and valuable works, to mention only George L. Mosse.[50] However, the authors of war publications met with strong and sometimes professional criticism actually during World War I. Pacifist Walther Schücking, who analysed the war publications of German professors, came to the conclusion that they not only lose scientific objectivity, but also fail to attract anyone (apart from the already convinced German chauvinists) due to their extremely aggressive and megalomaniac timbre.[51] There were also professional doubts concerning the practice of skull measurement, the methods of formulating scientific racial theories, etc.[52] However, it is doubtful if the 1915 thesis of

Friedrich Hertz, that "today, the racial theories in the style of Gobineau and Chamberlain are dead", fitted the reality of the war discourse.[53]

Karl Felix Wolff, whose article was previously quoted, described the vicious circle of sciences and ideologies that made the heterogeneous characterological literature practically immune to such general criticism, as by Friedrich Hertz or Friedrich Martius.[54] It might be construed as an example of *Denkkollektiv* defending its paradigm, but I would like to pinpoint another feature of the discourse illustrated above. It was the interdisciplinarity which allowed the use of scientific arguments derived from, say, research on prehistory in sociological or psychological debates. This also explains the methodological heterogeneity of scientific war literature. In fact, many authors combined specialist anthropological or other narratives with intellectual imports of various kinds. Quite often, even the "racial" publications include entire passages devoted to the "spirit of the nation", its philosophy, mentality, etc. Not incidentally, both genres of the war literature analysed above – psychologically and anthropologically oriented – frequently overlap in the selection of material and modes of argumentation. There were many ways of analysing the national character of war enemies. Even if we leave aside the early modern national characterologies or the phenomenon of "enlightened" stereotypes codified in lexicons and encyclopaedias, the *Völkerpsychologie* codified in the mid-19th century in Germany may be perceived as a forerunner and fellow traveller of racial war theories. *Völkerpsychologie* was designed, similar to *Rassenlehre* as described by Wolff, as a meta-science that used the material gathered by historiography, philosophy, theology, linguistic, ethnography, art history and other sciences to describe the spirit of the nation. The characterological war discourse analysed above did not substantially change this remaining programme by broadening the scope of sciences involved and involving racial theories.

2. "The Aryan Attitude of Mind" – Eastern Responses

The interdisciplinarity of German characterological discourse seemed of special importance for the interpretation of similar discourses in East-Central Europe. The "modern" tendencies present in the German discourse appeared in the East with a delay, as the scientific infrastructure was in general much less developed. Universities were less numerous and in many respects more conservative. Furthermore, the net of racial hygiene and anthropological organisations and institutes was not fully formed; however, analogous paths of thought can be traced there as well.[55] So far, I was not able to identify similar pairs of researchers as the French-German psychiatrist couple, Toulouse-Loewenfeld. Despite this, scientific war characterology was by no means a one-way road, the one-sided "orientalising" of the East. German scientific publications were known and commented on and, ironically, they served as arguments for the Russian intellectual counter-attack.[56] Semen L.

Frank analysed in the intellectual journal *Russkaya mysl* the variety of German sociological, historical and psychological publications to illustrate that the German science obviously gave honours to the *Leutnant*.[57] The racial theories have been analysed as proof of German bumptiousness.[58] The general timbre of the Russian symbolic answer was thus ethical rather than scientific: the nationalism of the German scientific publications was unmasked and served as a counterpart to Russian spirituality.[59]

Probably the most interesting critic of racial theories in the region was a Jewish merchant from the Galician city of Przemyśl (today in south-eastern Poland), Mateusz (using also the forename Matityahu and Matthias) Mieses. An amateur in many fields, he stubbornly demystified anti-Semitism and collected historical evidence of the Jewish impact on Polish and European culture. In 1912 Mieses published a book in Polish questioning all the pillars of Nordic theory, and ridiculed them mercilessly. In 1919 in Vienna he published the extended German version under the title *Zur Rassenfrage. Eine Stammes- und Kulturgeschichtliche Untersuchung* (On the Racial Question: A Study in Tribal and Cultural History). For Mieses, racial purity was a direct path to cultural decadence and fall. Nordic, blonde, dolichocephalic humans in fact were intellectually inferior to the more fruitful brunettes of Alpine, Mediterranean and Mongolic races. Germanic warriors did not possess even a pinch of culture and furthermore, in exact contrast to Woltmann's theory, all of the great European artists, scientists and politicians have been brunette and brachycephalic (ironically, Mieses draws from nearly the same roster of names as Woltmann, including Dante, Shakespeare and Goethe). Moreover, according to the Galician independent scholar, even in ancient Rome blonde hair was perceived as a sign of moral inferiority (which is why prostitutes wore blonde wigs). In general, as Mieses claimed, any true culture "feminises" and "weakens" the race; the "pure" Nordics would be doomed for a complete lack of civilisation.[60]

The impact of characterological discourse, including racial theories, during the war and immediately after in East-Central Europe was also visible in original literature by authors from the region. German arguments were transferred to publications by non-German authors. Polish and Ukrainian authors balanced their interpretation of war differently to the majority of Germans. According to them, it was Russia, not Britain or France, that was the main enemy and thus monopolised the criticism. One of the dominant motifs of these brochures was the Asiatic provenience of the Russian national character. There was a slight difference of opinion as to whether this Russian characteristic had been caused by history or biology. Some authors perceived it as a cultural heritage of Tartar despotism.[61] For the future leader of Ukrainian integral nationalism, Dmytro Doncov, Russian Asiatic passivity was inborn and proved distinctiveness of Ukrainians, characterised by "Western" European individualism, energy and "manly features".[62] A

German sympathiser of Ukrainian activists, Paul Ostwald, used the explicitly anthropological argument that "in any case we can prove that Ukrainians have only very limited common features with Russians. [...] Mongol influences as manifested in constitution of the Russian body cannot be found in the case of Ukrainians". The latter, according to Ostwald, represent the Dinaric race, whereas Poles and (interestingly) Russians – the *Weichselrasse*.[63] In a decisively more scientific publication by Ukrainian anthropologist Stepan Rudnyćkyj, from which Ostwald quoted extensively, Ukrainians were identified as a *Mischrasse*, though a completely different one than Russians or Poles. Ukrainians were taller than both Poles and Russians and rather dark-haired, and their faces were considerably longer than those of their neighbours (although their skulls remained brachycephalic). No traces of any Mongolic influences could be traced in the Ukrainian population.[64] According to Paul Rohrbach, editor of the journal *Das größere Deutschland*, Ukrainians are "of better blood than the Muscovites".[65]

Franz (Franciszek) Graf Kwilecki formulated an opinion on Russians in similar terms. According to this member of the Polish-Prussian aristocracy, ethnography proved that Russians were mixed people with prevalent influences of Tartaro-Mongolic blood: "The Mongolic nature of the Russian cannot be denied. Rarely does he enslave the others openly without hypocritical attempt to convince the observers about his innocence".[66]

Polish and Ukrainian authors were naturally of quite a common opinion about the Mongol, or Finno-Turanian racial status of Russians. Additionally, the racial difference between Poles and Ukrainians does not seem to have been a controversial issue, contrary to the older 19th-century statements of ethnic unity of all nationalities of the former Polish-Lithuanian Commonwealth. The problem appeared in the question of the racial background of Polish-Ukrainian differences. For Rudnyćkyj, Ukrainians have always been free from any Mongol influence, so evident in the Russian but also in the Polish case. Some Polish authors claimed the opposite. According to Julian Talko-Hryncewicz, the Galician Ukrainians were strongly mixed with Mongols, which should have been attested not only by their anthropological data, but also by their sanguine character, so distant from Finno-Slavic Belarusians.[67] In fact, the thesis of Ukrainian anthropologists was rather badly constructed logically. To believe that Ruthenians, with several centuries of Tartar rule behind them, contrary to all other East-Central European nationalities, never mixed with Tartars, seems to have been a matter of patriotism and feelings rather than conviction.

The identification of a "new" European race by Joseph Deniker in 1900 provided an elegant solution for the question of Ukrainian racial uniqueness. Deniker identified the Dinaric (of Adriatic) race, traced to South-Eastern Europe, Tyrol, north-eastern Bohemia, parts of Poland and Ukraine.[68]

The idea was quickly adapted by the representatives of nationalities inhabiting these regions. Niko Županić, for instance, claimed in 1912 that racial Serbs represented two groups: the Nordic of the Slavic character and the Dinaric. His description of the latter reads like a poem: "The falcon-eyes [...] shine with fire and announce a courageous, enduring heart. Those people are full of power and energy, with their expressive, intelligent faces". Generally, due to their strong psyche they do not suffer from any kind of addiction.[69] The picture of Ukrainians in the works of Rudnyćkyj refers to the same picture of an industrious, powerful and courageous race in constant struggle against racially different (and needless to add, inferior) neighbours.

In 1915, when the German army began to replace the Russian in East-Central Europe, many works devoted to Polish-Russian and Ukrainian-Russian relations were allowed to appear, due to the abolishment of Russian censorship. One of the most intriguing is the first volume of Feliks Koneczny's history of Russia.[70] Koneczny commented critically upon the current racial theories that in his concept of world history were replaced by the notion of civilisation.[71] According to this Polish historian, Russia built an ethnographic-civilisational mixture on a giant geographical space. In contrary to the West, where the ethnographic difference between the Romans and Germanic tribes was not fundamental, in Russia the ethnic ingredients were extremely different. Finno-Turanian tribes had nothing in common with Slavic colonisers, and in effect "this mixture of strange elements [...] composed a mechanical and not a chemical mixture – contrary to the western countries".[72]

Probably the most important Polish specialist in national characterology, Jan Karol Kochanowski, published his fundamental study, *Polska w świetle psychiki własnej i obcej. Rozważania* (Poland in the Light of Her Own and Foreign Psyche), during the Polish-Bolshevik war. It is impossible to verify which part of the international literature on his topic was known to Kochanowski, who followed the rule in quoting only his own publications. Nevertheless, his analysis of Polish, Russian and German national character belonged to the same discourse as the German publications quoted above. Kochanowski compared the Pole's psyche to the German, stating that the former was much more courageous and flexible due to both the genetic complexity and the meandering of the history of Poland. (Interestingly enough we can find the identical thesis in the first issue of the inter-war Czechoslovak journal *Anthropologie*, though in respect to the historically formed "better minds" of the Czechs).[73] According to Kochanowski, Poles represented the true Aryan Slavic race, in their psychological profile similar to the English and South Europeans.[74] Germans were described as a human mass possessing the psychological qualities of a crowd: deprived of any individuality and aggressive to the root. The industrious, creative and optimistic Pole had an intellectual potential that was lacking among the Germans, which indicated their tendency to conquer Poland. The author describes this mechanism in a thought-provoking, subversively sexualised manner: “The Prussian looks upon a Pole with a smile of an ‘adult’ gazing at a ‘child’: forbearing, tough and stupid, finally full of desire, like the North looks at the South, like Satan on God, seeing in him the unchallenged material for his apparently ‘perfect’ ‘creativity’”.[75] According to Kochanowski, Germans and Russians in many respects bare similar traces: lack of individualism, barbarian behaviour and aggression. However, Russians lacked even the tiny layer of civilisation that covers German barbarianism.[76] They simply did not belong to the European community of nations: “The nomadic Finns from the North, exactly as nomadic Asiatic Mongols from the East, Khazar nomads from the South and then the Cossacks – all those elements [...] are the Russian ‘yesterday’, if not ‘today’ manifesting itself in the agrarian structure of the country”.[77]

Interestingly enough, Kochanowski and other Polish authors,[78] with their entire critical attitude towards not only Russians but equally Germans, recognised some of the anthropological arguments of the Nordic idea. It was true – as Kochanowski wrote – that the conquerors had come from the North. However, the vitality and creativity was located in the European South, and only a combination of these two elements can conjure true humanity.[79] The rehabilitation of the national “femininity” present in Kochanowski's sexualised vision of the racial mixture was followed by other Polish authors. Philosopher Wincenty Lutosławski agreed with the thesis of a special status of women in Polish life and culture – and of a certain Polish characterological feminity. At the same time he refused to recognise it as a problem: Poles and other Slavs simply prove their masculinity in their attachment to perfect femininity. Lutosławski also claimed that Aryans originated in Poland, which gave the Polish population an exceptional merit of having “the Aryan attitude of mind” that “leads by mutual concessions towards a certain ideal of fairness. [...] The Poles are of the purest Aryan race, as they form that portion of the Aryans which remained in their original home. [...]. All the characteristic peculiarities of the Aryans are therefore found in the Poles”.[80] Therefore, had Poland not been suppressed by the German dynasties ruling its neighbours in the 18th century, it would have produced many more geniuses such as Marie Curie-Skłodowska.[81]

3. Conclusions

Kochanowski and Lutosławski used the motifs that were crucial in German works on national characterology during World War I. Like many German authors, they referred to the 19th- and 20th-century European gender discourse, trying to reinterpret the popular motif of national femininity. In fact, this discourse was formed long before the war, from the late 18th century onwards. As John Horne put it, the auto-stereotype of one’s own nation referred to the imagined masculinity, whereas the external and internal enemy was portrayed as either feminine or a feared type of masculinity.[82] The mechanism of identifying Germanness with manliness has been analysed in various aspects.[83] One of them is military and war history, stressing the links between the picture of the war enemy and gender discourse.[84] Putting the thesis of Karen Hagemann in a rather simplistic way, the initial German hidden agenda of combining the French with femininity was to replace fear of the enemy with a sense of superiority.[85] The male was characterised by creativity, activity, rationality and leadership; he was trustworthy and disciplined. The female, on the contrary, was emotional, passive, uncreative, non-individualist, psychically weak, superficial, etc. The association of the French, Slavs and Jews with femininity was present mostly in German-speaking culture of the 19th and 20th centuries, reaching its climax shortly before World War I. The famous book *Geschlecht und Charakter* by Otto Weininger, after its first publication in 1903, appeared in 18 German and Austrian editions, and in numerous translations by the end of war. Weininger, commonly acknowledged as a tragic genius, equated Jews with women, describing the latter as the pure spiritless materiality.

If we go back to the characterological publications in psychology, psychiatry and racial anthropology, we can easily find this kind of "feminisation" of the war enemy, especially in the psychological theories of suggestiveness, barbarianism, amorality, *psychopathia gallica*, neurasthenia, depression and hysteria. All of these were typically opposed to the "manliness" of one's own nation. Some Polish and Ukrainian authors seemed to seek "redemption" from the national feminisation by using the same argument against other nationalities, namely the Russians. Publications by Kochanowski and Lutosławski show another possibility of intellectual refuge from the gender trap – the reformulation of some aspects of femininity as a positive self-stereotype.

Another important mechanism of wartime publications could be summarised as exclusion from Europe. German authors quite naturally labelled Russia as an Asiatic state, whereas their Polish and Ukrainian colleagues pushed the argument further to discover the non-Slavic ethnicity of the Russians. As Lutosławski put it: "the Muscovites, who were Turanians, of the same race as the Finns, Hungarians, Bulgarians, Turks, and Tartars, now pretend to be Aryans, and even to belong to the oldest Aryans, called Slavs, merely because they use an Aryan name, and have accepted a Slavic language, displaying thereby an un-Aryan neglect of tradition. They conceal and deny their Turanian past, their tradition of slavery, tyranny and cruelty; and, with the name of Russians, obtained by false pretences […] they assume a leading place among the Slavs".[86] For the Polish philosopher even the Russian-Slavic identification was an argument in support of the thesis of their Turanian or Mongolic origin. The symbolical exclusion of Russia and

Russians appeared with help of both "modern" racial and anthropological arguments, and "old" references to the Asiatic nature of their cultural and psychological formation.

Here, a difference between the German on the one hand, and Polish and Ukrainian national characterology on the other may be partly understood as an outcome of an ethnic and political situation. In Germany, where during the war many authors did not hesitate to express anti-Slavic resentments, the internal differences between the East and East-Central European nationalities was of little importance. "Barbarianism" and racial inferiority were attributed generally to the whole Slavic "race". This situation outlived the armistice in revisionist and nationalist literature, directed against Poles, Czechs and Yugoslavs. Understandably enough, such an intellectual import could not satisfy Polish or Ukrainian public opinion. Here, the positive estimation of Slavic origin was a more convenient solution and, thus, the exclusion of Russians from the Slavic family quite a logical phenomenon. An interesting analogy was observed by Andrew D. Evans in his comparison of German and Austro-Hungarian anthropological discourses during the Great War: "Despite his overall claim that the peoples of the Russian Empire were a racial mixture, [Rudolf] Pöch [the chief Austrian anthropologist, the leader, with Josef Weninger, of the biometric research on prisoners in arms] sought to cast his subject, albeit to widely varying degrees, as Mongolian. Considering the political situation in Europe during World War I, it is hardly surprising that Pöch characterised Russians through racial characteristics traditionally viewed as alien to Central Europe. [...] Operating within the multi-ethnic context of Austria-Hungary, Pöch de-emphasised the potentially divisive notion of Germanness; instead, he implicitly contrasted Central Europeans to 'Asian' Russians".[87]

Though possibly derived from the quite practical needs of wartime propaganda, similar theses had been popularised in the Polish-Ukrainian context by mid-19th century ethnographers, with the most influential being Franciszek Duchiński.[88] Perhaps the most interesting point was precisely this influence of older, 19th-century modes of describing "the Other" with newly developed scientific disciplines. The newly introduced Ukrainian racial argumentation was a *novum* in methodological terms. Its ideological content remained comparable to Polish non-anthropological publications of the period. Psychiatry, anthropology and prehistory were evidently used as new tools in an old play.

Note

[1] Above all the predominant intellectual influence of Henri Tajfel should be mentioned here. See: Tajfel, 1981.

[2] For the actual state of the art in German and East-Central European historical research on stereotypes, see the volume: Hahn, Mannová, Scholz and Weger (eds.), 2007.
[3] See recently: Dyserinck, 2003, 8-28.
[4] For the beginnings of the modern national characterology in Europe see the volume: Stanzel (ed.), 1999.
[5] See the modern edition of programmatic texts of Völkerpsychologie: Lazarus and Steinthal, 1997.
[6] See the collection of paradigmatic texts from East-Central and Southeast Europe from the second half of the 19th century till the 1930s in: Ersoy, Górny and Kechriotis (eds.), 2010, 3-76.
[7] For the situation on the book market see: Potter, 2007, 12-24.
[8] See: Tollmien, 1993, 137-173. The fundamental works on the militarisation of German professoriate are: vom Bruch, 1980 and vom Bruch, 2006.
[9] Sombart, 1915.
[10] Rürup, 1984, 22-23.
[11] See: Schwabe, 1969, 28.
[12] vom Bruch, 2006, 121.
[13] See for example some of the studies in the volume: Kontler (ed.), 1995; especially: Diana Mishkova's "Friends Turned Foes: Bulgarian National Attitudes to Neighbours" and Tomás Maszerics' "Italy's Entry into the First World War and the Hungarian Press".
[14] See: Brakelmann, 1974.
[15] von Gruber, 1915, 14.
[16] Loewenfeld, 1917, 12.
[17] Ibid., 26-27.
[18] Loewenfeld, 1914, 7-10.
[19] Ibid., 33-36.
[20] Stransky, 1918, 8-9.
[21] Ibid., 10-11.
[22] See: Weindling, 1989, 283.
[23] Blaschko, 1915, 1; See also: Grabinski, 1917, 217-235, describes the immoral nature of women in France, England, Russia and Italy.
[24] Bloch, 1919, 497. Bloch refers to the yearly *Antropophyteia. Jahrbücher für folkloristische Erhebungen und Forschungen zur Entwicklungsgeschichte der geschlechtlichen Moral* edited by Friedrich S. Krauss in Leipzig. Krauss combined attention to sexual and scatologic vocabulary of East-Central European nationalities with popularisation of eugenics – see: Collius, 1912, 286-288.
[25] Füllkrug, 1919, 17.
[26] See for example: Haase, 1915, 84-85. The negative characteristics of the French are very similar to the female features defined by G. Heymans in his

influential *Die Psychologie der Frauen*, Heidelberg: Carl Winter's Universitätsbuchhandlung, 1910, especially 268-272.
[27] Giese, 1916, 14.
[28] Reichl, 1917.
[29] Ibid., 3-22.
[30] Ibid., 26.
[31] Günther, 1923, 7.
[32] Wilser, 1915, VI.
[33] Barelhako, 1915, 16-39.
[34] Woltmann, 1907; for the same argument see also: Schultze-Naumburg, 1928 and Schmid, 1915.
[35] See: Lanz-Liebenfels, 1918.
[36] Classen, 1918, 7.
[37] Fisher, 1923, 150.
[38] Stoddard, 1924, 13 and 16-17.
[39] See for example an analysis of the French racial composition: Coar, 1916, 6-7.
[40] Hauser, 1922, 111; See also: Baur, Fischer and Lenz, 1921, 293.
[41] Ibid., 113.
[42] Stoddard, 1924, 80-81.
[43] Grant, 1916, 167-168.
[44] Hilzheimer, 1919-1921, 205.
[45] Ibid., 206-207.
[46] Wilser, 1915, 30-38.
[47] Coar, 1916, 6.
[48] Hauser, 1922, 37-38.
[49] Wolff, 1927, 1.
[50] Mosse, 1998.
[51] Schücking, 1915, 2.
[52] See: Poniatowski, 1912. For general Polish critical comments on Aryan theory see: Krzywicki, 1902, 256-276.
[53] Hertz, 1915, IV.
[54] See: Weingart, Kroll and Bayertz, 1992, 311-312.
[55] Magdalena Gawin uses the term „pre-eugenic" describing the pre-1918 Polish discourses; see: Gawin, 2003, 12.
[56] See: Plotnikov and Kolesov, 2005, 31-38.
[57] Frank, 1916, 20-23; Frank, 1914.
[58] Ibid.
[59] See: Ivanov, 2005, 318-343.
[60] Mieses, 1919, 30-39.
[61] See for example: Verax, 1915, 8-12; Brückner, 1916; Brückner, 1915.
[62] Donzow, 1915, 63.

[63] Ostwald, 1916, 8.
[64] Rudnyćkyj, 1914, 12-13.
[65] Rohrbach, 1914; quote after: Remer, 1997, 290.
[66] Kwilecki, 1915, 26-33.
[67] Talko-Hryncewicz, 1913, 113 and 151.
[68] Deniker, 1900, 390-391.
[69] Županić, 1912, 41.
[70] Koneczny, 1917.
[71] See: Koneczny, 1935, 204; Koneczny, 1962, 208.
[72] Koneczny, 1917, 4.
[73] Hrdlička, 1923, 6.
[74] Kochanowski, 1925, 431.
[75] Ibid., 190.
[76] Ibid., 202.
[77] Ibid., 210.
[78] See: Lutosławski, 1917.
[79] Kochanowski, 1925, 61.
[80] Lutosławski, 1917, 43.
[81] Ibid., 53.
[82] Horne, 2004, 29.
[83] See: Florack, 2000, 319-337; Hausen, 1976; Iveković and Mostov (eds.), 2002.
[84] See: Hagemann, 2002; Hagemann and Schüler-Springorum (eds.), 2002; Horne, 2004; Frevert, 1997; Hagemann, 2007.
[85] Hagemann, 2002, 236.
[86] Lutosławski, 1917, 25-26.
[87] Evans, 2002, 235.
[88] See: Górny, 2005; Górny, 2004.

References

Barelhako [Ferdinand G. Faßhauer] (1915), *Der Weltkrieg 1914/15 als Siegeszug der germanischen Kultur. Der Aufstieg der deutschen Kultur*. Breslau: Anthropologischer Verlag,

Baur, E.; Fischer, E. and Lenz, F. (1921), *Menschliche Erblichkeitslehre*. München: J. F. Lehmann.

Blaschko, A. (1915), *Welche Aufgaben erwachsen dem Kampf gegen die Geschlechtskrankheiten aus dem Kriege? Nebst einem Anhang: Vorschlag einer neuen Organisation des Prostitutionswesens von Dr. D. Sarason*. Berlin/Leipzig: Johann Ambrosius Barth.

Bloch, I. (1919), *Das Sexualleben unserer Zeit in seinen Beziehungen zur modernen Kultur*. Berlin: Marcus (10th-12th edition).

Brakelmann, G. (1974), *Protestantische Kriegstheologie im Ersten Weltkrieg. Reinhold Seeberg als Theologe des deutschen Imperialismus*. Bielefeld: Luther.

Brückner, A. (1915), *Der Weltkrieg und die Slawen. Rede am 5. März 1915*. Berlin: Carl Heymann.

Brückner, A. (1916), *Die Slawen und der Weltkrieg. Lose Skizzen*. Tübingen: J. C. B. Mohr.

Classen, K. (1918), "Beiträge zum Indogermanenproblem" *Korrespondenz-Blatt der Deutschen Gesellschaft für Anthropologie, Ethnologie und Urgeschichte*, vol. XLIX, no.1/4 (January-April), 1-7.

Coar, F. (1916), *Über das Wesen französischer Macht*. Mönchengladbach: Sekretariat Sozialer Studentenarbeit.

Collius, A. (1912), "Sterilisation Minderwertiger und Sittlichkeitsverbrecher" *Anthropophyteia*, vol. IX,, 286-288.[es ist ein Jahrbuch]

Deniker, J. (1900), *Les races et les peuples de la terre. Éléments d'anthropologie et d'ethnographie*. Paris: Schleicher Frères.

Donzow, D. (1915), *Die ukrainische Staatsidee und der Krieg gegen Russland*. Berlin: Carl Kroll.

Dyserinck, H. (2003), "Komparatistische Imagologie und ethnische Identitätsproblematik" in: Schubert, G. and Dahmenn, W. (eds.), *Bilder vom Eigenen und Fremden aus dem Donau – Balkan Raum. Analysen literarischen und anderer Texte*. München: Südosteuropa-Gesellschaft, 8-28.

Ersoy, A.; Górny, M. and Kechriotis, V. (eds.) (2010), *Discourses of Collective Identity in Central and Southeast Europe (1770-1945). Texts and Commentaries, vol. III / 1 Modernism – Representations of National Culture*, Budapest/New York: CEU Press.

Evans, A. D. (2002), *Anthropology at War: World War I and the Science of Race in Germany*. Ann Arbor, Mich: UMI Diss. Services.

Fischer, E. (1923). "Spezielle Anthropologie: Rassenlehre" in: Fischer, E. (ed.), *Anthropologie*. Leipzig/Berlin: B. G. Teubner, 122-222.

Florack, R. (2000), "Weiber sind wie Franzosen geborene Weltleute. Zur Verschränkung von Geschlechter-Klischees und nationalen Wahrnehmungsmustern" in: Florack, R. (ed.), *Nation als Stereotyp. Fremdwahrnehmung und Identität in deutscher und französischer Literatur*. Tübingen: Max Niemeyer, 319-337.

Frank, S. (1914), "O poyskach smysla voyny [On the Search for the Sense of War]" *Russkaya mysl'*, vol. VII, no. 12, 2-4.

Frank, S. (1916), "Mobilizatsiya mysli v Germanii [The Mobilisation of Thought in Germany]" *Russkaya mysl'*, vol. IX, no. 37, 20-23.

Frevert, U. (1997), "Das Militär als "Schule der Männlichkeit". Erwartungen, Angebote, Erfahrungen im 19. Jahrhundert" in: Frevert, U. (ed.), *Militär und Gesellschaft im 19. und 20. Jahrhundert*. Stuttgart: Klett.

Füllkrug, G (1919), *Der Selbstmord. Eine moralstatistische und volkspsychologische Untersuchung*. Schwerin: Friedrich Bahn.

Gawin, M. (2003), *Rasa i nowoczesność. Historia polskiego ruchu eugenicznego (1880-1952)* [Race and Modernity. History of the Polish Eugenic Movement, 1880-1952]. Warszawa: Neriton.

Giese, F. (1916), *Die Idee einer Frauendienstpflicht. Tatsachen und Möglichkeiten*. Langenpfalza: Wendt & Klanwell.

Górny, M. (2004), "Argument z etnografii w polskich planach federacyjnych XIX wieku [The Ethnographic Argument in Polish Federation Plans of the 19th Century]" *Borussia*, vol 35, 139-150.

Górny, M. (2005), "Das ethnographische Motiv in den polnischen Föderationsplänen des 19. Jahrhunderts" in: Borodziej, W.; Duchhardt, H.; Morawiec, M. and Romsics, I. (eds.), *Option Europa. Deutsche, polnische und ungarische Europapläne des 19. und 20. Jahrhunderts*. Göttingen: Vanderhoeck & Rupprecht, vol. I, 187-204.

Grabinski, B. (1917), *Weltkrieg und Sittlichkeit. Beiträge zur Kulturgeschichte der Weltkriegsjahre*. Hildesheim: Franz Borgmeyer.

Grant, M. (1916), *The Passing of the Great Race or the Racial Basis of European History*. New York: Charles Scribner's Sons.

Günther, H. F. K. (1923), *Rassenkunde des deutschen Volkes*. München: J. F. Lehmann (3rd edition).

Haase, K. (1915), *Der weibliche Thypus als Problem der Psychologie und Pädagogik. Ein Beitrag zur künftigen Nationalerziehung*. Leipzig/Berlin: B. G. Teubner.

Hagemann, K. (2002), *"Männlicher Muth und Teutsche Ehre". Nation, Militär und Geschlecht zur Zeit des Antinapoleonischen Kriege Preußens*. Paderborn: Ferdinand Schöningh.

Hagemann, K. and Schüler-Springorum, S. (eds.) (2002), *Home/Front. The Military, War and Gender in Twentieth-Century Germany*. Oxford: Berg.

Hagemann, K. (2007), "Military, War, and the Mainstreams. Gendering Modern German Military History" in: Hagemann, K. and Quataert, J. H. (eds.), *Gendering Modern German History. Rewriting Historiography*. Oxford: Berghahn Books, 63-85.

Hahn, H. H.; Mannová, E.; Scholz, S. and Weger, T. (eds.) (2007), *Nationale Wahrnehmungen und ihre Stereotypisierung. Beiträge zur historischen Stereotypenforschung*. Frankfurt am Main: Peter Lang.

Hausen, K. (1976), "Die Polarisierung der "Geschlechtscharaktere" Eine Spiegelung der Dissoziation von Erwerbs- und Familienleben" in: Conze, W. (ed.) *Sozialgeschichte der Familie in der Neuzeit Europas. Neue Forschungen*. Stuttgart: Klett.

Hauser, O. (1922), *Rasse und Politik*. Weimar: Duncker.

Hertz, F. (1915), *Rasse und Kultur. Eine kritische Untersuchung der Rassentheorien*. Leipzig: Alfred Kröner (2nd edition).

Hilzheimer, M. (1919-1921), "Aphoristische Gedanken über einen Zusammenhang zwischen Erdgeschichte, Biologie, Menschheitsgeschichte und Kulturgeschichte" *Zeitschrift für Morphologie und Anthropologie*, vol. XXI, no. 2, 185-208.

Horne, J. (2004), "Masculinity in politics and war in the age of nation states and world wars, 1850-1950" in: Dudink, S.; Hagemann, K. and Tosh, J. (eds.), *Masculinities in Politics and War. Gendering Modern History*. Manchester & New York: Manchester University Press, 22-40.

Hrdlička, A. (1923), "Úlohy a potřeby anthropologie, zvláště v Československu [Tasks and needs of anthropology, especially in Czechoslovakia]" *Anthropologie*, vol 1, 3-8.

Ivanov, A. I. (2005), *Pervaya mirovaya voyna v russkoy literature 1914-1918 gg* [The First World War in the Russian literature, 1914-1918]. Tambov: TGU.

Iveković, R. and Mostov, J. (eds.) (2002), *From Gender to Nation*. Ravenna: A. Longo Editore.

Kochanowski, J. K. (1925), *Polska w świetle psychiki własnej o obcej* [Poland in the light of hers own and foreign psyche]. Częstochowa: A. Gmachowski.

Koneczny, F. (1917), *Dzieje Rosyi, t. I (do roku 1449)* [History of Russia, vol. I (up to 1449)]. Warszawa: Spółka Wydawnicza Warszawska.

Koneczny, F. (1935), *O wielości cywilizacji*. Kraków: Gebethner i Wolff.

Koneczny, F. (1962), *On the plurality of civilisations*. London: Polonia Publications.

Kontler, L. (ed.) (1995), *Pride and Prejudice. National Stereotypes in 19th and 20th Century Europe East to West*. Budapest: CEU Press.

Krzywicki, L. (1902), *Systematyczny kurs antropologji. Rasy psychiczne* [The Systematic Course of Anthropology. Psychical Races]. Warszawa: K. Kowalewski.

Kwilecki, F. (1915), *Polen und Deutsche gegen Russland*. Berlin: Germania.

Lanz-Liebenfels, J. (1918), *Rasse und Adel*. Mödling/Wien: „Ostara“ (=Ostara, Bücherei der Blonden, No. 71/1918).

Lazarus, M. and Steinhal, H. (1997), "Einleitende Gedanken über Völkerpsychologie, als Einladung zu einer Zeitschrift für Völkerpsychologie und Sprachwissenschaft" in: Eckardt, G. (ed.), *Völkerpsychologie – Versuch einer Neuentdeckung*. Weinheim: Psychologie Verlags Union.

Loewenfeld, L. (1914), *Über den National-Charakter der Franzosen und dessen krankhafte Auswüchse (die Psychopathia gallica) in ihren Beziehungen zum Weltkrieg*. Wiesbaden: J. F. Bergmann.

Loewenfeld, L. (1917), *Die Suggestion in ihrer Bedeutung für den Weltkrieg*. Wiesbaden: J. F. Bergmann.

Lutosławski, W. (1917), *The Polish Nation. A Lecture*. Paris: Boyreau et Chevillet.

Mieses, M. (1919), *Zur Rassenfrage. Eine Stammes- und Kulturgeschichtliche Untersuchung*. Wien/Leipzig: Wilhelm Braumüller.

Mosse, G. L. (1998), *The Crisis of the German Ideology. Intellectual Origins of the Third Reich*. New York: Howard Fertig (1st edition in 1964).

Ostwald, P. (1916), *Die Ukraine und die ukrainische Bewegung*. Essen: G. D. Baedeker.

Plotnikov, N. and Kolesov, M. (2005), ""Den inneren Deutschen besiegen". Nationalliberale Kriegsphilosophie in Rußland 1914-1917" in: Einermacher, K.; Volpert, A. and Bordjugov, G. (eds.), *Verführungen der Gewalt. Russen und Deutsche im Ersten und Zweiten Wetkrieg*. München: Wilhelm Fink, 31-70.

Poniatowski, S. (1912), *O wpływie błędów obserwacyjnych na wskaźniki antropologiczne* [On the observatory mistakes' influence upon anthropological indexes]. Warszawa: TNW.

Potter, J. (2007), "For Country, Conscience and Commerce: Publishers and Publishing 1914-18" in: Hammond, M. and Towheed, S. (eds.), *Publishing in the First World War. Essays in Book History*. London: Palgrave Macmillan, 11-26.

Reichl, H. N. (1917), *Die weibliche Kampfesart der Gegner Deutschlands im Weltkrieg*. Nürnberg: H. Reichl.

Remer, C. (1997), *Die Ukraine im Blickfeld deutscher Interessen. Ende des 19. Jahrhunderts bis 1917/18*. Frankfurt am Main: Peter Lang.

Rohrbach, P. (1914), *Der Krieg und die deutsche Politik*. Dresden: Verl. "Das Grössere Deutschland".

Rudnyćkyj, S. (1914), *Ukraina und die Ukrainer*. Wien: Verlag des Allgemeinen Ukrainischen Nationalrates.

Rürup, R. (1984), "Der "Geist von 1914" in Deutschland. Kriegsbegeisterung und Ideologisierung des Krieges im Ersten Weltkrieg" in: Hüppauf, B. (ed.), *Ansichten vom Krieg. Vergleichende Studien zum Ersten Weltkrieg in Literatur und Gesellschaft*. Königstein: Forum Academicum, 1-30.

Schmid, H. U. (1915), *Deutschtum und bildende Kunst. Rede am 22 März 1915*. Berlin: Carl Heymann.

Schücking, W. (1915), *Die deutschen Professoren und der Weltkrieg*. Berlin: Neues Vaterland.

Schultze-Naumburg, P. (1928), Kunst und Rasse, München/Berlin: J. F. Lehmann.

Schwabe, K. (1969), *Wissenschaft und Kriegsmoral. Die deutschen Hochschullehrer und die politischen Grundfragen des Ersten Weltkrieges*. Göttingen: Musterschmidt.

Stanzel, F. K. (ed.) (1999), *Europäischer Völkerspiegel. Imagologisch-ethnographische Studien zu den Völkertafeln des frühen 18 Jahrhunderts*. Heidelberg: C. Winter.

Stoddard, L. (1924), *Racial realities in Europe*. London: Charles Scribner's Sons.

Stransky, E. (1918), *Krieg und Geistesstörung. Feststellungen und Erwägungen zu diesem Thema vom Standpunkte angewandter Psychiatrie*. Wiesbaden: J. F. Bergmann.

Sombart, W. (1915), *Händler und Helden. Patriotische Besinnungen*. München: Duncker & Humblot.

Tajfel, H. (1981), *Human Groups and Social Categories. Studies in Social Psychology*. Cambridge: Cambridge University Press.

Talko-Hryncewicz, J. (1913), *Człowiek na ziemiach naszych* [The Man in our Lands]. Kraków: J. Mortkowicz.

Tollmien, C. (1993), "Der "Krieg der Geister" in der Provinz – das Beispiel der Universität Göttingen 1914-1919" *Göttinger Jahrbuch*, vol. 41, 137-173.

Verax. (1915), *Der Weltkrieg und das ukrainische Problem. Ein Beitrag zur Auflösung der gegenwärtigen politischen Lage*. Berlin: Carl Kroll.

vom Bruch, R. (1980), *Wissenschaft, Politik und öffentliche Meinung. Gelehrtenpolitik im Wilhelminischen Deutschland (1890-1914)*. Husum: Matthiesen.

vom Bruch, R. (2006), *Gelehrtenpolitik, Sozialwissenschaften und akademische Diskurse in Deutschland im 19. und 20. Jahrhundert*. Wiesbaden: Franz Steiner.

von Gruber, M. (1915), *Krieg, Frieden und Biologie. Rede am 28. Mai 1915*. Berlin: Carl Heymann.

Weindling, P. (1989), *Health, Race and German Politics between National Unification and Nazism, 1870-1945*. Cambridge: Cambridge University Press.

Weingart, P.; Kroll, J. and Bayertz, K. (1992), *Rasse, Blut und Gene. Geschichte der Eugenik und Rassenhygiene in Deutschland*. Frankfurt am Main: Suhrkamp.

Wilser, L. (1915), *Die Überlegenheit der germanischen Rasse. Zeitgemäße Betrachtungen*. Stuttgart: Strecker und Schröder.

Wolff, K. F. (1927), *Rassenlehre. Neue Gedanken zur Anthropologie, Politik, Wirtschaft, Volkspflege und Ethik*. Leipzig: Curt Kabitzsch.

Woltmann, L. (1907), *Die Germanen in Frankreich. Eine Untersuchung über den Einfluss der germanischen Rasse auf die Geschichte und Kultur Frankreichs. Mit 60 Bildnissen berühmter Franzosen*. Jena: Eugen Diederichs.

Županić, N. (1912), *Altserbien und die albanesische Frage*. Wien: Anzengrube

Soviet Eugenics and National Minorities: Eradication of Syphilis in Buriat-Mongolia as an Element of Social Modernisation of a Frontier Region 1923-1928

Vsevolod Bashkuev

Abstract

The anti-syphilis campaign carried out in Buriat-Mongolia in 1923-1928 is as yet an insufficiently studied historical aspect of the social and medical modernisation of national minorities in Soviet Russia. Eradicating syphilis in the context of Buriat-Mongolia's national autonomy was a state-supported, multi-staged effort aimed at the general improvement of the health, heredity and fertility of the Buriat-Mongol population. In essence, it was a peculiar example of Bolshevik eugenics as applied to autochthonic Siberian people with both short-term regional and long-term geopolitical goals in mind.

1. Introduction

This paper addresses a very important yet insufficiently studied historical aspect of the social modernisation of national minorities in Soviet Russia. The research focuses on the anti-syphilis campaign carried out in Buriat-Mongolia in 1923-1928. Eradicating syphilis in the context of Buriat-Mongolia's national autonomy was a state-supported, multi-staged effort aimed at the general improvement of the health, heredity and fertility of the Buriat-Mongol population. In essence, it was a peculiar example of Bolshevik eugenics as applied to autochthonic Siberian people with both short-term regional and long-term geopolitical goals in mind.

The Buriat-Mongolian Autonomous Soviet Socialist Republic was created on 30 May 1923. In line with Lenin's national self-determination policy, the Buriat-Mongols received a considerable degree of territorial and cultural autonomy. The period from 1923 to the mid-1930s was marked by a dynamic development of Buriat-Mongolian self-administration institutions, national culture and health care.[1]

Strategically located at the USSR's Inner Asian frontier, Buriat-Mongolia became "an outpost of socialism in the Buddhist Orient." It was believed that the Buriat-Mongolian nation would represent a model of successful social transformation for other Asian peoples and thereby assist in the exportation of the socialist revolution to neighbouring Asian states. Georgy Chicherin, the Soviet Union's People's Commissar of foreign affairs, wrote in a letter to Stalin on 12 June 1925 that the "...creation of the Buriat-Mongolian statehood within the autonomous Soviet republic had, first of all, a goal of the best influence of the USSR on the peoples of the Far East, their awakening and rapprochement with the USSR in a struggle against foreign imperialism".[2]

However, the fulfilment of such an important mission was impossible without colossal efforts to overcome numerous challenges related to underdevelopment. When it was founded, Buriat-Mongolia suffered from a virtual absence of economic infrastructure and chronic weaknesses in its social structures. Its population lived in abject poverty and was largely illiterate. Other problems included a pronounced lack of qualified staff for education and health care, rampant infectious diseases, and high mortality and low birth rates among the autochthones.[3]

Pitiful health and low birth rate in the native population were a constant headache for the Soviet authorities. They planned to empower the Buriat-Mongols to fulfil the ambitious Bolshevik geopolitical mission in the East, but this task required a strong and healthy nation. First, the Bolsheviks had to cure the Buriat-Mongols of infectious and social diseases, eliminate malnutrition and bad social habits, increase fertility and improve heredity. These tasks presupposed a radical improvement in the quality of human well-being through a highly calculated system of social, economic and ideological measures.[4]

Thus, in Buriat-Mongolia, the Soviets faced a great challenge of transforming what they called "unenlightened people" of a former imperial periphery into a true socialist nation over a short period of time. A person with a new socialist formation was envisaged as a thoroughly rational, civilised, modernised and politically conscious representative of *Homo sapiens*. He or she had to be physically fit for building socialism, mentally sound enough to understand and absorb the basics of Marxist ideology, and capable of producing healthy future generations. Eventually, this idealised human being was to become a pinnacle of a successful eugenics and social engineering programme under the auspices of Joseph Stalin's nationality policy.[5]

The gravity of this intention manifested itself in Bolshevik efforts to create a Commission for Healthcare Problems among the National Minorities in 1921. This authority (dubbed the National Minorities' Subdivision of the People's Commissariat of Public Health of the RSFSR) would be responsible for the comprehensive medical supervision of national minorities. This included sanitary and hygienic propaganda, and a study of living conditions and traditional lifestyles. Other functions included establishing special medical facilities for sick children and coordinating activities between the People's Commissariat of Public Health of the RSFSR, the People's Commissariat of Nationalities, and the public health commissariats of the autonomous republics of Soviet Russia.[6]

Though this project never materialised, the Soviet government did initiate a step-by-step programme of radical social modernisation for non-

Russian peoples of Siberia and Central Asia.[7] In doing so, the Bolsheviks were driven not only by their geopolitical and ideological goals, but increasingly by the aggravating health problems and a threat of extinction of some Asiatic national minorities. During the 1920s, Buriat-Mongolia represented one of the worst cases, especially concerning the so-called social diseases. Tuberculosis was widespread in the western districts of the republic, which was populated by Russians and semi-nomadic and sedentary Buriat-Mongols. Syphilis was rampant in the central and eastern parts, which was mostly inhabited by nomads.

These two diseases were generally considered indicators of social disorder and low living standards. The Bolsheviks ascribed their prevalence to the appalling social conditions of Siberian natives resulting from past discrimination by imperial authorities. Syphilis was an especially major concern. Highly adaptive to the given cultural environment, this disease was so widespread among the republic's nomadic native population that specialists at times spoke of 200,000 syphilitics out of a population of a little more than 500,000.[8] In some areas, the difference between registered syphilis cases among Buriat-Mongols and Russians was as high as 38 times.[9]

Beginning in the late 19th century, Russian venereologists drew a clear-cut borderline between the predominantly non-venereal syphilis in villages *(syphilis insontium)* and venereal syphilis in cities.[10] In remote areas such as Buriat-Mongolia, where practically no modern urban centres existed by the 1920s, syphilis was definitely considered as endemic (non-venereal). Soviet doctors described syphilis in Buriat-Mongolia as "the purely household infection." In their view, it originated in the traditional unkemptness of the Buriat-Mongols – their lack of bathing, habits of sharing smoking pipes, and unclean kitchen utensils, clothes and bed-sheets. In essence, the "cultural barbarianism" of nomadic life was viewed as the main contributing factor for the spread of syphilitic infection.

Without radically transforming the traditional lifestyle of the Buriat-Mongols by "acculturating" them, syphilis was seen as impossible to eradicate. Therefore a cultural revolution was to accompany the anti-syphilis campaign. The eradication of syphilis and the improvement of the human condition were impossible without a system of measures comprised of education, sanitary propaganda, cultural change and large-scale implementation of modern European medicine.

Officials of the Buriat-Mongolian Commissariat of Public Health were specifically anxious about the long-term consequences of syphilis on the future generations. Considering the low birth rates among Buriat-Mongols (about 17.6 per 1,000 compared to 45.5 per 1,000 among the Russians in 1924),[11] the high number of miscarriages, and widespread cases of congenital and tertiary syphilis, they feared the extinction of the Buriat-Mongols, or, at least, a dramatic reduction of their numbers.[12] This scenario would be absolutely contrary to the Bolshevik mission of a new socialist Buriat-Mongolian nation as a showcase of Communist nation-building. Undoubtedly, it would also gravely contradict with the grand Bolshevik plan of bringing the Buriat-Mongols (and other Asiatic peoples of Soviet Russia) from feudalism directly to socialism, bypassing the stage of capitalism.

To confront and solve this problem, the leadership of the Buriat-Mongolian republic, with full support from Moscow, embarked on a long and costly but, in the final analysis, effective anti-syphilis campaign. It comprised several stages, including thorough medical examinations of the population and painstaking research on the origins, characteristics, consequences and treatment of syphilis among Buriat-Mongols and Russians. The results were not purely scientific or clinical discoveries, but were a comprehensive analysis of the social construction of syphilis in different areas of Buriat-Mongolia.[13]

Furthermore, the campaign involved vigorous propaganda related to personal hygiene, sexual moderation, protection of maternity and infancy, basic sexual and hygienic education, and the rehabilitation of prostitutes, unemployed or criminal women and other suspected carriers of the disease. In addition, the Russian People's Commissariat of Public Health took advantage of foreign expertise. At the time, medical doctors from Germany were at the cutting edge of clinical research in syphilology.[14] Through the networks of the USSR's Academy of Sciences, as well as the personal connections of Professor Volf Moiseevich Bronner (director of the State Venereal Diseases Institute in Moscow), the Buriat-Mongolian case attracted considerable attention among German venereologists. The international transfer of knowledge was carried out though a joint Soviet-German medical expedition to study syphilis in Buriat-Mongolia in 1928.

The character of the anti-syphilis campaign in Buriat-Mongolia suggests that in the early Soviet social modernisation discourse, Bolshevik ideology was amalgamated with European eugenics concepts. The campaign itself was comprised of the politics of the body, social hygiene, pro-natalist and child-care policies, and ideological mobilising frames that portrayed syphilis as an enemy firmly associated with the dark legacy of Tsarism. The combination of geopolitical aspirations and ideological designs, with goals of social eugenics, points to the fact that the Bolsheviks theorised that a utopian, multipurpose eugenics programme would produce a totally harmonious human being in a matter of just a couple of decades. At the same time, despite its obvious overall utopianism, elements of this programme, once implemented, had far-reaching social, demographic and even eugenic effects. The anti-syphilis campaign in Buriat-Mongolia certainly was one of such elements and, as further study will demonstrate, had significant socio-historical importance for the emerging Buriat-Mongolian nation.

2. Endemic Syphilis in Buriat-Mongolia: The Background

Russian venereologists were certain neither about the origins of syphilis in Buriat-Mongolia nor the exact timing of its initial spread in the region. Dr. D. A. Lapyshev, whose venereologic detachment worked in the Troitskosavsk *aimak* (region) of Buriat-Mongolia in summer 1924 and in the Adon-Chelon *khoshun* (county) in summer 1925, came to the conclusion that syphilis among native populations dated back several centuries. This finding was based on the number of Buriat-Mongols afflicted with syphilis, the prevalence of tertiary and latent forms of the disease, and the high percentage of congenital syphilis.[15]

In a July 1930 article published in the local magazine *The Life of Buriatia*, Y. F. Zhorno stated that Russians brought syphilis to Buriat-Mongolia in the middle of the 18th century.[16] M. A. Kubantseva, in her article published in this magazine in autumn 1929, summarised several expert opinions. Though rather doubtful, the most interesting view was that syphilis came to Buriat-Mongolia from China, where an entire town allegedly was inhabited by syphilitics. However, based on 19th century archival data, Y. F. Zhorno refuted this hypothesis, stating that syphilis had spread from the West via Russian merchants travelling through Kyakhta to China.[17] Doctors L. G. Zaks and S. T. Ilyin found it difficult even to determine the approximate age of syphilis in Buriat-Mongolia and concluded that it could have been there for several decades.[18]

The speculation about the origins of syphilis, however, suggested that Buriat-Mongols certainly had suffered from the disease for quite some time, since it had a specific term in their language. Other venereal diseases such as gonorrhea were also reflected in the language and the terminology of local traditional medicine. *"Muu uboshen yara"* or *"Pal'kkhin uboshen"* – "evil" or "malicious" ulcer – was the term used by Buriat-Mongols to denote syphilis. *"Kyiton uboshen"* or "cold illness" was the traditional Buriat term for gonorrhea.[19]

Medico-anthropological observations done in the late 19th and early 20th century suggested that Buriat-Mongols could well differentiate between leprosy and syphilis.[20] They were very afraid of leprosy and strictly isolated lepers living in their midst. Horrible cases were recorded in which neighbours secretly killed lepers and burned their entire families alive inside yurts.[21] Nothing like this is recorded about syphilis. Medical examinations of Buriat-Mongolian families showed that in most cases they did not isolate syphilitics and that they continued to live under one roof with them. In a family, people with tertiary and congenital syphilis could coexist with completely healthy members. Seemingly the Buriat-Mongols viewed syphilis as a bad but common and familiar suffering. Contrastingly, leprosy was so terrifying that people would flee or become aggressive in protecting themselves from infection.

Over time, the Buriat-Mongols developed certain traditional ways of treating syphilis. Treatment among the illiterate Buriat-Mongols mostly was performed by elderly people who previously had suffered from the disease, or by *lamas* (Buddhist priests) trained in Tibetan medicine. A syphilitic patient was treated with mercury chloride, expensive herbs and cinnabar.[22] No doubt, the results of such therapy could hardly be compared with the effects of *Salvarsan*, an arsenic-based anti-syphilis drug of the pre-antibiotic epoch.

Statistics of the People's Commissariat of Public Health of the RSFSR indicate that out of 42,187 cases of infectious diseases registered in 1924 in Buriat-Mongolia, 10,926 of them were syphilis cases – more than a fourth of all cases. Of these, 6,588 cases were tertiary syphilis, 3,838 were secondary syphilis, and only 500 were recognised as recent cases.[23] The 1924-1925 report of the Buriat-Mongolian People's Commissariat of Public Health indicated that in heavily infected areas, the number of syphilis cases per 10,000 people was as high as 518, with an average of 246 cases per 10,000. By comparison, the average rate of syphilis contraction in pre-revolutionary European Russia was 72 cases per 10,000 (1914).[24]

These numbers demonstrate the scale of syphilis infection in Buriat-Mongolia. These figures, however, are almost certainly not 100 percent accurate, since some regions of the republic were too distant from urban centres to be reached by venereological detachments and lacked medical personnel capable of gathering more or less reliable data. As officials of the *Burnarkomzdrav* claimed, statistics presented in the reports must have represented only the very tip of the iceberg. In reality, the situation could have been far worse.

3. Social and Cultural Construction of Syphilis in Buriat-Mongolia: The Medical Viewpoint

Before directly turning to the construction of social and cultural origins of syphilis in Buriat-Mongolia as viewed by Soviet medical doctors, it is crucial to briefly address some influential viewpoints on syphilis and its social dimensions that formed among Russian doctors and scientists in the 19th and early 20th centuries. This background is important in the context of our study, due to the conceptual continuity in how pre-revolutionary Russian and Soviet specialists perceived syphilis, explained its causes, and developed strategies and methods to eradicate it.

A very detailed analysis of views that were prevalent among the Russian doctors on syphilis and its social nature in the 19th century was conducted by Laura Engelstein. She argues that unlike their European colleagues, Russian medical doctors tended to look for epidemiological clues not in the clinical picture of the disease, but in its social and cultural context. Many of them were *zemstvo* physicians working in the countryside. They shared the populist commitment to act in the interest of the common people, educating them and raising their cultural level, following their "distinctive medico-political ethos."[25]

Russian medical observers unanimously characterised rural syphilis as endemic and non-venereal in origin. They saw it "not as the result of sexual promiscuity, the egotistic search for private pleasure in disregard of the collective norm, but as the result of social promiscuity, a reflection of collective tyranny and the weakness of self." They viewed the peasant community as a deeply traditional milieu in which there was hardly a place for sexuality and, therefore, in Engelstein's words, "they persistently rejected evidence of sexual misbehaviour that testified to the crumbling of traditional bonds."[26]

Poverty and ignorance were perceived as the main contributing factors in the spread of syphilis. The crowding of children in village huts, the impossibility of observing proper hygiene standards under conditions of poverty, carelessness stemming from general ignorance, and unfamiliarity with the specific nature of the disease were listed as the causes of rural syphilis. In fact, pre-revolutionary physicians called syphilis "the Russian people's everyday disease."[27] Twenty years later in Buriat-Mongolia, Soviet physicians would refer to it as "the purely household infection." These characteristics share similar connotations. They both imply that backwardness was the main cause of syphilis in a Russian wooden hut and in a Buriat-Mongolian felt yurt alike.

Another important circumstance was that Russian physicians characterised syphilis as an invisible, but dangerous enemy threatening national honour and security.[28] Soviet doctors effortlessly borrowed this image and built it into their own agenda, adding to it the Bolshevik rhetoric of class struggle. Overall, they quite comfortably appropriated the conceptual framework developed by their predecessors in constructing social and cultural causes of syphilis, defining backwardness, ignorance and superstition as the main culprits. At the same time, Soviet physicians generally adopted the pre-revolutionary epidemiological approach in studying rural syphilis. They painstakingly gathered demographic, anthropological, cultural and socio-economic data to expose the social roots of the disease. Timely exposure and removal of these roots were viewed as the necessary condition of success in the cultural context of Buriat-Mongolia.

Syphilis in Buriat-Mongolia was unanimously defined as the disease of the native population. This viewpoint was statistically well grounded. In his 1926 report based on earlier research trips to the most affected districts, Vice People's Commissar of Public Health V. N. Zhinkin indicated that in the Khorinskii *aimak*, Russians had 61 syphilitics per 10,000 people compared to 627 for Buriat-Mongols; and rates of 13 and 503 for Russians and Buriat-Mongols, respectively, in the Bokhanskii *aimak*. Similar figures were typical in other districts except in urban areas, where the gap between Buriat-Mongol and Russian syphilitics was smaller.

Doctors also highlighted a clear cultural pattern in the geographical distribution of syphilis in Buriat-Mongolia. The situation in western Buriat-Mongolian *aimaks* was better than in eastern ones. In their reports based on data gathered during expeditions by venereological detachments west and east of Lake Baikal, Soviet physicians noticed that western Buriat-Mongols were generally more "civilised" than their eastern counterparts. Most people in the west were sedentary or semi-sedentary, whereas eastern Buriat-Mongols were nomadic herders. Western Buriat-Mongols were at least partially aware of the destructive character of the "malicious ulcer." They took some measures to isolate syphilitic settlements, imposing a sort of improvised "quarantine" on them by not marrying people from these areas, minimising contacts with infected people, and so on. In the words of Dr. G. Nagibin, "This instinctive method of isolation is, in fact, preventive since it does not leave syphilis a chance to refresh. However, it is possible only in the western *aimaks*, where the population is relatively cultured, willingly receiving treatment, if needed, and informed about the outcomes of the disease."[29]

Eastern Buriat-Mongols, with their nomadic lifestyle, perpetual shabbiness and, in the words of one medical observer, "endless patience", presented in the eyes of the European-educated physicians a population uniformly afflicted with a wide variety of venereal and skin diseases, ranging from syphilis to lice infestation. Their suffering in many cases lasted for generations, since syphilis in some families was traceable to very distant ancestors. Dr. D. A. Lapyshev, who headed a venereal detachment in Adon-Chelon county in 1925, characterised everyday life and nomadic practices of eastern Buriat-Mongols as follows:

> The archetypal aboriginals of this land are the Buriats and rarely the Tungus, typically roamers and herders, scattered here and there. The Russian population is non-existent here. The original circular white yurts with the typical fireplace and miserable belongings are liberally scattered in the vast steppe…, only rarely concentrating into large settlements – *uluses* along the rivers.[30]
>
> …Bed-sheets and blankets are not typical here; people sleep side by side on the same felt pad they seat in the daytime, covering themselves with the same clothes they always wear. Bloomers and a shirt – worn by both men and women – are never washed. They are worn by the owner until they tear apart and disintegrate right on him. There are no baths whatsoever, and a normal Buriat rejects washing. The soap is unknown

> here. We will mention as a joke the fact that during our stays in the *uluses* crowds of people would come to stare at our morning washes. ... Our frequent appeals to the patients to come bathe themselves in the river even without soap always met energetic refusals. It is noticeable that explanations the Buriats gave for their refusal to bathe are not devoid of some specific pragmatism: 'If I wash away the dirt, the lice will eat me up...'[31]

Doctor S.T. Ilyin, head of another venereological detachment in the Khorinsk *aimak*, emphasised the fact that Buriat-Mongols frequently shared the same dirty bowls, which typically were cleaned by licking instead of washing. Among the practices that contributed to the quick spread of syphilis and other infectious diseases, he mentioned smoking of the same pipe by several people in a row, exchanging clothes, using animal hides instead of blankets, and sleeping together of men, women and children.[32]

The sexual habits of Buriat-Mongols represented another important research aspect. Physicians pointed out that while the sexuality of rural Russians had been sufficiently researched, the sex life of Soviet national minorities remained largely unstudied. For social venereology, knowledge of sexual habits was extremely important to develop preventive measures against venereal diseases.[33] The staffs of all venereological detachments in Buriat-Mongolia took care to investigate the sexual practices of native people that could spread syphilis. During field trips they interviewed dozens of patients about their sex lives, habits and preferences. Some of the data was published. This gave us the chance to look closer upon this side of life of Buriat-Mongols.

Several doctors highlighted the fact that in the shabbiness and overcrowding of traditional Buriat-Mongolian dwellings, children of a very young age would regularly witness the sexual relations of their parents, growing accustomed to this aspect of life. Moreover, being herders, they constantly observed and closely studied the sexual behaviour of domesticated animals, not finding anything disgusting or shameful about it.

Dr. A. M. Pesterev, who worked in venereological detachments in the remote Khorinskii and Eravniniskii districts in 1925-27 and in Zakamenskii district in 1930, studied the characteristics of the sex life of Buriat-Mongols. He calculated that 55.7 percent of Buriat-Mongolian men and 78.7 percent of women began sexual activity very early. The most common ages were 14, 15 and 16.[34] Pesterev stated that in 45 percent of cases, women began having sexual relations well before their first menstruation. He attributed this to the fact that regardless of their age, Buriat-Mongolian women were often sexually abused by men.[35]

Among the factors contributing to the early beginning of sex life and lax sexual behaviour, Pesterev singled out the following:

- Overcrowding in yurts;
- The almost exclusively meat and dairy diet of Buriat-Mongols, which induced sexual arousal;
- Alcohol abuse, a rather widespread addiction among many indigenous peoples;
- Horse riding as the most common way of transportation, which stimulates erogenous zones from the very early age;
- Absence of cultural needs, which makes sexual intercourse a sort of entertainment;
- High infant mortality in Buriat families, which made children the main purpose of marriage. A childless man would abandon his sterile wife and look for another woman until he had children; women were very afraid not to bear children, since sterility was considered a social stigma;
- Ignorance about venereal disease. People would not stop their sexual activity even when infected, spreading the disease. In most cases, eastern Buriat-Mongols did not associate syphilis and gonorrhoea with sexual activity. In fact, many people thought gonorrhoea was a normal physiological process with men, much like menstruation with women;
- No tradition of sexual abstention. Because family and religion did not regulate the sex life of Buriat-Mongols, people had many sex partners, which caused the spread of syphilis and gonorrhoea.[36]

While some of these assumptions look naïve or speculative today, others suggest that like his pre-revolutionary Russian colleagues, Pesterev found the roots of syphilis in socio-cultural conditions rather than sexual superfluity. He clearly understood that nation-building finally gave the indigenous peoples of Siberia a long-awaited chance to eliminate syphilis. Therefore, Pesterev concluded:

> The task of social venereology in the course of building up of the curative and preventive measures should be the urgent sanitation of the sexual life and family life in general in order to paralyse infections like syphilis and gonorrhoea, whose further spread may lead to the extinction of the Buriat nationality. The ethnic perspectives of the Buriat nationality greatly depend on how early and intensively the fight against the old life and sexual habits will start.[37]

In this way, there was a definite continuity of ideas between the pre-revolutionary Russian physicians who studied rural syphilis, and the venereologists of the early Soviet period who struggled with syphilis within national minorities. Meanwhile, the prevalence of syphilis in the Buriat-Mongolian population continued to worry even the most experienced specialists. Dr. Lapyshev's 1924-1925 expeditions found only 49 syphilis-free families among the 155 families examined: in the other 106 families either all or some members had some form of syphilis. His examinations revealed that 75 percent of the described cases were among people aged 20 to 60, which encompassed people of reproductive age. Latent forms were found in 68.8 percent of all cases. Active forms were found in 40 cases, out of which only eight cases were fresh forms.

Syphilis among Buriat-Mongols was predominantly tertiary with a broad variety of symptoms. Patients with latent syphilis had skin lesions (65.4 percent of cases), affection of facial bones (54.5 percent) and skeletal deformities (26.3 percent). There were also cases of degenerative syphilis going deep into familial ancestry.[38] These cases showed that syphilis manifested its destructive consequences in following generations.

Another alarming tendency was the hybridisation of syphilis and tuberculosis, which venereologists described in a number of cases. Dr. Lapyshev stated that the danger of degeneration and extinction of Buriat-Mongols was quite imminent if the socio-cultural conditions in which syphilis flourished were not radically improved in the future.This viewpoint was shared by Y. F. Zhorno, who wrote:

> The setting of a whole range of the key issues in the fight against syphilis of the Buriat-Mongols...requires an immediate implementation. ... We witness colossal mortality of the Buriats and, possibly, their extinction. ... If to consider the paucity of the Buriat population multiplied by the early mortality of children and cultural backwardness, every Buriat syphilitic and, especially, a syphilitic pregnant woman fully treated within two years will be a large and noble victory in the battle against venereal disease in the Buriat republic.[39]

In this way, Soviet medical observers in Buriat-Mongolia considered syphilis to be a great social evil, menacing the population with extinction. Among its primary causes they listed the profound "cultural and social backwardness" of the Buriat-Mongols, the almost total ignorance of the transmission methods and consequences of syphilis, and the unsanitary conditions of their traditional lifestyle. Among the contributing factors were primitive sexual habits and the widespread sexual abuse and exploitation of women, which enabled the spread of venereal syphilis. The absence of social hygiene, modern obstetrics and European medicine called for urgent action.

4. Anti-syphilis Campaign and Social Eugenics in Buriat-Mongolia, 1923-1928

Fortunately for Buriat-Mongols, by the 1920s there were specialists in the People's Commissariat of Public Health of the RSFSR with first-hand experience of health problems in Buriat-Mongolia. Their support helped begin the tedious work of eliminating endemic syphilis. Primarily, there was Dr. Volf Moiseevich Bronner, head of the Venereological Sub-Commission of the People's Commissariat of Public Health of the RSFSR, an influential and highly professional venereologist. Bronner was born in Buriat-Mongolia in 1876, finished high school in Chita and commenced studies at Tomsk University. However, Bronner would soon get involved in revolutionary activity and, in 1899 after the 10th semester at the medical faculty, he was expelled. He graduated from Berlin University's faculty of medicine in 1900. In March 1900 Bronner successfully defended his doctoral thesis and was conferred the degree of medical doctor. From 1900 to autumn 1901 he served as the municipal doctor in Verkhneudinsk. From 1906 to 1913 he worked in Paris, first at the urological clinic of Professor Guillion, then at the Pasteur Institute. Bronner edited the French *Journal Clinique d'Urologie* and authored a number of influential works in the fields of urology and venereology.[40]

From 1915 Bronner lived in Moscow, where he headed the venereological hospital No. 9871. In October 1918 he began working at the Commissariat of Public Health as the secretary of the venereal disease section. From 1920 he worked as the head of the Venereological Sub-Commission. In 1922 Bronner organised the State Venereological Institute in Moscow, which was named after him. He was characterised as one of the few public figures in the field of Soviet venereology with a solid scientific background, knowledge of European medical practices and a broad scientific approach to problems of venereal disease in the USSR.[41]

The experience of living in Buriat-Mongolia must have enlightened Bronner to the health problems of indigenous people. His lobbying for the anti-syphilis campaign there was very strong. Buriat-Mongols could not have hoped for a better advocate of their vital interests – an influential scholar and medical doctor with wide international connections, a unique specialist in venereology, a high state official in Moscow, and an old Bolshevik with access to the highest-ranking party functionaries.

Bronner was the initiator of the First All-Union Convention against venereal disease, which was held in Moscow in 1923. It was a turning point in the history of anti-venereal medicine in the USSR: after this convention began a rapid growth of venereological dispensaries, mobile venereological detachments and rural venereological stations. In the six years from 1923 to 1929, the number of venereological dispensaries in cities grew from 29 (located only in large regional capitals) to 172 (101 of which were located in smaller district and county centres). The same dynamic of growth was displayed in the rural venereological stations: from 12 in 1924 to 243 in 1929 (101 located in county centres).[42]

With centrally coordinated efforts to organise a united front against venereal disease, the government of the newly created Buriat-Mongolian republic launched a local anti-syphilis campaign. Backed by Professor Bronner, it was carried out simultaneously in several directions. Its medical component involved rebuilding the medical network, opening specialised venereological dispensaries, conducting scientific research on syphilis among Buriat-Mongols, and carrying out mass medical examinations by venereological detachments. Special attention was given to remote areas.

The second direction encompassed a vigorous education campaign to promote personal and social hygiene, European medicine and healthy living. Much attention was to be paid to informing people about the destructive consequences of syphilis and its means of transmission. The educational component also involved creating a strong sentiment against the old prejudices and those who, in Bolshevik eyes, personified and perpetrated them: the Lamaist clergy, Shamans, traditional healers and such "rudiments of the past."

Thorough medical examinations of the population began in 1923, and serious efforts to examine Buriat-Mongolian herders en masse were undertaken in 1924-1925, when three venereological detachments worked simultaneously in the Aginskii, Khorinskii, and Barguzinskii districts, which were the worst syphilis-affected areas of Buriat-Mongolia. It must be stressed that the efficiency of such venereological detachments was rather high. At first, Buriat-Mongols were very suspicious of Russian doctors, who not only looked unfamiliar but also performed mysterious and seemingly magical operations with strange instruments and devices. Superstitious Buriat-Mongols were afraid of them, which required some effort to persuade them to allow medical checks and treatment.[43]

However, once these doctors earned the nomads' trust, the popularity of anti-syphilis treatment rose so high that hundreds of sick Buriat-Mongols would arrive at medical stations from all directions. Lapyshev wrote that, as a rule, the worsening symptoms of untreated syphilis quickly disappeared after a few infusions of *Neosalvarsan*. People who previously had looked like walking corpses were resuscitated and could even return home on horseback.[44]

The effect of the venereological dispensaries was as positive as that of the venereological detachments. After the Russian Civil War, the hospital network in Buriat-Mongolia literally lay in ruins. In 1924 the average number of patients per hospital cot was 1,870, and the number of patients per one doctor equalled 19,590.[45] In 1924-1925 the number of venereological dispensaries in the republic increased to five, but only three were operational since there were no doctors to staff the other two.[46]

In addition to treating syphilis, venereological dispensaries registered and traced cases, which helped gather information about patterns of the disease in Buriat-Mongolia. According to the main regional newspaper *Buriat-Mongol'skaya Pravda,* in three years there were 95,870 visits to the venereological dispensary in Verkhneudinsk alone. Not only townsfolk but people from neighbouring villages also went to the dispensary in search of medical assistance.[47]

Soviet authorities in Buriat-Mongolia were certain that success in the fight against syphilis and other venereal diseases could not be attained solely by medical measures. A broad propaganda campaign was required, too. Therefore, each year local newspapers announced "three days of fight against venereal diseases," during which doctors gave public lectures about the dangers of venereal diseases, exhibitions were held in the capital and regional centres, and communist activists gave speeches against superstitions and promoted healthy ways of life.

Functionaries of the *Burnarkomzdrav* understood well that the shamans, lamas and traditional healers presented a formidable obstacle to sanitising the population and eradicating venereal disease. The 1925 report stated: "Traditional healing is rampant in the Buriat republic. Its cause generally lies in the rusticity of the population, especially of the Buriat-Mongols. The Buriat way of life, their isolation from qualified medical service as well as their century-old customs and religious traditions facilitate the spread of traditional healing and hamper the fight against it."[48] Another obstacle was Tibetan medicine, which the Soviet physicians viewed as no less harmful a practice. The report commented: "There are huge numbers of medical lamas in the Buriat republic calling themselves 'Tibetan medics'. The bulk of them are absolutely unenlightened and ignorant in medicine. Making use of the rusticity of the masses, they are immensely popular with the culturally backward Buriat population."[49]

Mass media played an especially important role in the anti-clerical and anti-healer propaganda. Newspapers published articles about the harmful influence of lamas, shamans and traditional healers. One article told of a healer in a Russian village who treated different diseases without the slightest knowledge of medicine. The healer's career ended up badly: he fell ill and died – from syphilis. He infected all of his family members and, possibly, many people who came for treatment. Through such simple, hyperbolised, articles, Soviet authorities persuaded the population that healers, shamans and lamas were not qualified to treat dangerous social diseases such as syphilis.[50] Only qualified medical personnel could help people suffering from venereal diseases, and such help was only available at state medical institutions.

Much information was distributed through village reading rooms and yurts *(izba-chital'nia, yurta-chital'nia)*. Propaganda materials were translated into the Buriat-Mongolian language and disseminated. However, the occasional lack of qualified translators led to odd incidents. In February 1928, Soviet authorities issued many anti-syphilis slogans, published in both Russian and Buriat languages. The Buriat translation of the Russian slogans was so inaccurate that its meaning was totally contrary to the original. If the Russian original slogan said, "Venereal diseases, especially syphilis, are a dangerous scourge of the human organism", its wrongly translated Buriat equivalent would announce "Ulcer, syphilis is not dangerous for humans".[51] Of course, one can have difficulty believing such profound professional incompetence, but taking into account a chronic lack of professionals in the Soviet bureaucratic system, such mishaps seem possible.

The campaign against "social [sexually transmitted] diseases" also encompassed measures to eradicate social ills such as prostitution in the cities. A special rehabilitation facility for prostitutes and unemployed single women was opened in Verkhneudinsk, where women were supposed to learn cutting and sewing for future employment. They were also regularly examined by doctors.[52]

All in all, measures to curb the spread of social diseases and to sanitise and enlighten the population went in line with Bolshevik ideology. Social diseases were portrayed as the heavy legacy of the past, a result of the Tsarist government's oppressive policy of keeping the indigenous people of Buriat-Mongolia dependent and exploited. Social classes alien to the proletarian dictatorship, such as the Lamaist clergy, shamans, and traditional healers, were viewed as perpetrators of old superstitions that were placing obstacles on the path to progress. The effectiveness of the anti-syphilis work was attributed to modern European medicine. Much effort was directed at improving the health of the population, eradicating old traditional practices and introducing new, European-style hygiene. The Soviet government went so far as to accept foreign scientific assistance. With the help of Professor Bronner and after three years of international negotiations, a joint Soviet-German expedition to study endemic syphilis in Buriat-Mongolia was organised in 1928.

5. International Transfer of Knowledge in the Fight against Syphilis: The Soviet-German Medical Expedition of 1928

Negotiations between Soviet and German officials about a possible joint expedition to Buriat-Mongolia began in 1925. On the German side, the talks were carried out by chief officers of the *Notgemeinschaft der Deutschen Wissenschaft*, officials of the German Ministry of Foreign Affairs, and the

Embassy in Moscow. On the Soviet side, the main negotiators were officials of the People's Commissariat of Public Health, People's Commissariat of Education, and VOKS (the All-Union Society for Cultural Connections with Foreign Countries).

The importance of the expedition was so great that in the course of a preliminary meeting held in Berlin in January 1926, it was decided that a group of specialists should be sent to Buriat-Mongolia for a preparatory observation. From the Soviet side the group included Professor Bronner. Germany was represented by dermatologist A. Stümer and psychoneurologist Karl Wilmanns. The preparatory observation revealed tensions between Soviet and German doctors: the Soviets were offended by the self-assuredness of the Germans, and the Germans complained about the inadequate level of scientific and professional training of their Soviet colleagues and the primacy of politics in their relationships with foreign colleagues.[53]

Despite these minor tensions, the expedition's goals quickly became the major concern. In her article, Canadian researcher Susan Gross Solomon states that the goals of German and Russian teams were so profoundly different that it may be plausible to speak about not one, but two expeditions. She explains her thesis through the different scientific agendas of the German and Russian doctors. For the Germans, who lost colonies after the defeat in World War I and with them access to "human laboratories" for the study of infectious diseases, the Buriat-Mongolian population presented a treasury of actual syphilis cases. German specialists could study clinical characteristics of the disease in a population that largely had been unexposed to European medicine. It provided unique access to a "human laboratory" that German doctors otherwise would have lacked in Europe, where untreated syphilis on such a scale was already eradicated by that time.[54]

The Germans were interested in testing Professor Wilmanns' theory about a connection between the *Salvarsan* treatment and the development of tertiary syphilis. Wilmanns' hypothesis maintained that under the effects of *Salvarsan*, the *treponema pallida* mutated and transformed from a dermotrope, a pathogen that affected skin, into a neurotrope, a pathogen that affected the nervous system.[55] To confirm his theory, Wilmanns needed access to a large number of untreated syphilitics. Members of the expedition would also study the peculiarities of the *Salvarsan* treatment under real conditions.

The Soviet doctors did not match the scientific training of their German colleagues, but they were superior in the organisation of preventive measures. Moreover, they could easily access the affected regions of the USSR to test new medicines and treatment methods, which they expected to borrow from the Germans. Therefore, contact with more experienced and modernised German medical specialists was a crucial element of their agenda. Transfer of knowledge in the field of syphilis treatment was the main objective of the Soviet team.[56]

At the same time, Soviet physicians portrayed syphilis as a great social evil whose eradication was a crucial prerequisite of modernising the region. Aware of the after-effects of *Salvarsan* treatments, Soviet doctors were concerned about the possibilities of its widespread use, but not so much about the mutations of the microbe that caused syphilis. Moreover, in their struggle against venereal diseases, Soviet medical authorities paid more attention to eradicating their social causes than to purely medical aspects. Syphilis in small indigenous nationalities in Siberia seriously threatened their future existence. The Buriat-Mongols, who had recently proclaimed national-territorial autonomy, were the largest of these populations, and their continuous degeneration at the very beginning of the Soviet experiment would be detrimental to the international image of the USSR. This is why for the Soviet research team, something larger and more important than merely testing a medical theory was at stake.

The expedition arrived in Buriat-Mongolia in June 1928. Its work was thoroughly reported in the local press. The arrival was filmed by *Sovkino* for the newsreels. Articles about the expedition's progress regularly appeared in the local daily, *The Buriat-Mongolian Pravda*.

The expedition began its work in the village of Kulsk on 4 June 1928. Active research was conducted throughout the summer. Doctors examined about 200 patients a day, mostly Buriat-Mongols. Interestingly, local Russians viewed the predominance of Buriat-Mongols among the patients of Kulsk hospital as discrimination against them. Alarmed by growing nationalist sentiments, the doctors began examining Russian patients.

At first, the local lamas actively agitated against the expedition. They feared that the European doctors would win over their loyal clientele. Lamas spread rumours that the German and Russian doctors took people's blood to perform magic rituals that would harm patients. But as soon as the first effects of *Salvarsan* treatments were observed by the Buriat-Mongol patients, they spread the word, and hundreds of people from isolated settlements hundreds of kilometres away came to Kulsk. The lamas then began singing a different tune. They concocted a story that German and Russian doctors came to Kulsk to learn Tibetan medicine from them. In their own anti-syphilis therapy, some lamas would include recommendations to visit the European doctors.[57]

The Soviet specialists worked in a different direction. While the Germans conducted tests and research, Soviet doctors painstakingly investigated the social origins of the disease. They carried out so-called "sincere interviews", asking patients about their sex life and habits, and filling out special questionnaires. To ensure maximum sincerity and precision of answers, they started treatment with *Salvarsan* only after the patients had thoroughly and completely answered all of the questions in the questionnaire.[58]

The Soviet-German expedition ended its work in September 1928. Both German and Russian specialists were highly satisfied with its results. In a series of articles published in *The Buriat-Mongolian Pravda*, scientists gave interviews about the major outcomes of the three-month effort. Dr. E. Radnaev of *Burnarkomzdrav* told a correspondent that the Germans achieved all of their planned objectives. He noted that the German members concentrated exclusively on gathering scientific materials and that their team was well diversified: venereologists studied symptoms of the disease, neurologists observed the effects of syphilis on the nervous system, and so on.[59]

The expedition made several important scientific findings. First, latent syphilis turned out to be the prevalent form of the disease in the population, as opposed to fresh cases, of which only a handful were registered. Neurosyphilis constituted only 15 percent of more than 1,300 patients whose spinal fluid was tested. Contrary to expectations, there were relatively few cases of congenital syphilis. The head of Russian team, Professor Rossiianskii, suggested that this was due to early mortality of congenital syphilitics. The fact that the expedition found rather few cases of gummous syphilis and fresh syphilis was, in his view, a direct result of earlier medical and preventive work conducted by the venereological detachments.[60]

The expedition highlighted the need to continue studying hereditary syphilis and its effects on future generations. The number of patients that the expedition could examine during the limited period was also rather small (3,996 Buriat-Mongols and 971 Russians), and the successful campaign against the disease required continuous efforts to diagnose more people. Moreover, the Soviet-German expedition found that diagnosing cases with the Wassermann blood serology test was too complex to be conducted in the field, and that simpler but equally effective methods were required.[61]

In the final analysis, the expedition was an example of a successful transfer of state-of-the-art medical knowledge across national and ideological borders. Both sides achieved their goals. The results confirmed that a complex effort to research and fight social diseases was required in Buriat-Mongolia. A plan was devised to establish an institute of social hygiene on the basis of a biochemical laboratory in Verkhneudinsk in 1929-1930. This institute would organise research work in the following directions: 1) determining causes of the low birth rate and researching the physique of Buriat-Mongols; 2) studying social diseases, primarily syphilis and tuberculosis; and 3) special research projects necessitated by the settling down of nomadic Buriat-Mongols.[62]

Unfortunately, not all of the plans materialised. But in 1928-1931, the number of venereological ambulatory clinics in Buriat-Mongolia increased from six to nine. Therapeutic assistance was provided by two houses of sanitary education in Verkhneudinsk and Troitskosavsk. In 1928-1930 there were three more expeditions to study gonorrhoea and tuberculosis, and a group of eight doctors studying labour and everyday life conditions as well as social diseases in three communities of Buriat-Mongolia.[63] As Bronner mentioned in his article on the results of the Third All-Union convention to fight venereal diseases held in 1929 in Moscow, the data from Buriat-Mongolia revealed a decrease in the number of syphilis cases. He said this was a direct result of the vigorous anti-syphilis campaign of the late 1920s, which significantly reduced syphilis cases, especially in villages, across the entire USSR.[64]

6. Conclusions

The anti-syphilis campaign of late 1920s in Buriat-Mongolia stands out as a successful example of eradicating social diseases in the national autonomous republics of the USSR. The catastrophic scale of the problem and efforts to understand its malignant consequences necessitated measures on both the national and international levels. The campaign involved the highest medical authorities from the RSFSR and the Soviet Union. Powerful support from Moscow, from the People's Commissar of Health N. A. Semashko and V. M. Bronner, greatly assisted the newly formed Buriat-Mongolian republic to solve its century-long problem of rampant syphilis.

The problem of social diseases among national minorities in the USSR was primarily viewed in the context of social and cultural modernisation. In the long run, Buriat-Mongols were expected to fulfil a thoroughly geopolitical Bolshevik mission of revolutionising the East by providing a successful case of modernisation. However, in the short-term perspective, huge challenges of underdevelopment had to be overcome in very little time. This dissonance between ambitious goals and the more than problematic reality necessitated urgent, resolute and coordinated action by the Soviet authorities in Moscow and Buriat-Mongolia.

In the Bolsheviks' eyes, nationalities such as the Buriat-Mongols appeared "uncultured" and "backward." This vision was thoroughly in line with their highly pragmatic, European-based ideology. However, this understanding of backwardness profoundly differed from the colonial views of the imperial period. Surprisingly, of all its castigation of traditional culture as being primitive, the reasoning of Soviet physicians generally lacked racist connotations. On the contrary, backwardness, superstitions, dirt and rampant social diseases were constructed as the heavy legacy of the past. The new Soviet lifestyle had to rectify these conditions as quickly as possible.

In order to become a truly socialist, advanced nation, the Buriat-Mongols were to be enlightened, educated and cured. A healthy body emerged as a natural prerequisite of a successful socialist construction. Therefore, Soviet specialists paid special attention to propaganda related to sanitation, social and personal hygiene, a healthy way of life, moderation, physical culture and so on. In the appalling living conditions of Buriat-Mongolian nomadic settlements, it was necessary to teach people some elementary principles of European hygiene and preventive medicine. Without such basic activities, all efforts to eliminate social diseases among large populations of uneducated, superstitious and poor people would be futile.

The anti-syphilis campaign in Buriat-Mongolia comprised politics of the body, social hygiene, pro-natalist and child-care policies, as well as ideological mobilising frames portraying syphilis as an enemy and firmly associating it with the dark legacy of Tsarism. This suggests that the early Soviet social modernisation discourse was heavily influenced by European eugenics concepts. A combination of geopolitical aspirations and ideological designs with goals of social eugenics reveals a utopian multipurpose eugenics programme theorised by the Bolsheviks. They hoped this would produce totally harmonious "socialist" people in just a couple of decades. At the same time, despite its obvious overall utopianism, elements of this programme, once implemented, had far-reaching social, demographic and even eugenic effects. The anti-syphilis campaign in Buriat-Mongolia, for example, certainly had significant socio-historical importance for the emerging "socialist Buriat-Mongolian people."

In an effort to curb social diseases, Soviet doctors welcomed foreign expertise. For political and geopolitical reasons, their natural ally was Germany, which had been isolated and humiliated by the terms of the 1919 Versailles Treaty. At the time, German venereologists and eugenicists were working actively in social venereology, social hygiene and racial research. Co-operation with them in the fight against social diseases was mutually beneficial. German specialists would gain access to a large "human laboratory" of untreated syphilitics and a chance to test the validity of their medical hypotheses. Meanwhile, Soviet doctors would use German medical expertise to determine effective and quick methods to treat syphilis with state-of-the-art drugs, such as *Salvarsan*. Efforts were made to begin producing the drug of the same effectiveness and high quality in the USSR, and specialists were sent to German pharmaceutical enterprises to learn the technology. Based on V. M. Bronner's 1929 report, the quality of USSR-produced *Salvarsan* approached the German version, and during a short period of time completely replaced it.[65]

Soviet doctors clearly understood that the eradication of syphilis, tuberculosis, trachoma and other social diseases was vitally important for the development of Buriat-Mongols. They painstakingly reconstructed the mechanisms of infection, and investigated the sexual habits of Buriat-Mongols in an effort to identify the main social causes of the syphilis epidemic. A victory meant a guarantee of success towards totally eradicating venereal diseases in national autonomies and union republics of the USSR. Doctors perceived the anti-syphilis campaign as directed not against one infectious disease, but as a complex fight against prejudices, superstitions, backwardness and sexual exploitation. To generalise, it was a fight against the old way of life and its deeply entrenched social and physical ills. In this sense, forming a new socialist person – healthy, rational and productive – could only be attained through a deep social modernisation of the entire ethnos and the destruction of its traditional culture and religion.

Notes

[1] There is a wealth of historical literature published in the USSR and Russia on the creation of the Buriat-Mongolian ASSR, national statehood and new national elites of the Buriat-Mongols. See: Erbanov, 1926; Okladnikov, 1951; Haptaev et al. (eds.), 1954; Haptaev et al. (eds.), 1959; Zateev, 1961; Soktoev, 1961; Chimitdorzhiev, 1991; Elaev, 2000.

[2] Quoted in: Elaev, 2000, 186.

[3] *Buriat-Mongol'skaia Avtonomnaia Sovetskaia Sotsialisticheskaia Respublika (Materialy k otchetu IV s'ezda sovetov) 1926/27 – 1927/28 g.g.* [The Buriat-Mongolian Autonomous Soviet Socialist Republic (Materials for the Report at the Fourth Convention of the Soviets) 1926/27 – 1927/28] (1929). Verkhenudinsk: Burgosizdat, 191.

[4] Parallel developments took place in other nations in the making. In my view, a good comparison can be made with a social eugenics program in early Republican Turkey in the 1930s. Despite clearly different political agendas, aims, discourses, methods and practices of implementation of social eugenics programs in Buriat-Mongolia and Kemalist Turkey closely resembled each other. For information on the social eugenic discourse in Turkey see: Alemdaroglu, 2005, 68-70.

[5] Michaels, 2003, 46-47.

[6] Project of the Commission for Health Care Problems among the National Minorities, 1921: Gosudarstvennyi Arkhiv Rossiskoi Federatsii (State Archive of the Russian Federation – further: GARF), A-482/19/44/3.

[7] For a detailed account of the Soviet medical policies in Kazakhstan see: Michaels, 2003.

[8] Toward the question of the spread of syphilis in Buriatia. A report by Dr. Zhinkin, 1926: Natsional'nyi Arkhiv Respubliki Buriatia (National Archive of the Republic of Buriatia – further: NARB), R-661/ 1b/1/153.

[9] Ibid., 150 reverse.

[10] Engelstein, 1992, 167-169.
[11] Report by the representative of the Buriat-Mongolian ASSR at the All-Union Central Executive Committee Chenkirov and deputy People's Commissar of Health of the BMASSR Zhinkin to the Department of Maternity and Childhood Protection of the People's Commissariat of Health of the RSFSR, 10 March 1926, ibid., 154.
[12] Report by Zhinkin: NARB, R-661/ 1b/1/152 reverse, 153.
[13] Hirsch, 2005, 245.
[14] Solomon, 2006.
[15] Lapyshev, 1929, 550.
[16] Zhorno, 1930, 99.
[17] Batoev, Demkova and Batoev, 2009, 244.
[18] Kubantseva, 1929, 59.
[19] Batorov, 1926, 35-36.
[20] Plishkin, 1930, 100.
[21] Ibid., 101, 103.
[22] Batorov, 1926, 36.
[23] Monthly statistical information on the spread of highly infectious diseases in the Buriat-Mongolian ASSR in 1924 by the month, 1924: GARF. A-482/10/431/17.
[24] Report of the Peoples' Commissariat of Health of the Buriat-Mongolian ASSR prepared for the conference of the Peoples' Commissariats of Health of autonomous republics, 1925: NARB, R-661/1b/1/184, 184 reverse.
[25] Engelstein, 1992, 175.
[26] Ibid., 177.
[27] Ibid., 179.
[28] Ibid., 176.
[29] Nagibin, 1928, 126-127.
[30] Toward the question of the spread of syphilis: NARB, R-661/ 1b/1/150.
[31] Lapyshev, 1929, 548.
[32] Zaks and Ilyin, 1927, 858.
[33] Pesterev, 1930, 89.
[34] Ibid., 90-91.
[35] Ibid., 93.
[36] Ibid., 97.
[37] Ibid., 98.
[38] Lapyshev, 1929, 552.
[39] Zhorno, 1930, 99.
[40] Biography of the Head of Venereological Department of the People's Commissariat of Health V.M. Bronner: GARF, A-482/41/418/3.
[41] Ibid., 4.
[42] Bronner, 1929a, 4-5.

[43] Extremely interesting anthropological comments can be found in the reports of the Russian doctors on the traditional nomadic way of life, customs, and religious beliefs of the Buriat-Mongols. For instance, see: Toward the question of the spread of syphilis…: NARB, R-661/ 1b/1/151 reverse, 152; Lapyshev, 1929, 548; Zhinkin, 1927, 55-64.
[44] Lapyshev, 1929, 552.
[45] Report of the Peoples' Commissariat of Health, 1925: NARB, R-661/1b/1/172 reverse.
[46] Ibid., 184.
[47] "Na bor'bu s venericheskimi bolezniami. Zdorov'e trudiaschikhsia – delo samikh trudiaschikhsia [Against the Venereal Disease. The Health of Proletariat is the Concern of the Proletariat]" *Buriat-Mongol'skaya Pravda* [The Buriat-Mongolian Pravda], no. 27 (2 February), 1928.
[48] Report of the Peoples' Commissariat of Health, 1925: NARB, R-661/1b/1/171.
[49] Ibid.
[50] "Sharlatanstvo [Quackery]" *Buriat-Mongol'skaya Pravda* [The Buriat-Mongolian Pravda], no.88 (16 December), 1923.
[51] "Chto skazal Narkomzdrav o sifilise [What the Peoples' Commissariat of Health Said about Syphilis]" *Buriat-Mongol'skaya Pravda* [The Buriat-Mongolian Pravda], no. 29 (4 February), 1929.
[52] Report of the Peoples' Commissariat of Health, 1925: NARB, R-661/1b/1/184 reverse.
[53] Wilmanns, 1995; Solomon, 1993, 209.
[54] Solomon, 212
[55] Ibid., 213.
[56] The State Venereological Institute in Moscow delegated V.M. Bronner and S.E. Galperin to Germany in December 1927 to observe the biologic and clinical control of German-made *neosalvarsan*. Such trips that transferred scientific knowledge across the borders were rather common in the mid-1920s, when Soviet physicians strove to establish domestic production of high quality drugs and draw on the German clinical experience of fighting infectious diseases. See: Memorandum of the Director of the State Venereological Institute V.M. Bronner to the Peoples' Commissar of Health, 1927: GARF, A-482/35/201/95.
[57] "Tseli i zadachi russko-germanskoi ekspeditsii [Goals and Objectives of the Russian-German Expedition]" *Buriat-Mongol'skaia Pravda* [The Buriat-Mongolian Pravda], no. 161 (17 July), 1928.
[58] Solomon, 1993, 217.
[59] "Rezul'taty raboty russko-germanskoi ekspeditsii [Results of the Russian-German Expedition]" *Buriat-Mongol'skaia Pravda* [The Buriat-Mongolian Pravda], no. 198 (29 August), 1928.

[60] Podgoornyi, 1928, 116.
[61] Ibid., 117.
[62] *Buriat-Mongol'skaya Avtonomnaya Sovietskaya Sotsialisticheskaya Respublika*, 192-193.
[63] *Otchet pravitel'stva Buriato-Mongol'skoi ASSR 1928-1930. Izdanie TSIK i SNK BMASSR.* [The Report of the Government of the Buriat-Mongolian ASSR, 1928-1930. The Publication of TSIK and SNK of the BMASSR] (1930). Verkhneudinsk: Burgosizdat, 82-83.
[64] Bronner, 1929b, 8.
[65] Ibid., 12-13.

References

Alemdaroglu, A. (2005), "Politics of the Body and Eugenic Discourse in Early Republican Turkey" *Body & Society*, vol. 11, no. 3, 61-76.

Batoev D. B., Demkova N. P. and Batoev S. D. (2009), *Pervyi narkom Andrei Timofeevich Trubacheev v istorii zdravookhraneniia Buriatii* [The First People's Commissar Andrei Timofeevich Trubacheev in the History of Healthcare in Buriatia]. Ulan-Ude: GUZ RCMP MZ RB.

Batorov, P. P. (1926), "Materialy po narodnoi meditsine Alarskikh buriat [Materials on the Traditional Medicine of the Alar Buriats]" *Buriatovedcheskii sbornik* [The Collection on Buriat Studies], no. 1, 33-41.

Bronner, V. M. (1929a), "K tret'emu vsesoyuznomu s'ezdu po bor'be s venericheskimi boleznyami [Towards the Third All-Union Convention on the Fight against Venereal Diseases]" *Zdravookhranenie* [Health Care], no. 5, 4-11.

Bronner, V. M. (1929b), "Itogi Tret'ego Vsesoyuznogo S'ezda po bor'be s venericheskimi boleznyami [The Results of the Third All-Union Convention on the Fight against Venereal Diseases]" *Zdravookhranenie* [Healthcare], no. 7, 5-10.

Chimitdorzhiev, Sh. B. (1991), *Kto my – buriat-mongoly?* [Who Are We Buriat-Mongols?]. Ulan-Ude: Assotsiatsiia literatorov Buriatii.

Elaev, A. A. (2000), *Buriatskii narod: stanovlenie, razvitie, samoopredelenie* [The Buriat Nation: Formation, Development, Self-Determination]. Moscow: RAGS.

Engelstein, L. (1992), *The Keys to Happiness: Sex and the Search for Modernity in Fin-de Siècle Russia*. Ithaca,NC: Cornell University Press.

Erbanov, M. N. (1926), *Vorposy natsional'no-kul'turnogo stroitel'stva v Buriatii* [The Questions of National-Cultural Construction in Buriatia]. Verkhneudinsk: Burmongiz.

Haptaev, P. T. et al. (eds.) (1954), *Istoriia Buriat-Mongol'skoi ASSR* [History of the Buriat-Mongolian ASSR]. Ulan-Ude: Buriatskoe knizhnoe izdatel'stvo, vol.1.

Haptaev, P. T. et al. (eds.) (1959), *Istoriia Buriatskoi ASSR* [History of the Buriat ASSR]. Ulan-Ude: Buriatskoe knizhnoe izdatel'stvo, vol. 2.

Hirsch, F. (2005), *Empire of Nations: Ethnographic Knowledge and the Making of the Soviet Union*. Ithaca,NC: Cornell University Press.

Kubantseva, M. A. (1929), "Partiino-sovetskii aktiv v bor'be s venericheskimi bolezniami [The Party-Soviet Activists in the Fight against Venereal Diseases]" *Zhizn' Buryatii* [The Life of Buriatia], no. 5, 57-69.

Lapyshev, D. A. (1929), "K kharakteristike sifilisa sredi buriat [Towards the Characteristics of Syphilis among the Buriats]" *Vrachebnaya Gazeta* [Physician's Newspaper], no. 8, 546-554.

Michaels, P. A. (2003), *Curative Powers: Medicine and Empire in Stalin's Central Asia*. Pittsburgh, PA: University of Pittsburgh Press.

Nagibin, G. (1928), "Tuberkulez i sifilis sredi buriat i russkikh Alarskogo aimaka [Tuberculosis and Syphilis among Buriats and Russians of the Alar District]" *Zhizn' Buryatii*, no. 4-6, 126-136.

Okladnikov, A.P. (ed.) (1951), *Istoriia Buriat-Mongol'skoi ASSR* [History of the Buriat-Mongolian ASSR]. Ulan-Ude: Buriatskoe knizhnoe izdatel'stvo.

Pesterev, A. M. (1930), "Polovoi byt buriat [The Sexual Life of the Buriats]" *Zhizn' Buryatii*, no. 7, 89-98.

Plishkin, D. N. (1930), "Nekotorye istoricheskie dannye o prokaze v Buriatii [Some Historical Data about Leprosy in Buriatia]" *Zhizn' Buriatii*, no. 4, 100-104.

Podgoornyi, I. D. (1928), "Predvaritel'nye itogi rabot russko-germanskoi ekspeditsii [The Preliminary Results of the Russian-German Expedition]" *Zhizn' Buriatii*, no. 10-12, 115-117.

Soktoev, I. A. (1961), *Formirovanie sotsialisticheskoi intelligentsia v Buriatii* [Formation of the Socialist Intelligentsia in Buriatia]. Ulan-Ude: Buriatskoe knizhnoe izdatel'stvo

Solomon, S. G. (1993), "The Soviet-German Syphilis Expedition to Buriat-Mongolia, 1928: Scientific Research on National Minorities" *Slavic Review*, vol. 52, no. 2, 204-232.

Solomon, S. G. (ed.) (2006), *Doing Medicine Together: Germany and Russia between the Wars*. Toronto: University of Toronto Press.

Wilmanns, K. (1995), *Lues, Lamas, Leninisten: Tagebuch einer Reise durch Russland in die Burjatische Republik im Sommer 1926*. Pfaffenweiler: Centaurus Verlag.

Zaks, I. G. and Ilyin, S. T. (1927), "Opyt izucheniya zabolevaemosti venboleznyami v Buriato-Mongolii [The Experience of the Study of Venereal Diseases in Buriat-Mongolia]" *Venerologiya i dermatologiya* [Venereology and Dermatology], no. 9, 857-875.

Zateev, V. I. (1961), *Formirovanie i rastsvet buriatskoi sotsialisticheskoi natsii* [The Formation and Blossoming of the Buriat Socialist Nation]. Ulan-Ude: Buriatskoe knizhnoe izdatel'stvo.

Zhinkin, V. N (1927), "Opyt izucheniya sotsial'no-bytovykh zabolevanii sredi kochuyuschikh buriat Aginskogo aimaka [The Experience of the Study of Social Diseases among the Nomadic Buriats of the Aginski District]" *Zhizn' Buryatii*, no. 1-3, 55-64.

Zhorno, Y.F. (1930), "Neotlozhnye zadachi v bor'be s sifilisom v Burrespublike [Urgent Tasks in the Fight against Syphilis in the Buriat Republic]" *Zhizn' Buryatii*, no. 7, 98-104.

Sterilisation in the Swedish Welfare State: A Gender Issue?

Maija Runcis

Abstract

When the first moves towards the introduction of sterilisation laws were made in 1922, the public discourse on racial hygiene was at its height in Sweden. Political parties, doctors, philanthropists and social workers alike were agreed on the potential social advantages of eugenics. On the question of sterilisation, however, representatives of both the conservative and liberal factions accepted state surgical intervention whenever it was motivated on the grounds of race biology and/or eugenics. The political left, for its part, felt that sterilisation surgery was essential to the common good for social reasons as well. Focussing on gender division in Swedish sterilisation policy until the mid 1970ies the article demonstrates that biological indications became less important after World War II as social behaviour became the most important argument for sterilisation. The author stresses the point that female "deviance" was a male construction and targeted socially weak women, at first hand. Further the author argues that this form of forced social homogenization by bio-political means as an integral part the Swedish welfare state.

1. Introduction

The traditional image of Swedish sterilisation policy has emphasised nationalism, conservatism, notions of popular degeneration and dreams of racial purity. Some observers and eugenicists have drawn parallels with Nazi sterilisation policies of the 1930s and 1940s, which resulted in the forced sterilisation of 400,000 people, men and women in equal numbers. But this was not the case in Sweden. Swedish zoologist Nils von Hofsten, who acted as the expert on hereditary biology and eugenics at the Swedish National Medical Board, declared after World War II that Sweden was the "best in the world" in terms of sterilising "feeble-minded" individuals.[1] He rejected all comparisons to Germany and the Nazi sterilisation programmes – though the Swedish sterilisation practice did not stop after World War II, but continued until 1976. Swedish eugenicists used the sterilisation programme as a brick in constructing the new welfare state – *folkhemmet* – a new and progressive "people's home".

Nils von Hofsten's viewpoint is one example of an apparent inability among contemporary Swedish medical doctors and physicians to see parallels between Nazi racial policy and other kinds of racial policies, such as that found in Sweden. The medical doctors and eugenicists were convinced that sterilisations of so-called anti-social and feeble-minded individuals were made in the nation's best interest and for the best of the individual.[2] The strong belief in eugenics united both the conservative middle-class and the radical socialists, who allowed themselves to be convinced that certain categories of people should be stopped from procreating and that the occurrence of social problems could be prevented by applied eugenics.

As a whole, sterilisations appeared to have been a non-controversial subject on the political and societal agenda –until 1997, when the Swedish daily newspaper Dagens Nyheter reported about it. The Swedish sterilisation laws were adopted in 1934 and 1941 on the basis of a broad consensus. It was a sophisticated circle of medical experts and radical politicians (the social engineers), with great influence and power in society, who were the driving force behind the issue of sterilisation itself. They were driven by a strong confidence in the possibilities of medical science, and their ideas were influenced by a belief in modernity, and future progress, rationalism and science.[3] In the 1930s eugenic sterilisations fit within a preventive social and health policy designed by the Social Democratic reformers in co-operation with other parties.[4] But it was mainly Social Democrats who, as members of the government, bore the responsibility for the political decisions and the practical implementation of the sterilisation laws.

2. The Early Swedish Welfare State – a Short Introduction

In 1928, Swedish Social Democrat leader Per Albin Hansson gave a famous speech proclaiming his vision of a Swedish *folkhem*, a modern and progressive "people's home" that would accommodate and reconcile conflicting political traditions, and set in place a welfare system to redress the iniquities of unemployment and poverty. Significantly, he used a concept earlier associated with conservative ideology, duly recast and co-opted into a left-wing ideology of modernity that welcomed close collaboration with large-scale industry and firmly believed in state initiative, social engineering and science. In the early 1930s we saw the ascendancy of liberal, social ideals, and their widespread support amongst the educated elite.[5] There was considerable continuity in the social-liberal thinking favoured by the Social Democrats, with a premium on morality, conscientiousness and preventive social policy. Both Liberals and Social Democrats in Sweden ascribed great importance to the state, and their policies were formed in concert with Sweden's great popular movements of that time. The Social Democrats were keen for all citizens to participate in the construction of the new *folkhem*.

The interests of the collective society were placed above the individual's, which meant that the individual had to submit to decisions for the common good of society. The idea was that public welfare would benefit all members of society, not just certain groups.[6] The new social reforms introduced during the 1930s would help unfit and poor people to participate in society and, at the same time, purify the collective "people's home" of undesired participants. If these measures did not succeed, society could use other measures to exclude or include undesired individuals. In fact, not everyone had the physical and mental strength to take an active part in this collective construction of society. What would happen to them – would they be excluded or supported by the state?

The entire period during which the welfare state emerged can best be described as the age of utilitarian-rational functionalism – when old mores were overturned and social engineers and medics polished the squalid corners of the human temple. Collectivism was all. And it was the state, with the help of experts, which would accomplish the modernisation that the nation was thought to need and from which everyone would benefit.[7] The late 1930s saw an increase in public debate, not at least in the wake of the Myrdals' controversial 1934 book *Kris i befolkningsfrågan* ("Crisis in the Population Question").[8] The Social Democrats had come to power and the all-important Population Commission had been appointed. Now, both individual and society would be transformed, and for this a master plan was needed. This master plan included the formation of a new, healthy individual able to actively participate in building the "people home". One way to achieve this was by preventing hereditary mental disease from being passed on to future generations. Mental disease and epilepsy were already criteria and an impediment to marriage, with the aim of preventing hereditary ill persons from giving birth. The questions here are – who was diagnosed as feeble-minded and deviant, and how could the society prevent the expansion of feeble-mindedness in society? The strong belief in science and eugenics became the new, powerful solution to the problem, which became transformed into the sterilisation laws.

3. The Introduction to a Swedish Sterilisation Programme in 1930s and 1940s

Social engineering was a modern, national project included in programmes about public health, hygiene, eugenics and/or racial biology that was set in a broader socio-political context. Racial biology emerged as the paramount scientific tool for social policy, to the extent that biological inheritance became more important than the right of ownership in determining one's class, because it was seen as affecting all classes equally.[9] Its most radical proponents argued that it could scientifically prove that class affiliation had no significance in terms of the qualities of the individual.

When the first moves toward introducing sterilisation laws were made in 1922, the public discourse on racial hygiene was at its height in Sweden. The Swedish Institute for Racial Biology became a government institution with state support. The "issue of the feeble-minded" was reviewed by a government-appointed commission of inquiry, and legislation had already been introduced to ban marriages involving certain categories of mentally ill and deviant people. Political parties, doctors, philanthropists and social workers alike agreed on the potential social advantages of eugenics. The political parties could not agree, however, on the fundamental position of the individual in society.[10] In the traditional liberal view the individual was autonomous, and the state's role therefore should be restricted. Laws existed to protect the individual from official abuse and safeguard his or her right to life and property. On the question of sterilisation, however, representatives of both the conservative and liberal factions accepted state surgical intervention whenever it was motivated on the grounds of race biology and/or eugenics. The political left, for its part, felt that sterilisation surgery was essential to the common good for social reasons as well.[11]

Early arguments in favour of a sterilisation law originally emphasised the motives of eugenics. One problem, however, was that at this time, during the 1920s, scientific findings with respect to hereditary links were still very vague in the fields of eugenics and race biology. In purely practical terms, this would have meant for sterilisation purposes that only in a very few properly verified cases could people receive surgery. This is why medical doctors and eugenicists urged that social motives for sterilisation receive legislative support.[12]

On the day that Parliament passed the first sterilisation law (8 May 1934), a new commission was established to regulate and facilitate conducting additional surgical procedures on individuals. On this occasion, doctors and politicians agreed on the social and population policy aims involved. Women's sexuality and reproduction now entered the picture, as some of those who became pregnant lacked the means to look after their children. Sterilisation laws would help society prevent such problems from arising. When discussing and drawing up their legislative proposals, scientific experts took into account the socio-political motives. This is evident, for instance, from the clear use of gender distinctions in the preparatory work and later in the application of the law.[13] Terms such as "uninhibited sexuality", "unprincipled", "unreliable" and "unfit to care for children" were used primarily to describe "feeble-minded" women and single mothers. Unrestrained female sexuality in such cases presented a "threat to society", as there was a risk that these women would bring into the world "unwished for" children who had no means of support. In this context, sterilisation could be seen as a kind of "correction" of women's sexual behaviour and "immorality", which were said to have negative ramifications for society.

When the law was expanded in 1941, "anti-social" behaviour became an additional ground for sterilisation. Another new feature – one that would prove to be of particular importance – was that the law would no longer apply to women and men in the same measure. In line with social reform work in other areas of Swedish society, women were made the centre of attention in the new sterilisation law. Added to the sterilisation "indicators" that applied specifically to women were marked departures from the moral and mental code of the times, failing health and general weakness in connection with pregnancy.[14] From a medical viewpoint, surgical sterilisation at that time was a riskier operation for women than for men. Despite this, the operations often were performed at the same time as abortions, when the risks were even greater, or when the woman was so weak that her physical health itself might present a risk. However, the equivalent operations on men – out of consideration for their wives' health – were not permitted by the law. This means that a woman could be sterilised even if there was a risk she might die from the operation, but it was not allowed to sterilise their husbands instead, in order to save their wife's life.

When Sweden's "population crisis" of the 1930s became a central issue (initiated by Alva and Gunnar Myrdal), demands were voiced for the ban on birth control to be lifted. At that time, there was also increasing public demand for sex education among poor families and in schools. The Association for Sex Education (RFSU) was launched, and its activities included fitting women with contraceptives and providing sex counselling for the general public. The organisation, however, did not recommend sterilisation as a method for birth control following the passage of the sterilisation law. It was clear that the wording of this law and the areas for which it was intended did not apply to the same type of people as those seeking help and support from the RFSU.[15] On the whole, it was "right-living" and well-behaved women, and well-educated women from the middle class who "got themselves into trouble" who received help and support from the organisation. The women exposed to sterilisation operations during the first two decades of the law, on the other hand, were those giving birth to children who were subsequently taken into care by the authorities.[16]

4. Eugenics in Practice

The chief motive for introducing the sterilisation laws was to prevent mentally ill and "feeble-minded" individuals from reproducing, mainly for population policy and social reasons. When the "feeble-minded" were set apart and classified as different or deviant, this was done on medical, eugenic and moral grounds, and also to some extent on social grounds. Eugenicists argued that social and cultural deviations and the overstepping of moral boundaries stemmed from mental illness and abnormality. Use of the law peaked in the mid-1940s, when some 75% of the operations registered in the official statistics were performed for eugenic reasons.[17] The statistics show that eugenics in practice was synonymous both with deviant behaviour and normality. The diagnosis lay in the eyes of the observer.

Many sterilisation operations took place in a context of compulsion or coercion, as it would be defined today. These non-voluntary operations were performed for different reasons, including as a precondition for being discharged from an institution, as a condition for eugenic abortion, or in some borderline cases with the influence of persuasion. Any operation performed on legally competent persons required a written consent. This consent was mostly given following more or less pressure from local authorities.[18] They could coerce a woman with different threats, such as: "If you don't sign the application, we will take your child away from you", or "If you don't sign the application, we will keep your child allowances", and so on.

In Sweden, some 63,000 people were sterilised during the 40 years that the sterilisation laws were in force. The share of sterilised women grew increasingly from 63% in 1941 to 98% in 1976. How is it possible to explain that the majority of the sterilised persons were women if eugenics was regarded as a gender-neutral science? Searching in contemporary medical dictionaries for the concept of "feeble-minded" and "unfit", we find that feeble-mindedness was a more typical diagnosis for boys and men. So again, how can we explain the fact that most of the sterilised people were women? My answer to these questions is that we must shed light on the implementation of social reforms in the Swedish welfare state. The main question must be: to what extent did eugenics influence the welfare state policy?

As a rule, the concept of "feeble-mindedness" has been defined synonymously as what is usually referred to nowadays as "mental retardation" or "mental disability". Exploring the sterilisation files and the diagnoses, it is difficult to give "feeble-minded" any precise definition. The principal component in such diagnoses related to the person's inability to master his or her behaviour and emotions. Usually, such factors derived from the person's upbringing at home, in a foster home, at school, in church (if the person went through confirmation), in the workplace or in a public institution, and how the person reacted to these environments. Deviant behaviour in the home, for instance, might mean that children were disobedient, unkempt and dirty. In school and also in church, where children received compulsory confirmation lessons, deviant behaviour might involve a lack of attention and poor concentration ("daydreaming"), liveliness, or a lack of respect for the teacher. In the workplace or work-home/institutions, people could be considered "feeble-minded" if they were deemed so apathetic that they lacked a will of their own, or if their will was so strong that they were rebellious and defied their superiors. Because uninhibited behaviour could be an obstacle to learning and discipline, those classed as "feeble-minded" were also regarded as "inferior", with defective intellectual powers. In other words, "feeble-minded" was developed as a generic term for people who in one way or another disturbed the culture of conformity.

The concept of feeble-mindedness in the application of the sterilisation law also involved a power relationship – based primarily on relations between persons of superior and inferior standing – and the relationship between the sexes. In the 1930s and 1940s, doctors and social engineers claimed that feeble-mindedness was becoming one of the foremost problems in Swedish society, as deviant behaviour and appearance were at odds with the ideal of diligence and proper conduct in the nascent "people's home". Parallel to this development, professional experts strengthened their position by assuming responsibility for solving these problems.

By general agreement, public monitoring of the "diligence ideal" was the responsibility of teachers, clergymen, heads of institutions and locally elected politicians. Applications for sterilisation included issues related to confirmation classes, conduct and school grades, and whether people recommended for sterilisation had a debt to society in the form of children taken into care or the need for social handouts, for instance, if they themselves had been taken into public care in one form or another.[19] During the 1930s until the end of 1950s, surgical action was initiated primarily by elected representatives on municipal child care committees and poor relief committees, and by doctors and staff at approved schools for young offenders. Committee reports were directed mainly at the weakest and most vulnerable individuals in the community – women, children, teenagers and young adults. Local authorities often collaborated with doctors to make sure that an application was actually submitted.[20] The official right of the poor relief system to apply for the sterilisation of individual members of the community reinforces the impression of the sterilisation laws as being both gender- and class-related.

Individual citizens named in the applications were often persuaded – especially after the directives issued by the National Medical Board in 1939 – to sign the sterilisation forms.[21] From the 1950s onwards, therefore, it is difficult to see who actually initiated the applications.

The sterilisation law of 1934 allowed operations on the "feeble-minded" when two doctors agreed on their necessity after joint consultation, while other decisions were to be made by the National Medical Board. During the initial period of the law's existence, however, some doctors were found to be more inclined to sterilise than others. This meant there was a greater risk of the arbitrary processing of sterilisation applications and of the violation of an individual's legal rights. Therefore in 1941, the system was reorganised so that all decisions were taken at central level, except in the case of "physical defect, deformity and emergency". The Board's committee of forensic psychiatry, renamed the committee of social psychiatry in 1947, became in principle the sole decision-making body. These two committees made their sterilisation decisions based on certificates of age or testimonials from relatives, from social workers or "others" as mentioned on the application form, or on the basis of medical certificates.

Overall, these documents were intended to give a composite picture of the person being proposed for sterilisation and provide a basis for an assessment and a final decision. When such cases were transferred to the social psychiatry committee in 1947, two new members were brought in, having been appointed by the government to four-year terms. In certain ways, this reorganisation of the decision-making procedure marked a shift in the orientation of the sterilisation laws toward a more untrammelled instrument of social welfare policy – i.e., the position advocated mainly by Social Democratic MPs in the public debate. The committee of forensic psychiatry was left to conduct mental examinations on criminals, and beginning in 1944 to make decisions in castration cases.

Thus, the central decision-making was done by the National Medical Board, which consisted of politically appointed representatives (MPs) and a small scientific elite comprised of eugenicists and doctors. Legal experts, however, were left without any say in the matter, and indeed were not allowed to take part in the decision-making process. Their only task was to comment on laws as they were being prepared.[22] Familiarity with social problems and a person's social situation, which was required for the law to work, was based at the local level. It was provided mainly by locally elected politicians and municipal social workers. As these people had no knowledge of the degree of heredity in certain mental diseases, they took a far more crass view of things, believing they could reduce local government spending while also "cleansing" the community of "bad" social environments and overcrowding. It is important to keep in mind, however, that Swedish society at that time was very poor and the tax base was very small. Society could not afford to increase its social expenditures.[23] Society had a common interest to limit such expenditures and prevent social problems from arising.

During the late post-war period, the implementation changed in a number of respects. Eugenics and concerns about "anti-social behaviour" as major motives for sterilisation were replaced by a concern for individual social and medical problems, and to some extent, so-called family planning. But in these cases it was always state authorities who made the final decisions. Beginning in the 1960s, implementation of the law and the strong focus on the interests of society shifted toward a focus on the interests and preferences of the individual.

5. Sterilisations and the Swedish Welfare State – an Ill-matched Couple

As mentioned, the aim of sterilisation policies from 1934 to the early 1950s was to sterilise "unfits", both inside and outside various institutions. It was the "unfit" from poor social environments, and those who for various reasons were regarded as unsuitable parents, who the state wanted to prevent from giving birth to who they predicted would be maladjusted, vicious or simply unsupported children. Most of the women who were sterilised during this period were single mothers whose children had been taken into public care. Decisions to remove children from parental custody were usually made based on the woman's financial circumstances or her alleged neglect of the children. Thus, these women were all dependent on public support and considered to be, or likely to become, "bad mothers". In this way, they did not live up to the expected role for a woman to be the bond holding the family together and to be a moral example for all. Consequently, women alone had to bear the responsibility for birth control, and if for one reason or another they were considered unfit to do so, they risked being sterilised.

As the sterilisation law was implemented, women were convicted on vaguer grounds than men for their own misconduct, unreliability, immoral living, or physical and mental failings, which were duly reported by employers, child care officials, neighbours and others around them. Likewise, doctors' judgments were often based on other people's interpretations of women's behaviour and their social situation. Loose rumours and information from relatives, teachers and vicars could form a general foundation for assessing whether a woman should be sterilised and appeared unchallenged in doctors' certificates. Further, in these cases, the conceptions of deviant behaviour differed for men and women: public life was reserved for men, while women were associated with the home and the private sphere. The sexually unreliable woman was positioned against the "normal" idea of a woman who was natural, intuitive, a self-controlled sensual elemental being, and a good moral example, all at the same time.[24]

To a greater degree than women, men who were proposed for sterilisation were inmates in work institutions and mental hospitals. Because they had committed criminal acts, the reports on them were often based on court records and police reports, or previously conducted studies of their mental condition. Thus, men brought into the sterilisation programme often had already been convicted or locked away, and sterilisation usually was proposed in connection with their released. It might be said that women were tried and convicted of crimes against the moral standards of the day in a court of "popular discourse", while men, in most cases, were convicted of crimes against laws in a proper court and punished on legal grounds.[25]

Applications from poor relief and child-care authorities were mainly aimed at society's very weakest and defenceless individuals – women, children and juveniles. The applications reflect strong social economic motives. Local authorities often collaborated with doctors so that an application would actually come about, and in most cases it was women who were singled out because of socially unsatisfactory conditions and inadequate parental responsibility. The official right of poor relief authorities to apply for the sterilisation of individual inhabitants of municipalities confirms the picture of the sterilisation laws as being class-related legislation.

Various types of sexual behaviour were interpreted as innate personal characteristics. Uninhibited sexuality was associated with "feeble-minded" women, and immoral behaviour or other sexual abuses were said to characterise homosexual or criminal men. The morals of the day dictated that social problems associated with sexuality, morality and negligent child care were primarily a "female" problem. For their part, deviant men were taken to task for economic rather than moral reasons. Men were associated with work, supporting a family and shouldering responsibility. Were they to fail on any one of these counts, they too risked being considered "feeble-minded". Such men could be used for labour (workforce in the community) if they were sterilised and released from whatever institution they were in, whereas women in similar cases were thought to likely become even more "depraved", because after sterilisation they would be able to engage in sex even more promiscuously – without the risk of unwanted pregnancies.

The modern Swedish welfare state provided for and supported many new scientific and other professional experts in the social field, and these experts obtained significant power to make decisions about the ordinary daily lives of people, and the state (the social workers and social engineers) tried to educate them for society's good and conformity. Well-borne and good members of the community were to be helped and encouraged in various ways so they could be better equipped to contribute to the country's material production, as well as in the reproduction of healthy children. "The others", those targeted by the sterilisation law, represented the flip-side of the social coin -- the opposite of the norm.

On a societal local level (child care institutions), the authorities worried that modern society could bring about a series of negative consequences regarding the quality of the population. The risk of social reforms and relief work was quite simply that "the worst off individuals" lived longer, which could result in increased fertility. Therefore, sterilisations were considered to be "more needed as a counterbalance against what in some cases was an unavoidable unlucky consequence of albeit good and significant relief work".[26] When Sweden introduced a new social reform in 1947 – the child allowance only for mothers – many of the "social engineers" and reformers demanded the mass sterilisation of poor and unfit women.[27] Otherwise, society would be forced to support or take care of many unwanted children.

Swedish law prohibited the use of any physical force, and in fact some 90-95% of the persons sterilised from 1935-1975 signed their own applications. From a legal point of view, these are considered voluntary operations. On the other hand, free will actually did not exist for persons who were sterilised; they could hardly influence the decision regarding the operation because institutions and physicians used different techniques of persuasion and indirect coercion. Former Prime Minister Olof Palme spoke in Parliament about the sterilisation law being an encroachment of authority, because women – not alone, but with their husband's consent – were allowed to decide whether they wanted to be sterilised. Moreover, certain criteria of the Medical Board were to be fulfilled in order to obtain their permission. It was also very common for women who requested but were refused abortions to be recommended for sterilisation instead. Women would give birth to children, be exhausted and let themselves be sterilised because it was not legal to sterilise men – an operation that today is considerably easier to carry out on men.

With the sexual liberation movement of the 1960s, women took on the right to regulate their sexuality themselves, which in some cases meant refraining from sex in a marriage because contraceptives were not possible. The sterilisation files from the 1960s to 1970s state that abstinence occurred because of a woman's fear of pregnancy. This was seen, mostly by social welfare workers, as a threat to the unity of marriage. It was women's own need for guidance that led social welfare workers to recommend sterilisation for women with pregnancy anxiety, in order for "marriage happiness" to be spared from abstinence. These operations were officially defined as medical, and eugenics was less important here, if at all. Another purpose was to avoid physical danger and lighten the burden of repeated pregnancies for mothers living in socially distressful circumstances. Very often, these women were also under pressure because of their social situation, and most likely they were influenced by the opinions of medical and social workers.

In this way the practice of the law was changed, but without leading to a change in the sterilisation law during this period. On the other hand, a new reason and a new social category for sterilisation were created in practice – the conscientious, anxious housewife.

Notes

[1] *Statens offentliga utredningar* (The Official State Report – further: SOU), 2000, vol. 22, 206.

[2] Runcis, 1998, 63.

[3] Hirdman, 1989.

[4] Runcis, 1998, 157.

[5] Hedin, 2002.

[6] Hallberg and Lernestedt (eds.), 2002.

[7] Hallberg and Lernestedt (eds.), 2002.

[8] Myrdal and Myrdal, 1934.

[9] Ibid.; This was a main argument of the Social Democrats and the Myrdals in support of the sterilisation laws.
[10] Broberg and Tydén, 1991.
[11] Runcis, 1998,157.
[12] Ibid.; Hatje, 1974, 186.
[13] Medicinalstyrelsens arkiv 1935-1946 [The Archives of National Medical Board] preserved in the National Archives.
[14] *SOU*, 1929, vol. 14; *SOU*, 1933, vol. 22; *SOU*, 1941, vol. 25.
[15] See as example: Lennerhed, 2000.
[16] Runcis, 2007.
[17] *Statens officiella statistik* [Official Reports of Statistics], 1947.
[18] *Medicinalstyrelsens tillämpningsföreskrifter*, 1939 and 1947. (Printed Instructions from the National Medical Board, 1939 and 1947).
[19] Runcis, 1998, Attachment: Sterilisation application.
[20] Runcis, 2007.
[21] *Medicinalstyrelsens tillämpningsföreskrifter*, 1939. (Printed Instructions from the National Medical Board, 1939)
[22] In late 1940s and 1950s there was a major debate over the so-called "administrative deprivation of liberty" (See: *SOU*, 1960, vol. 19).
[23] *SOU*, 2000, vol. 22.
[24] Runcis, 1998, 185.
[25] Ibid.
[26] *SOU*, 1936, vol. 46, 16.
[27] Myrdal, 1946, 57.

References

Betänkande med förslag till steriliseringslag: Avgivet av tillkallade sakkunniga den 30 april 1929 [State report with law proposal on sterilisation. Reported by called in expert advise 30 April 1929]. Stockholm: Beckman in: *SOU*, 1929, vol. 14.

Förslag till lag om sterilisering av vissa sinnessjuka, sinnesslöa eller av annan rubbning av själsverksamheten lidande personer [Law Proposal on Sterilisation on some mentally retarded, feebleminded or persons from other kind of mentally illness]. Stockholm: Marcus in: *SOU*, 1933, vol. 22.

Betänkande angående sterilisering [State report on Sterilisation]. Stockholm: Marcus in: *SOU*, 1936, vol. 46.

Strafflagberedningens promemoria med förslag till lag om kastrering m. m. Stockholm [Memorandum from penal preparation with law proposal on castration and so on] in: *SOU*, 1941, vol. 25.

Rättssäkerheten vid administrativa frihetsberövanden: sakkunnigbetänkande med förslag till lag om socialdomstol [Law and order within administrative detention: State report by "called in expert advise" with law proposal on social court]. Stockholm: Idun in: *SOU*, 1960, vol. 19.

Broberg, G. and Tydén, M. (1991), *Oönskade i Folkhemmet: Rashygien och sterilisering i Sverige* [Unwished in People's Home: Eugenics and Sterilisation in Sweden]. Stockholm: Gidlund.

Hallberg, P. and Claes L. (eds.) (2002), *Svenska värderingar? Att se och ompröva det invanda* [Swedish Values? To See, and Reconsider Ingrained Opinions]. Stockholm: Carlsson.

Hatje, A.-K. (1974), *Befolkningsfrågan och välfärden. Debatten om familjepolitik och nativitetsökning under 1930- och 40-talen* [The Population Problem and Welfare. The Debate about Family Politics and Nativity during the 1930s and 40s], Stockholm: Allmänna Förl.

Hedin, M. (2002), *Ett liberalt dilemma. Ernst Beckman, Emilia Broomé, GH von Koch och den sociala frågan* [A Liberal Dilemma. Ernst Beckman, Emilia Broomé, GH von Koch and the Social Issue] Dissertation. Stockholm/Stehag: Brutus Östlings Bokförl. Symposion

Hirdman, Y. (1989), *Att lägga livet tillrätta. Studier i Svensk folkhemspolitik* [Constructing the Everyday Life. Studies in Swedish Politics of the 'People's Home']. Stockholm: Carlsson.

Lennerhed, L. (2000), "Steriliseringar och sexuell reform. Om Elise Ottesen-Jensen och Riksförbundet för sexuell upplysning [Sterilisations and Sex-reforms]" in: Broberg, G. (ed.), Fem uppsatser om steriliseringen i Sverige (Five essays on the sterilisation in Sweden) Lund: Univ., 22-46.

Myrdal, A. and Myrdal, G. (1934), *Kris i befolkningsfrågan* [Crisis in the Population Question]. Stockholm: Bonnier.

Myrdal, A. (1946), "Kontanta barnbidrag kräver skärpt steriliseringslag? [Child Allowances in Cash Demands a Stricter Sterilisation Law?]" *Tidskrift för barnavård och ungdomsskydd*, no. 2, 57.

Runcis, M. (1998), *Steriliseringar i Folkhemmet* [Sterilisations in the 'People's Home'], Dissertation. Stockholm: Ordfront förl.

Runcis, M. (2007), *Makten över barnen. Tvångsomhändertagande av barn 1928-1968* [The Power over the Children]. Stockholm: Atlas.

Tydén, M. (2000), Från politik till praktik. De svenska sterikliseringslagarna 1935-1975 [From Politics to Practice. The Swedish Sterilisation Laws]. Stockholm: Fritzes in: *SOU*, 2000, vol. 22.

Eugenic Concerns, Scientific Practices: International Relations and National Adaptations in the Establishment of Psychiatric Genetics in Germany, Britain, the US and Scandinavia 1910-1960

Volker Roelcke

Abstract
The chapter uses the international institutionalization of psychiatric genetics as an exemplary case on two levels: It illustrates the inseparable relations between eugenics and medical genetics until the 1960s; and it addresses some core issues in the trans-national history of eugenics. On the empirical level, the chapter focuses on the emergence of research programmes, institutions, and the activities of historical protagonists in four national contexts: Ernst Rüdin and the German Research Institute of Psychiatry (Deutsche Forschungsanstalt für Psychiatrie) in Munich, Franz Kallmann and the department of psychiatric genetics at the Institute of Psychiatry in New York, Eliot Slater and the MRC Centre for Psychiatric Genetics in London, and Erik Essen-Möller with the programme in psychiatric genetics at the University of Lund. It is shown that there was a continuous mutual exchange of ideas and practices. Each of the historical actors followed both scientific and political agendas.

The early decades of the institutionalisation of psychiatric genetics exemplify the inseparability of the history of eugenics and medical genetics, at least in the first half of the 20th century.[1] In addition, they may also be seen as an empirical case for elaborating on some issues relevant to the transnational history of eugenics.[2]

For the purpose of some conceptual considerations derived from the historical case, it is useful to start with a provisional definition: eugenics may be defined as a social movement characterised by a specific configuration of aims, means, actors and justifications. Its central aim is to improve the biological "quality" of a population. The methods to achieve this aim are interventions – educational, legal, as well as physical – in human reproduction. Both the aim and the applied methods are justified by scientific expertise. Finally, the main actors who pursue eugenics programmes are professionals such as physicians, biologists, statisticians, lawyers and teachers. At all levels – local, regional and national – conditions have had an impact on the specific and continuously changing configurations of problem perceptions, scientific practices and interpretations.[3]

In the following, the emergence of research programmes and institutions in psychiatric genetics will be focused upon in four national contexts by examining the concerns, activities and interrelations of protagonists in this field: Ernst Rüdin and his group at the German Research Institute for Psychiatry (*Deutsche Forschungsanstalt für Psychiatrie*, or DFA) in Germany, Franz Kallmann in the United States, Eliot Slater in the United Kingdom, and Erik Essen-Möller in Sweden. Today, the latter three protagonists are considered to be the founding fathers of psychiatric genetics in their countries.[4] All of them, however, had been research fellows at the DFA in Munich in the mid-1930s, which at that time was directed by Rüdin.

These cases, with their mutual interrelations as well as their specific national contexts, provide insights into the impact of transfer and adaptation processes in the development of eugenics concerns as well as scientific activities. They also contribute to a historiographical approach that focuses on the integral role of eugenics in the history of human genetics, and vice versa.[5] This is of particular interest, since human geneticists today claim that medical genetics only had connections – if at all – to what Daniel Kevles has termed "reform eugenics", but that it had little or nothing to do with the ideologically contaminated "mainline eugenics" of the 1920s and early 1930s. According to Kevles, and also according to the self-image of many geneticists today (including the authors of historical accounts of genetics, or heredity), "mainline eugenics" was racist, had a strong class bias, lacked scientific credibility, and was associated with the application of force. In contrast, "reform eugenics", which according to this picture emerged in the second half of the 1930s, was scientifically up-to-date and supposedly compatible with liberal and socialist political ideas. Conventionally, German racial hygiene is placed unequivocally on the side of "mainline eugenics", and indeed is seen as a particularly radical manifestation of this variant.[6]

As a number of historians have pointed out, such a dichotomous image of the history of eugenics and its relation to medical genetics ignores the complex relationships and indeed the continuities which existed between representatives of the purported two strands of eugenics since the mid-1930s.[7] Such continuities may be identified in respect of the relations, transfer and adaptation processes between the Munich group around Rüdin, and the protagonists of psychiatric genetics in the US, the UK and Sweden.

1. Ernst Rüdin and the Department of Genealogy and Demography at the German Research Institute for Psychiatry

Ernst Rüdin (1874-1952), a Swiss citizen, was a pupil of Emil Kraepelin, the dominant figure in German psychiatry during the first quarter of the 20th century.[8] Around 1910, both Kraepelin and Rüdin proposed the idea of a central German research institute to investigate the spread and causes of psychiatric disorders and social deviance, with explicit reference to the need for preserving or even improving the hereditary health of the German population – that means, with a eugenics-based argument.[9] Kraepelin and a few other university psychiatry professors lobbied for this research institute[10] and eventually were successful. The DFA was inaugurated in 1917 in Munich, backed by funds from both state authorities and private philanthropic sources.[11] Among the new institute's five departments was one for genealogy and demography, which had the explicit aim to investigate the epidemiology and genetics of psychiatric disorders for eugenics purposes. Rüdin was appointed director of this genetic-epidemiological department and remained in this function until the end of World War II.

In 1911 Rüdin formulated an elaborate research programme to study the genetics of major psychiatric conditions, drawing on the most advanced methodologies of contemporary sciences, including psychiatry itself, as well as anthropology, genealogy, statistics and various laboratory sciences.[12] In the following years, he turned this programme into practice, focussing on one distinct disease entity, schizophrenia. Rüdin built on Wilhelm Weinberg's methodology of population genetics, in particular his "sibling method". Weinberg was himself a social hygienist and a physician for the poor (*Armenarzt*), though also eugenically oriented.[13] For his empirical study, Rüdin studied all of the approximately 700 patients suffering from dementia praecox/schizophrenia in Bavarian asylums, as well as 4,800 of their siblings and 1,400 parents. This was in contrast to previous investigations that had selectively looked at families with more than one "case".[14] Applying further methods of correction, in particular those formulated by Weinberg, Rüdin arrived at results which considerably diverged from previous studies such as that by Jolly, and by Charles Davenport and Aaron Rosanoff at the Eugenics Record Office in Cold Spring Harbor on Long Island, New York.[15] Rüdin's findings suggested that the incidence of dementia praecox amongst the offspring of one affected parent was around 6%, and among the siblings of an affected patient around 4.5%, in contrast to the much higher rates expected by a recessive mode of inheritance according to Mendelian laws. The problems of interpreting these data in terms of Mendelian inheritance led Rüdin and his team to extend the database and methods applied to this material. The aim was to calculate the probability of the manifestation of a mental disorder in an individual on the basis of the manifestation ratio in his/her immediate relatives (*empirische Erbprognose*), which was highly relevant information in the context of eugenics counselling on human reproduction.[16]

Rüdin's 1916 study received rapid international recognition and remained for decades a standard text in psychiatric genetics. Rüdin and his group, amongst them Hans Luxenburger and Bruno Schulz, refined their methodologies, including the application of the then-highly promising twin-method of medical genetics to hereditary studies in psychiatry, and built up a broad and internationally acclaimed programme in the epidemiology and genetics of psychiatric and neurological disorders.[17] In 1934 the methodology and practices by Rüdin and his team in Munich were described as being at the forefront of international efforts in the field and possessing paradigmatic character by Aubrey Lewis (in the mid-century the leading British psychiatrist), in the first British handbook on human genetics.[18]

After the Nazi takeover in 1933, Rüdin and Eugen Fischer, director of the Kaiser Wilhelm Institute for Anthropology, Human Heredity, and Eugenics, became members of the government's Expert Committee for Health and Racial Policy, which in conjunction with state administrators modified an existing draft version of a sterilisation law. It is amply documented that the definitive law which allowed compulsory sterilisation was enacted in January 1934.[19] Rüdin was one of three authors of the official commentary. This commentary explicitly referred to the method of *empirische Erbprognose* which, as pointed out above, had been developed at the Munich institute. At the end of 1934, the Rockefeller Foundation stopped the institutional funding for the Munich department. To make up for this, Rüdin turned to authorities of the Nazi state – in particular the Ministry of the Interior and the Chancellory of the "Führer" – and he repeatedly received large sums of money for his research. This research was explicitly designed to "contribute to the scientific foundations" of the regime's population and race policies.[20]

In spite of this orientation towards the regime's political aims, the research by the DFA was still considered to lead the field internationally. The Rockefeller Foundation continued to fund fellowships for young, promising scientists from abroad to work at the Munich institute, such as Eliot Slater, who in 1934 followed Erik Essen-Möller on a one-year grant. Staff members at the DFA, in particular Hans Luxenburger and Bruno Schulz, were quoted in the international literature of the field until the 1950s and 1960s. Immediately before World War II began, in spring 1939, Gunnar Dahlberg, the director of the State Institute for Human Genetics and Racial Biology at the University of Uppsala, invited Rüdin to present one of four plenary lectures in the Human Genetics Section at the Seventh International Congress of Genetics, for which Dahlberg served on the programme committee.[21]

Thus, in the case of Rüdin, an explicit eugenics and (with it) political motivation for his genetic research had already existed more than two decades before the Nazi takeover. And after 1933, Rüdin's department continued to attract scientists internationally, despite its close co-operation with Nazi authorities. After the war, Rüdin was suspended from office; he died in 1952 at age 78. His department within the DFA was downsized to a one-man endeavour, run by Bruno Schulz until the mid-1950s.

2. Franz Kallmann and Psychiatric Genetics in the US

Franz Kallmann (1897-1965) was born in Silesia, with a Jewish family background.[22] After graduating in medicine, he specialised in psychiatry and neurology and trained under Karl Bonhoeffer in Berlin. As part of the requirements to work in Bonhoeffer's department, Kallmann went through a short training at the Berlin psychoanalytic institute, studying under Franz Alexander, Sandor Rado and Otto Fenichel.[23]

In 1928 Kallmann was appointed chief of the neuropathology department at the large asylum of Berlin-Herzberge. Here, he began to collect files on patients suffering from schizophrenia, dating from the late 19th century until his day. It was his aim to extend Rüdin's earlier studies on the genetics of this disorder, as well as to analyse the reproduction rates of schizophrenic patients and their relatives. He did all of this with an explicit eugenics motivation, and indeed he was a member of the German Society for Racial Hygiene since very early in his career. To update his methodological knowledge and intensify his research, Kallmann took a leave to continue his work at Rüdin's department in Munich in 1931. Until 1936, he regularly returned to the DFA for a few weeks each year to develop his close co-operation with colleagues there.

Due to his Jewish family background (although he was a baptised Protestant), Kallmann lost his position in Berlin in 1935. For the same reason (not because of his work in psychiatric genetics, which was completely compatible with the accepted views of the regime), in 1936 he was also excluded from the German Society for Racial Hygiene, a measure about which he was explicitly disappointed.[24] That he accepted the Nazi sterilisation law is documented by the conclusions of a paper in which he presented preliminary results of his research to the International Congress of Population Science held in Berlin in 1935. Here, he argued that due to the differential fertility rates amongst various groups of patients suffering from schizophrenia, "in regard to the recessivity and belated manifestation of the schizophrenic disposition, it is desirable to extend prevention of reproduction to relatives of schizophrenics who stand out because of minor anomalies, and, above all, to realise this intervention for those undesirable from the eugenic point of view at the beginning of their reproductive years".[25] Thus he not only accepted the sterilisation law, but he even recommended that the law be extended to healthy, heterozygous carriers of the assumed gene for schizophrenia, for which he postulated recessive inheritance. Kallmann thought that the healthy carriers of this gene might be identified by thorough clinical investigations eliciting minor psychological anomalies.[26]

Again because of his Jewish origins, he was not allowed to personally present a paper on his research in November 1935 at the annual congress of the association of German neurologists and psychiatrists in Dresden. Instead, the paper was read by Bruno Schulz, a colleague from the

DFA. That the contents of the paper – on the fertility of various groups of schizophrenic patients – passed censorship is evidenced by the fact that it was published under Kallmann's name in the proceedings of the congress.[27] Later hagiography (e.g. in the obituary by Slater) turned this single instance of being prevented from presenting a paper into the plural, and wrongly claimed that his statistics could only get into print "by quotation marks in papers of anothers' authorship", thereby presenting Kallmann as opposed to the regime's hereditary health policies.[28]

At the end of 1936, Kallmann managed to emigrate to New York. Anthropologist Franz Boas, himself of German-Jewish origins and an explicit critic of German racial laws as well as racist interpretations of eugenics, helped him find a provisional position and initial funding. To Boas, Kallmann presented Rüdin as a victim of the Nazi regime, since his psychiatric-genetic research programme had allegedly been restricted by political pressure and censorship. In a letter to Boas written immediately after his arrival in New York, Kallmann also constructed a dichotomous image of the situation in Germany according to which the various political institutions of the regime were arguing with "pseudo-scientific dogmas" (*pseudo-wissenschaftlichen Irrlehren*), whereas he and his colleagues in the field of psychiatric genetics had been doing proper science and fighting against such ideologies.[29] The two arguments – the image of a suppressed scientific field of medical genetics, created by Kallmann himself, together with the fact that as a Jew he was forced to leave Germany – were later used to construct the legend that Kallmann had been opposed to the German programme of racial hygiene and the Nazi sterilization law,[30] which he definitively was not.

In New York, Kallmann spent a few years on short-term grants, all devoted to projects on psychiatric and broader medical genetics. From his research in Berlin and Munich, he had concluded that the disposition for manic-depressive disorder was inherited by a dominant gene, and the disposition for schizophrenia by a homozygous recessive gene. In 1938, two years after his emigration to the US, he published an English translation of the original German manuscript of his book on the genetics of schizophrenia.[31] It was a family study of schizophrenia of 13,851 relatives of 1,087 patients admitted to a Berlin hospital during a 10-year period. He calculated the prevalence of the different forms of schizophrenia in relatives of patients compared with the general population. Whereas the expected prevalence in the general population was 0.85%, in the relatives of patients it was 16.4% (children) and 11.5% (siblings). This implied the hereditary nature of schizophrenia, although socio-educational transmission of the disorder was not ruled out. In the preface of the book, Kallmann explicitly acknowledged Rüdin and the Munich group for their scientific inspiration and support. Building on this work, and supplementing the method of family-studies by the twin-method, Kallmann then published a number of articles on the genetic component of schizophrenia in the 1940s.[32] In a final monograph on this issue, he made use of a combined twin-adoption method drawing on concepts formulated by Hans Luxenburger (a member of the Munich team) in the late 1920s and 1930s.[33] For many years these studies were the major evidence for genetic factors in manic depressive illness and schizophrenia.[34] All of these results were widely received and discussed in the field of medical genetics. Thus, some of Kallmann's as well as Luxenburger's results and interpretations, for example, were quoted and discussed in the *Principles of Human Genetics*, a standard textbook published in 1960 by Curt Stern, one of the leading American geneticists and himself an émigré from Germany.[35] Even in the late 1960s and early 1970s, the results of Kallmann's studies on the genetics of schizophrenia commenced in Munich were cause for a large empirical study to – now critically – re-evaluate his hypotheses.[36] Kallmann also applied the twin-method to the study of tuberculosis, homosexuality and involutional psychosis.

In the first years following his emigration, Kallmann was in a state of uncertainty regarding his employment. Initially, he received funding from the Emergency Committee in Aid of Displaced Foreign Physicians (an organisation that supported Jewish physicians from Germany and the occupied territories), the Carnegie Corporation of New York (1938-1941) and the Rockefeller Foundation (1945-1951).[37] Finally, he managed to get a tenured position that was combined with a full professorship in psychiatry (from 1954 onwards) at Columbia University. His research programme developed into the first department of psychiatric genetics in the US, and additionally it was the second programme in the broader field of medical genetics (preceded only by the programme in medical genetics at Ohio State University under Laurence H. Snyder).[38] In terms of methods, he continued working with concepts and analytical tools from the Munich school, albeit supplemented by further approaches. In 1962 he commenced cytogenetic studies.

In 1948 Kallmann co-founded the American Society of Human Genetics with Nobel Prize winner Hermann J. Muller and Laurence Snyder, and from 1952 onwards he acted as its president.[39] He was also a founding editor of the *American Journal of Human Genetics* and served as co-editor of the *Journal of Heredity* and the *Acta Geneticae Medicae et Gemellologiae*, a journal of twin studies and medical genetics founded in 1952 by the Catholic eugenicist and medical geneticist Luigi Gedda, a friend of Kallmann.[40] Kallmann was also a member of the organisation committee of the first and second International Congress of Human Genetics in Copenhagen and Rome, in 1956 and 1961, respectively.[41] From 1962-65 he was chair of the Permanent Committee for International Conferences of Human Genetics. Thus, in the 1950s and early 1960s, he was among the most distinguished figures not only in psychiatric genetics, but also in the broader field of human genetics. He was also awarded honorary degrees from the universities of Torino and Salerno. Upon his death in 1965, the *New York Times* honoured Kallmann as the foremost representative of psychiatric genetics in the US.[42] His colleague and friend Slater said that psychiatry had lost "one of its most notable pioneers".[43]

While serving as president of the American Society of Human Genetics in the 1950s, Kallmann was also a member of the Board of the American Eugenics Society. In fact, as pointed out by Diane Paul, five of the first six presidents of the American Society of Human Genetics shared these roles, with Hermann J. Muller being the only exception.[44] Throughout his career, Kallmann maintained his interest in eugenics. When applying for funding for his projects in the US, he regularly pointed to the importance of the new medical knowledge to be gained in terms of the genetic constitution as well as the health of the population. Shortly after his emigration, he delivered a paper to the Eugenics Research Association where he repeated the results he had previously presented at the Berlin congress of population science. Here, he talked of the "danger of the development of new schizophrenic cases" arising from the "unions" of heterozygous carriers of the supposed genes. "From a eugenic point of view, it is particularly disastrous that these patients not only continue to crowd mental hospitals all over the world, but also afford, to society as a whole, an unceasing source of maladjusted cranks, asocial eccentrics and the lowest types of criminal offenders. Even the faithful believer in the predominance of individual liberty will admit that mankind would be much happier without those numerous adventurers, fanatics and pseudo-saviours of the world who are found again and again to come from the schizophrenic genotype".[45] He concluded that "there should be legal power to intervene, in addition to the general eugenic program of the biological education of all adolescents, marriage counsel, obligatory health certificates for all couples applying for a marriage license, and the employment of birth control measures".[46]

Following the same rationale, he recommended in a 1938 paper the sterilisation of mentally disabled family members of schizophrenic patients if they were inaccessible to genetic counselling.[47] A year later he wrote: "In democratic countries with high standards of ethics and moral discipline, the general public should be educated rather than forced to realize the advisability of submitting to reasonable eugenic measures adopted for the perpetuation of common happiness and security".[48] Here, the reference to democracy is used as a rhetorical means to justify intervention in human reproduction to achieve "common happiness and security", whereas *individual* well-being is not on the agenda. Also, the use of force is not completely rejected, but educational measures are preferred to achieve what is defined by *experts* as eugenic aims, whereas the perspective of *those concerned* is neglected.

What Kallmann dropped after his move to the United States, however, were explicit references to terms such as race and the biological quality of the (own) national body. Thus, his talk of populations could now be interpreted as referring to humankind in general. He also stopped arguing explicitly for laws to enforce regulations on reproduction, in particular against the will of those concerned, and he substituted this by calling for educational measures, the nationwide provision of genetic counselling, and voluntary sterilisation. Further, he stressed that his genetic interpretation of schizophrenia was compatible with social and even psychoanalytical approaches which, according to him, only referred to the *manifestation* of the already pre-existing, genetically caused disorder.[49]

Despite the new availability of therapeutic interventions such as psychopharmaceuticals, eugenic recommendations made as preventive measures did not disappear, though they lost their prime place among Kallmann's considerations and recommendations.[50] In the context of a conference on the state of psychiatric genetics at Columbia University held in 1961, which celebrated the 25th anniversary of his research programme and department, Kallmann elaborated on the dangers of "bad mutations" for the collective gene pool, and said that medical geneticists should act as "expert guides" for measures to improve the biological quality of the population. All available scientific and technical means should be employed to direct the evolution of the human species towards "unimpaired health potentials". To accomplish this, Kallmann argued, genetic selection had to be controlled. Since the potentially dangerous effects of "bad" mutations could only be partially compensated by a supporting environment, an adequate population policy also had to encompass prenatal selection.[51]

The sick or suffering individual was absent in this programmatic statement. Instead, Kallmann referred to one of the basic ideas of early eugenics, namely that of counter-selection: in contrast to "natural" evolution where "bad" hereditary traits were supposedly eliminated by the "struggle for existence", the social institutions and health-care provisions of modern societies preclude this process of selection and thus contribute to a potential deterioration of the collective gene pool.[52] The definition of what is considered "bad" remained with the medical geneticist, whereas the individuals concerned (the supposed carriers of "bad" genes) had no voice.

In conclusion, it may be stated that Kallmann kept his overall eugenics concerns and aims. But by "sanitising" his rhetoric, excluding references to race and the national body politic, and by propagating educational measures as well as genetic counselling, he adapted to the values and rhetoric of the surrounding American society. For his scientific activities, he continuously developed the methodological reservoir he had acquired at the DFA while at the same time including newly available concepts and procedures.

3. Eliot Slater and the Institutionalization of Psychiatric Genetics in the United Kingdom

Eliot Slater (1904-1983) is seen today as the "doyen and founder of psychiatric genetics in Great Britain".[53] After earning a degree in medicine, he joined the staff of the Maudsley Hospital in London in 1931, where within a few years he met a group of German émigré psychiatrists, notably Willi Meyer-Gross, Eric Guttmann and Alfred Meyer. Aubrey Lewis, then chief-of-staff of Maudsley and later the dominant figure in British psychiatry, had himself developed an interest in genetic issues.[54] The suggestions and guidance of Lewis, as well as contact with statistician and population geneticist Ronald A. Fisher, led Slater to choose genetics in psychiatry as the focus of his further scientific endeavours. Both Fisher and Lewis were protagonists of the British Eugenics Society. In the early 1930s Fisher in particular lobbied for the introduction of a sterilisation law in the UK, after the Wood Report of 1929 had documented the high and apparently increasing incidence of feeble-mindedness, which was seen as both a genetic and a social condition. In order to obtain data to support the sterilisation campaign, the Eugenics Society turned to Rüdin, who was pleased to supply information including a comprehensive list of references on the inheritance of feeble-mindedness and a number of reprints.[55] Lewis, however, was critical of the new Nazi law on forced sterilisation. In August 1933 he published an anonymous comment on the law in *The Lancet*, in which he spoke of an "irresponsible extremist" implementation of the ideas of eugenics in Germany, a "gross overstatement of our present knowledge of heredity", and a "disregard for the individual human being" whose interests were subordinated to those of the society.[56]

Following the suggestion and support of Lewis, Slater received a Rockefeller fellowship for 1934-35 to become acquainted with modern methods in psychiatric genetics and epidemiology. During this year Slater worked for five months at the DFA in Munich. He also went to Heidelberg and Berlin, where he became acquainted with the principles of mutation genetics at the genetic department of the Kaiser-Wilhelm-Institute for Brain Research under geneticist Nicolai Timofeeff-Ressovsky. In addition, he visited the newly established Institute for Hereditary Health and Racial Hygiene at Frankfurt University directed by Otmar von Verschuer, who was regarded as a pioneer in the twin-methodology.

In Munich, Slater began a study of three generations of manic-depressive families. The study consisted of 315 probands with recurrent affective disorders, their parents and children, using the material on probands collected by Rüdin and his team, and supplemented by his own interviews with children of the probands.[57] As a methodical innovation, he applied three different methods of age-correcting to the same data sets. He also was the first to use the analysis of variance in psychiatric research.[58] One result of this study was that manic-depressive disorder was not connected with downward social mobility for the family, an important result in the context of eugenics-related considerations, which confirmed previous results of Luxenburger in this direction.

After he returned to Britain, Slater, encouraged by Lewis, prepared the way for later genetic studies by first carrying out an epidemiological study of the incidence of mental disorders in England and Wales.[59] This approach had been established by the Munich group around 1930 and was also followed in various Scandinavian contexts.[60] The rationale was to enable comparisons between the frequency of various mental disorders in the average population and in the families of psychiatric patients, which would be the starting point for genetic investigations. Slater made direct use of the methods of Schulz and Luxenburger, and compared his findings from Britain with the results from Germany.[61]

For his further work, Slater received funding from the Mental Disorders Committee of the Medical Research Council (MRC), which until then mainly had supported Lionel Penrose's study on the heredity of the mentally defective in Colchester. The committee now sought to extend the scope of genetic work in psychiatry, and found Slater to be a very suitable person to do this. In 1936 Slater began setting up a genealogical databank of psychiatric patients in Greater London, with a particular focus on twins. This project was closely modelled after the work done by Rüdin and his colleagues at the Munich GDA, which Slater again visited in 1937.[62] When he delivered a lecture on "The Inheritance of Mental Disorder" in 1936, more than two-thirds of his references for valid evidence on various disorders were from authors affiliated with the DFA. At the same time, he clearly criticised the methods applied in an earlier publication on the inheritance of epilepsy by Rosanoff from the Eugenic Record Office at Cold Spring Harbor.[63]

It was apparently no problem for the MRC that Slater had received his training at German research institutions, in particular the DFA, which were known for their close co-operation with the Nazi regime. In 1936, a year after the MRC's decision to fund Slater, he published a report and critique of the German sterilisation law, which in later hagiography was used as supposed proof of his opposition to the political "abuse" of genetic knowledge.[64] However, a closer look shows that beyond heavy attacks against the *Führer*'s policies in general and the spirit of compulsion pervasive in German society and culture, Slater did not criticise the law in principle but rather a number of its technical details, such as the difficulties of making exact diagnoses and shortcomings in the practical implementation of the law. As is apparent, he had very detailed knowledge of the various dimensions of the implementation and the reactions towards the law, both amongst the population and physicians. He also discussed the question if voluntary sterilisation might be preferred over enforced measures: "Against this (i.e., the idea of voluntary sterilization) one must put the opinion of such a well-informed authority as Sjögren of Sweden, who told me that in his opinion the reverse would be the case. Apparently the experience in Sweden with voluntary sterilization has been very disappointing, only a few hundred having been sterilized since the introduction of the law a few years ago".[65] He described the elaborate, although according to his judgment still insufficient, efforts on the indications for sterilisation, and pointed to mistaken assumptions abroad: "There is no foundation in the idea, sometimes found outside Germany, that the law is administered in a partial way as a punishment for political offenders".[66] Further, Slater discussed possible expansions of the law, such as including children of two recessive parents, and sterilising the monozygotic twin of a person already ordered to be sterilised. He argued that "this could be taken as an improvement in the justice of the law", since "the phenotypically healthy twin is at least as dangerous eugenically as the sick twin".[67] Finally, he goes to some length to discuss the 1935 "law for conjugal health". Here, he regards mandatory marriage certificates as problematic due to difficulties in practicalities, but finds the newly adjusted German income tax system "from a eugenics point of view" much better than the English system.[68]

Slater's moderate criticism of the law in 1936 caused some irritation to Rüdin, but did not prevent Slater from turning to Rüdin to support him (Slater) in the publication of two lengthy articles in German psychiatric journals.[69] He also contributed a methodologically interesting article (on the probability of manifestation of genetic dispositions) to the *Festschrift* on the occasion of Rüdin's 65th birthday. The birthday celebrations were held in Munich, where members of various political institutions of the regime met with researchers from the DFA and the international scientific community. The contributions were published in a special issue entitled "Rüdin-Festschrift" of the *Allgemeine Zeitschrift für Psychiatrie*, the journal of the German Association of Neurologists and Psychiatrists, of which Rüdin was the president.[70] After an introduction by Luxenburger, Herbert Linden, physician and head of the *Medizinalabteilung* of the Ministry of the Interior, presented a paper on the "fight against sexual delinquency with medical means". Kurt Pohlisch, a close personal friend of Rüdin and professor of psychiatry at Bonn University, as well as director of the second-largest research institute on psychiatric genetics in Germany, contributed considerations on the family method. Rüdin's and Pohlisch's common pupil Friedrich Panse presented a paper about a new "genetic order of human hereditary syndromes".[71] The international community was represented by Slater, Essen-Möller and the Swiss pupil of Rüdin, Carl Brugger.[72] Slater made it clear that his own work started out from the concept of *empirische Erbprognose* developed at the Munich institute, and concluded that it was Rüdin's special achievement to have insisted on the necessity of the recurring analyses of thoroughly collected empirical "material" with ever-revised methods.[73] This was a paper written in German in 1939, immediately before the beginning of World War II. At the same time, in mid-1939, the Mental Disorders Committee of the MRC decided that work on genetics in psychiatry was of growing importance for the future, and in particular Slater's twin-studies should be regularly supported the following year.[74]

After the war, Slater was appointed physician in psychological medicine at the National Hospital, Queen Square, where he assembled a group of neuropsychiatrists. Here, he observed a number of long-standing paranoid psychoses resembling paranoid schizophrenia in patients suffering from epilepsy. In a combined genetic and clinical follow-up study, the evidence suggested that these psychotic states, although closely resembling schizophrenia, were biologically completely distinct and rather related to an inherited disposition for temporal lobe epilepsy.[75] This finding reconfirmed the biological heterogeneity of clinical cases of schizophrenia, and the necessity already proclaimed by Rüdin of both very careful clinical "case"-identification and differentiation of genetic studies, as well as co-morbidity as a potential clue for further investigations into the inheritance of the disorder.

Slater also continued his work on the genetics of neurotic and psychotic twins, and in the 1950s he formulated an influential theory about the monogenetic origins of schizophrenia. This theory remained a theoretical challenge and reference point in the field for several decades. Together with Meyer-Gross and Martin Roth, he published *Clinical Psychiatry*, a textbook that was reprinted many times and translated into several languages. *The Genetics of Mental Disorders*, which Slater published with Valerie Cowie in 1971, remained a standard book on the subject for two decades.[76] Even in the 1950s and 1960s, his publications include references to the work of members of the Munich Institute, in particular Hans Luxenburger and Bruno Schulz.[77] In 1959 the Medical Research Council decided to establish a centre for psychiatric genetics. This was the first research institution in the field in the UK. Slater acted as its director until his retirement in 1969. From 1961-1972 he was also the editor-in-chief of the *British Journal of Psychiatry*, one of the most prestigious positions in the field in the UK. He was a member of the Royal Commission on Capital Punishment and held honorary fellowships at the Royal Society of Medicine, Royal College of Psychiatrists and American Psychiatric Association. His high esteem in Britain was sealed by receiving the 1966 award of the Commander of the Order of the British Empire. In his later life, he joined the Voluntary Euthanasia Society.[78] He died in 1983.

Parallel and intrinsic to his scientific activities, Slater continued to be a member of the Eugenics Society, which he joined in 1937 and for which he served as a member of the Council (in part in the office of Vice President) from 1944 onwards for nearly 30 years. He regularly contributed to the

Eugenics Review, as well as to the *Annals of Eugenics*, the latter later renamed as *Annals of Human Genetics*. In a short article "Modern Tendencies in Eugenics", which appeared immediately after the war in 1945, Slater gave an interesting evaluation of the Nazi sterilisation programme: though he did not reject negative eugenic measures in general, he argued that in Nazi Germany negative eugenics "was taken to its logical limit, and a good bit beyond", since individuals suffering from hereditary conditions could not themselves decide to be sterilised, but the decision was forced upon them. Further, he argued "the Nazis were in too much of a hurry [...]. We need to know a great deal more about human heredity than we yet do to lay down any absolute rules".[79] The implication is that the sterilisation law was imposed by "the Nazis" (suggesting a homogeneous group) in spite of scientific evidence. No mention was made by him that the law was drafted, publicly justified, and the official commentary authored by eugenically motivated medical geneticists, with Rüdin as a protagonist. Thus, similar to Kallmann in his correspondence with Boas, Slater after the end of the war rhetorically disconnected the Nazi sterilisation policies from proper science and juxtaposed them, thereby completely ignoring the close co-operation of scientists with the regime, and the fact that in many respects the sterilisation law as an initiative came from the side of the scientists.[80]

In the article, Slater also repeated the core eugenics argument of counter-selection as a threat to the genetic health of the population: in contrast to "a barbaric state of society [...] in a civilised society with social agencies to protect the least fortunate children, the mentally and bodily enfeebled may outbreed the others". In view of this differential birth rate, and since "it seems probable that important qualities such as good health and intelligence [...] have a basis in heredity", he concluded that social measures such as "good feeding and upbringing, good education [...] could be great forces for good; but they could not compensate for a consistent deterioration of innate endowment".[81]

In another paper read to the Eugenics Society in December 1947, Slater characterised the Nazi sterilisation law as "draconian" and the "strictest" of all implemented eugenic laws, but not as qualitatively different from other such measures.[82] He also referred to an overview and evaluation of 15 years of the Danish sterilisation law (implemented in 1932) by geneticist Tage Kemp. According to Slater, Kemp's "informative paper" had shown that the eugenics legislation was a "humanitarian measure", but that it had also illustrated the "limited scope and effect of such measures when we think of the community as a whole".[83] For the time being, he was alarmed that "we are confronted with what appears to us a ruinous rate of loss of good genes and so a corresponding rate of increase of bad ones. This is the eugenic problem that far outweighs all others," including that of mental disorders.[84] He concluded that "the more intelligent half of the community" should be encouraged "to have more, not fewer, children than the less intelligent" in order to avoid the imminent deterioration of the genetic preconditions for intelligence in the population.[85]

In the Galton-Lecture delivered by Slater in 1960, he rejected compulsory interventions into human reproduction, but still urged improving the genetic quality of human populations through biological methods.[86]

4. Erik Essen-Möller and the Programme in Psychiatric Genetics and Epidemiology at the University of Lund

After earning a B.A. degree in genetics and statistics at the University of Lund in 1926, Erik Essen-Möller (1901-1992) received a degree in medicine in 1931.[87] Inspired by geneticist Herman Nilsson-Ehle, who also was a fervent racial hygienist, Essen-Möller decided to focus his career on medical and in particular behavioural genetics. With a two-year fellowship from the Rockefeller Foundation (1931-1933) and additional contributions from Swedish sources (Royal Travel Fellowship of the University of Lund, Kungl. Fysiografiska Säklapet I Lund[88]), he studied the fertility of the mentally ill at the DFA in Munich. As Essen-Möller stated, this topic was suggested by Rüdin, who also supported him in all respects. Essen-Möller admired the "generous organisation" of the DFA and its director, whom he perceived as interested both in technical details as well as the "big issues" (*große Linien*).[89] He also made it explicit that the topic was first of all motivated by "eugenic-social" goals, but that it should also contribute to clinical knowledge.[90] At the DFA he met Luxenburger, Kallmann and Schulz, who strongly influenced his methodology,[91] and with all of whom he established close and enduring co-operation. He also spent short periods in 1936 researching methods of proving paternity at the Kaiser Wilhelm Institute of Anthropology, Human Heredity, and Eugenics in Berlin under the directorship of Eugen Fischer, and at the Anthropological Institute of the University of Vienna under Josef Weninger.[92]

Essen-Möller published the findings of his investigation into the fertility of some groups of patients with mental disorders conducted in Munich in 1935 (in German) as his doctoral thesis in medicine. The central outcome of this study was that the fertility of schizophrenics was diminished compared to the healthy population, which thus was of minor concern to eugenicists.[93] At the International Congress of Population Science in Berlin in 1935, presided over by Eugen Fischer, he presented a paper on some implications of this work for population policies. He opened his contribution by stating that for eugenic aims, the knowledge of the fertility of psychiatric patients with hereditary conditions was essential; that his research into the related questions had been initiated by Rüdin; and that the empirical "material" had been supplied by the Munich institute. In principle, the fertility rates of most groups of mental patients (in particular of patients suffering from schizophrenia and epilepsy) was lower than that of the general population, but the relationship between the (declining) fertility rate in the general population and in defined groups was changing over time. By sterilising all known "cases", he argued, the birth of individuals suffering from manic-depressive disorder might be only very slightly reduced, whereas the eradication (*Ausschaltung*[94]) of the number of schizophrenic patients and epileptics in the population might be considerably accelerated. Thus, the systematic sterilisation in the case of the latter two diseases would enhance the process of "natural" reduction due to "low" fertility rates, in particular regarding schizophrenia.[95]

Here, we find the motivation and core of one of Essen-Möller's several research programmes: his eugenic concerns and goals, in particular the effective control of the reproduction of patients suffering from supposedly hereditary conditions, lead to elaborated theoretical work that in retrospect could well be framed as "basic science". Thus, Essen-Möller began working on what he called "differential reproduction" of psychiatric patients and their family members. This research led him to important methodological developments such as elaborated methods for correcting statistical biases related to the concept of empirical prognosis of heredity (*empirische Erbprognose*) developed by the Munich group. This work also led to important contributions to theoretical debates on the mode of inheritance of psychopathological traits and disorders (dominant, recessive and, since the 1960s, polygenic), and the epidemiology of psychiatric conditions.

Like Slater, Essen-Möller participated in the celebration of Rüdin's 65th birthday in Munich in 1939. Essen-Möller read a paper about the potential of the concept of personality diagnosis coined by his Swedish teacher Henrik Sjöbring for the investigation of genetics in psychiatry. Reconfirming the programmatic statement Rüdin had formulated from the beginning of his genetic studies, Essen-Möller pointed to the necessity of careful clinical delineation of psychopathological states and syndromes as a prerequisite of reliable genetic studies.[96]

After returning to Sweden, he began a twin study of schizophrenic patients based on 10,000 consecutive admissions to four Swedish mental hospitals. Sixty-nine pairs of twins could be identified and closely investigated both in respect of their psychopathology and family relations, and after differentiating monozygotic from dizygotic twins. This study closely followed Luxenburger's approach of analysing an unselected series of twins in regards to the genetic and environmental factors going into various traits or symptoms. Similar to the approach used in Munich, this sample also was intended to be something like the empirical "raw material" for further studies of other researchers. The methods of recruitment, the sample selection, and all the cases therefore were described in considerable detail, and indeed used by a number of later authors. The first results of an analysis were published in 1941 (again in German), but this was only the beginning of a unique 30-year longitudinal study.[97]

Next to the follow-up twin study outlined above, he also inaugurated and carried out in 1947 another unique community-based longitudinal investigation into the epidemiology and genetics of mental disorder. With his colleagues, he examined 2,550 individuals, including 600 children, with the aim to describe and quantify their psychopathology and personality variants, and to analyze their family relationships. The last follow-up analyses were published in 1986 and 2005.[98] Like Kallmann and Slater, Essen-Möller continued to use concepts and methods of the Munich school developed in the 1930s and early 1940s well into the 1950s and 1960s, and he also explicitly appreciated the conceptual tools and results he had received from the DFA. Thus, in 1955, he devoted a paper to an update of the concept of *empirische Erbprognose*. In 1958 he wrote a lengthy obituary for Bruno Schulz in which he sketched the importance of Schulz's work for the next generation of psychiatric geneticists.[99]

Essen-Möller's first academic appointment was that of an assistant professor of medical genetics at the University of Lund in 1935, which was turned into an assistant professorship of psychiatry under his teacher Sjöbring in 1939. In 1943 he was appointed chairman of the Psychiatry Department at the Karolinska Institute in Stockholm, but he returned to the University of Lund in 1944 to become the successor of his teacher and head of the Department of Psychiatry, where he stayed until his retirement in 1967. In Lund he built up a strong programme in psychiatric genetics and epidemiology with a high international profile, and with close co-operation in particular with Slater and his group in London.[100] He served as an expert advisor for psychiatric epidemiology and genetics to the WHO from 1959 to 1967 and again from 1981 to 1985, and was appointed an honorary member of the Royal Medico-Psychological Association in Britain. In 1985 he was also awarded the prestigious Paul Hoch Medal of the American Psychopathological Association for his work in psychiatric genetics and epidemiology.

Like Slater and Kallmann, Essen-Möller throughout his career stuck to his eugenic convictions. Thus, at the First International Congress of Human Genetics in Copenhagen in 1956, he delivered the most extensive statement during the final panel discussion, which was devoted to the topic, "Epidemiological Control of Hereditary Diseases. Genetic-Hygienic Registration". The terminology here is interesting: the term racial hygienic, or eugenic, is replaced by genetic-hygienic. Essen-Möller's contribution argued that genetic expertise needed to be applied for improving the quality of the population in order to lower the number of criminals and increase the proportion of people with socially desirable qualities and behaviour. There is no word on the interests or well-being of individual patients or their suffering.[101]

5. Conclusions

The case of the four protagonists of psychiatric genetics in Germany, the US, UK and Sweden illustrates the international scientific credibility and authority of their respective research programmes and practices, all of which were motivated by eugenic-oriented ideas and which intended to contribute to eugenics policies in the various national contexts. All of them followed both scientific *and* political agendas, though they were geared to the prevailing political contexts of the respective national settings, and, of course, they depended on each actor's political and moral convictions. Each continuously adapted his research agenda to problem-constellations perceived as intellectually as well as politically interesting and promising in view of broader concerns, which in turn was dependent on the available financial resources and specific political configurations.

After his emigration to the US in 1936, Kallmann maintained his overall eugenic concerns and aims. But by sanitizing his rhetoric, excluding references to race and the national body politic, and propagating educational measures as well as genetic counselling, he adapted to the values and language of American society. Thus, his talk of populations could now be interpreted as referring to mankind in general. In the next few years, he softened his convictions and finally stopped arguing for laws to enforce regulations on reproduction, in particular against the will of those concerned, and substituted this by calls for educational measures, the nationwide provision of genetic counselling, and voluntary sterilisation. Further, he stressed that his genetic interpretation of schizophrenia was compatible with social and even psychoanalytical approaches. In order to justify his scientific work before his emigration, and that of the Munich group so essential to him, Kallmann presented Rüdin as a victim of the Nazi regime by constructing an image according to which the psychiatric-genetic research at the DFA had allegedly been restricted by political pressure and censorship. Thus he created a dichotomous image of the situation in Germany, according to which the political authorities were arguing with "pseudo-scientific dogmas" (*pseudowissenschaftlichen Irrlehren*), whereas he and his colleagues in the field of psychiatric genetics had been doing proper science and fighting against such ideologies.

Similarly, Slater argued during the post-war period that the German sterilisation law had been imposed by "the Nazis" (Slater in the original) in spite of scientific evidence. The participation of Rüdin and other medical geneticists in drafting and implementing the law was bracketed out, and instead Slater suggested a contradiction between political authorities ("the Nazis") and scientific expertise. In contrast to his 1936 comments on the sterilisation law, in which he did not criticise the possibility of involuntary enforcement but only the exact indications and the practicalities of the implementation, he weakened his statements in the post-war period by characterising the Danish sterilisation law, for example, as "humanitarian" (in contrast to the "draconian" German law). But on the other side he also pointed to its inefficacy.

On another level of analysis, the case of the international institutionalisation of psychiatric genetics illustrates that the specific developments in Germany after 1933 may not be explained by the fact that the kind of eugenics and medical genetic practices there remained in a scientifically backward state because of their political "contamination". It was only Slater who, at a relatively early 1936, criticised (albeit moderately) the authoritarian application of the results of genetic research to eugenic policies. In the case of Kallmann, as well as Scandinavian medical geneticists such as Essen-Möller, we find such an explicit critique only developing considerably later, after the defeat of Nazi Germany, and when working in the context of democratic societies, relying on the resources and gratifications provided by the corresponding institutions. There existed, however, a pre-World War II tradition of a political as well as a scientific critique of the German version of medical genetics and eugenics, albeit outside of the community of medical geneticists.[102]

Thus, a simple dichotomous image of eugenics as formulated by Kevles does not fit the historical evidence. Instead, an interpretation is needed that takes into account three historical facts: first, the comparatively early onset of the institutionalisation of eugenically motivated human genetics in Germany, which had already peaked by the late 1920s during the democratic Weimar Republic; second, its high standing in the international scientific community until well into the second half of the 1930s; and third, the continuation of its research agendas and methodologies in the UK, the US and Scandinavia well into the 1950s. With these historical facts in mind, a different interpretation for the divergence of German eugenics is more plausible. Such an interpretation may be built around two analytical concepts: first, the authority of scientific expertise in the public sphere, and second, the readiness of scientists as well as politicians to co-operate with each other – in both respects the constellation in Germany may well be seen as diverging from other national contexts.

The degree of authority ascribed to scientific expertise in relation to other kinds of expertise in public life and politics may be seen as a differentiating factor between the German situation in the 1920s and 1930s, and that in other national contexts. For the German situation, a particularly high degree of authority of scientific expertise may be observed, as indicated by the early, extensive and in many respects successful establishment of eugenic research institutions and their close co-operation with the state.

Regarding the second factor, recent historiography has documented that among scientists, the degree of readiness to co-operate with all types of political groups and institutions was remarkably high, independent of the scientific standing and quality of work. This does not automatically imply that all of the scientists who co-operated with state agencies accepted *all* political presuppositions, or that state agencies accepted *all* scientific presuppositions or implications of the other side, as long as the profit for one's own projects and aims was seen to prevail.[103] On various levels of state and party organisation, Nazi authorities had a considerable interest in co-operating with scientists, both for a scientific underpinning and justification of their policies, and in the sense of a genuine interest to improve the health and efficiency of the population, or body social. However, this readiness for co-operation, and the provision of heavy funding, was conditional on the direct involvement in the health and population policies of the regime. Most scientists in the realm of eugenics and medical genetics, such as Rüdin, Eugen Fischer and Otmar von Verschuer, did not at all mind accepting this condition.

Finally, it may be argued that the two explanatory factors emerging from the analysis of this case – the authority of scientific expertise in the public sphere, and the readiness of scientists to follow the temptations of those in power – are crucial factors not only for an adequate historical analysis of eugenics, but also for broader issues related to the integral political dimension in the history of science, as well as for today's public debate on the role of science in a civil and humane society.

Notes

---

[1] Roelcke, 2007, 173-190. The following chapter is as considerably revised and extended version of this publication.

[2] An outline of an international history of eugenics is given in: Weindling, 1999, 179-197; a broad international perspective is also taken in: Turda, 2010.

[3] For the outline of a conceptual framework of international transfer and adaptation processes in medical history, and their impact on national and local developments exemplified by the case of psychiatry and mental health care, see the introduction in: Roelcke, Weindling and Westwood, 2010.

[4] For this evaluation of Kallmann, Slater and Essen-Möller, see e.g.: Gershon, 1981, 273-276; Gottesman and McGuffin, 1996, 537-548; Gottesman, 1993, 4-5; as well as Strömgren, 1994, 405-410.

[5] Although Stefan Kühl in his book *The Nazi Connection* claims to look into the international history of eugenics, racial hygiene, and human genetics, he in fact ignores the scientific activities, and the retrospective strategies (after

1945) of his historical actors to separate politics from scientific human genetics, as well as the plausibility which scientific authority gave to eugenicists' claims: Kühl, 1994, xvii.

[6] Kevles, 1985. For accounts of the history of genetics depicting eugenics as a historical backdrop and ideological obstacle, see: Dunn, 1991; Carlson, 2004; for examples of medical geneticists propagating such a discontinuity, see e.g.: Propping, 1989, 1; Propping and Heuer, 1991, 78-93; for more examples, see: Massin, 1999, 12-64.

[7] See, for example: Paul, 1995; Weindling, 1999; Roelcke, 2007; as well as Roelcke, 2002, 21-55; Koch, 2004, 315-331; Schwerin, 2005.

[8] The historical knowledge on Rüdin is summarized in: Roelcke, 2000, 112-150; and Roelcke, 2006, 73-87.

[9] Kraepelin, 1908, 745-751; as well as the positive comments made in: Rüdin, 1909, 254-257.

[10] Sommer, 1910, 295-298; Alzheimer, 1911, 242-246.

[11] Weber, 1991, 74-89.

[12] Rüdin, 1911, 487-585.

[13] Weinberg, 1908/09, 377-392 and 440-460; Weinberg, 1910, 34-43; Weinberg, 1912a, 646-649 and 697-701; on the "sibling method", see: Weinberg, 1912b, 165-174.

[14] Rüdin, 1916.

[15] For a detailed analysis of the differences between the approach by Rüdin and by Rosanoff/Davenport, see: Cottebrune, 2009a, 35-54.

[16] See e.g.: Schulz, 1926, 1-37.

[17] See: Roelcke, 2002.

[18] Lewis, 1934, 87.

[19] Bock, 1986; Weindling, 1989.

[20] Roelcke, 2006.

[21] However, due to illness, Rüdin finally did not participate in the congress and was substituted by his pupil Kurt Pohlisch, chair of psychiatry at Bonn University. See: *Archiv des Arbeitskreis Psychiatriegeschichte Bonn*: letter from Rüdin to Pohlisch (28 June 1939); as well as a letter from Eugen Fischer to Pohlisch (23 June 1939); I wish to thank Linda Orth (Bonn) for calling my attention to these documents.

[22] An outline of Kallmann's biography with a focus on the time up to his emigration from Germany to the States is given in: Mildenberger, 2002, 183-200; for a more detailed reconstruction, with focus on the period of emigration and afterwards, see: Cottebrune, 2009b, 296-324.

[23] Rado, 1962, 259-260; Slater, 1965, 1440.

[24] Letter of Kallmann to Rüdin, dated 13 May 1936, quoted in: Mildenberger, 2002, 191.

[25] Kallmann, 1936a, 725-729; in the historical literature, this quote was first referred to (however wrongly translated for the English version of the book) in: Müller-Hill, 1988, 11.
[26] This has already been pointed out by: Müller-Hill, 1988, 28-29.
[27] Kallmann, 1936b, 119-124.
[28] Slater, 1965, 1440; similarly in: Gershon, 1981, 273-276.
[29] Quoted in: Cottebrune, 2009b, 303-304.
[30] This idealised image of Kallmann may be found e.g. in: Rainer, John 1989, 1361-1362; and as recent as in: Strous and Edelman, 2007, 212.
[31] Kallmann, 1938a.
[32] E.g.: Kallmann, 1946, 309-322.
[33] Kallmann, 1953; this represents a continuation of methods and concepts from: Luxenburger, 1928, 297-326.
[34] Gershon, 1981, 274.
[35] Stern, 1960, 581-584.
[36] Shields, Gottesman and Slater, 1967, 385-396; see also: Rainer, 1973, 40-46.
[37] Cottebrune, 2009b, 312.
[38] Herndon, 1962, 245.
[39] Paul, 1995, 121.
[40] Hurst, 1992, (no pagination).
[41] Herndon, 1962, 247.
[42] Quoted in: Cottebrune, 2009b, 297.
[43] Slater, 1965, 1440.
[44] Paul, 1995, 121.
[45] Kallmann, 1938b, 105.
[46] Kallmann, 1938b, 113.
[47] Kallmann, 1938, 195-199.
[48] Kallmann, 1939, 143-145.
[49] Kallmann, 1946, 320.
[50] Kallmann, 1948, 339-344; see also: Cottebrune, 2009b.
[51] Kallmann, 1962, 250-255, in particular 251.
[52] This idea is constitutive for the early development of eugenics/racial hygiene and closely related to theories of „degeneration" and a pessimistic outlook on modern civilization. See: Weindling, 1989, chapter 1 and 2, and for the broader development of cultural pessimism related to fears of degeneration and the emerging movement of eugenics: Roelcke, 1999, chapter 5.
[53] Gottesman and McGuffin, 1996, 537.
[54] See e.g.: Lewis, 1931, 119-125.
[55] Mazumdar, 1992, 196-210.
[56] Anonymous [Lewis, Aubrey], 1933, 297-298.

[57] Slater, 1936a, 981-990; Slater, 1936b, 626-634.
[58] Gottesman and McGuffin, 1996, 543.
[59] For the following passages on Slater, I draw on material presented by: Cottebrune, 2013 (publication in preparation).
[60] See e.g.: Schulz, 1931, 386-411.
[61] Slater, 1935, 172-186.
[62] This is evident from a letter of Slater to Rüdin, dated 8 September 1937, in: *Historical Archives of the Max-Planck-Institut für Psychiatrie München*: GDA 132.
[63] Slater, 1936c, 277-284.
[64] Gottesman and McGuffin, 1996, 543.
[65] Slater, 1936d, 285-295.
[66] Ibid., 290.
[67] Ibid.
[68] Ibid., 294.
[69] Slater, 1938, 1-47; Shrimpton and Slater, 1939, 715-718.
[70] See the Table of Contents (1939). *Allgemeine Zeitschrift für Psychiatrie* 112 („Rüdin-Festschrift").
[71] It has to be noted here that Linden was a central administrative figure in the implementation of the Nazi programme of patient killings ("euthanasia"), and that both Pohlisch and Panse belonged to the expert advisors (*Gutachter*) for the initial phase of this programme (*Aktion T4*); the close relation between Pohlisch and Rüdin is also documented by the fact that Pohlisch held the funeral eulogy after Rüdin's death in 1952.
[72] On Brugger's programme in psychiatric genetics at the University of Basel and his relation to the Munich group, see: Ritter and Roelcke, 2005, 263-288.
[73] Slater, 1939, 148-152.
[74] Cottebrune, 2009c; an article on this is in preparation.
[75] Beard and Slater, 1962, 311-316.
[76] Slater and Cowie, 1971.
[77] See e.g.: Slater, 1953/1971, 84-87, where Slater refers positively to Rüdin, Luxenburger, and Kallmann.
[78] Roth, 1983, 1906.
[79] Slater, 1945, 182-185, quotations on 182, 183.
[80] For a synthesis of historical knowledge on medicine during the Nazi time, and the active part physicians and biomedical scientists had in Nazi health and population policies, see: Roelcke, 2010a, 17-28.
[81] Slater, 1945, 184.
[82] Slater, 1948, 64.
[83] Ibid., 66.
[84] Ibid., 67.
[85] Ibid., 68.

[86] Slater, 1960, 91-103.
[87] A short outline of Essen-Möller's biography may be found in: Gottesman, 1993; as well as Hummel, 1981, 7-14.
[88] Essen-Möller, 1935, 3.
[89] Essen-Möller, 1935, 3-4.
[90] Ibid., 7: "Die vorliegende Arbeit stellt sich als Aufgabe, die Fruchtbarkeit gewisser Gruppen von Geisteskranken zu untersuchen und mit der entsprechenden Fruchtbarkeit der Gesamtbevölkerung zu vergleichen. Es sollen damit in erster Linie eugenisch-soziale, in zweiter Linie auch klinische Ziele verfolgt werden."
[91] Gottesman, 1993, 4; see e.g.: Essen-Möller, 1941, 7-10, where Essen-Möller formulates his strong tribute to Luxenburger; see also: Essen-Möller, 1958, 51-59.
[92] Hummel, 1981, 8.
[93] Essen-Möller, 1935.
[94] Ibid., 721, 722.
[95] Essen-Möller, 1936, 719-724.
[96] Essen-Möller, 1939, 25-31.
[97] Essen-Möller, 1970, 315-317.
[98] Hagnell, 1986, 89-110; Nettelbladt, Bogren and Mattisson et al., 2005, 444-452.
[99] Essen-Möller, 1955, 334-342; Essen-Möller, 1958, 51-59.
[100] See: Gottesman, 1993; as well as Munk-Jörgensen, 2007, 6-10.
[101] Essen-Möller, 1957, 464-468.
[102] See: Weindling, 2006.
[103] For broader considerations of "science" and "politics" as resources for each other, see: Ash, 2002, 32-51; for a critique of this approach, see: Roelcke, 2010b, 176-192.

References

Alzheimer, A. (1911), "Ist die Einrichtung einer psychiatrischen Abteilung im Reichsgesundheitsamt erstrebenswert?" *Zeitschrift für Neurologie*, vol. 6, 242-246.

Anonymous [Lewis, Aubrey] (1933), [Leading article] "Eugenics in Germany" *The Lancet*, vol. II, 297-298.

Ash, M. (2002), "Wissenschaft und Politik als Ressourcen füreinander" in: vom Bruch, R. and Kaderas, B. (eds.), *Wissenschaften und Wissenschaftspolitik. Bestandsaufnahmen zur Formationen, Brüchen und Kontinuitäten im Deutschland des 20. Jahrhunderts.* Stuttgart: Franz Steiner, 32-51.

Beard, A. W. and Slater, E. (1962), "The schizophrenia-like psychoses of epilepsy" *Proceedings of the Royal Society of Medicine*, vol. 55, 311-316.

Bock, G. (1986), *Zwangssterilisation im Nationalsozialismus. Studien zur Rassenpolitik und Frauenpolitik.* Opladen: Westdt. Verl.

Carlson, E. A. (2004), *Mendel's Legacy: The Origin of Classical Genetics.* Cold Spring Harbor, NY: Cold Spring Harbor Laboratory Press.

Cottebrune, A. (2009a), "Zwischen Theorie und Deutung der Vererbung psychischer Störungen. Zur Übertragung des Mendelismus auf die Psychiatrie in Deutschland und in den USA, 1911-1930" *NTM – Zeitschrift für Geschichte der Wissenschaften, Technik und Medizin*, vol. 17, no. 1, 35-54.

Cottebrune, A. (2009b), "Franz Josef Kallmann (1897-1965) und der Transfer psychiatrisch-genetischer Wissenschaftskonzepte vom NS-Deutschland in die USA: Wissenschaftstransfer im Kontext nordamerikanischer Eugenik" *Medizinhistorisches Journal*, vol. 44, no. 3/4, 296-324.

Cottebrune, A. (2013), "Eliot Slater and the Roots of Psychiatric Genetics in Great Britain" [paper presented at the workshop *Eugenics, Race and Psychiatry in the Baltic States. A Transnational Perspective*, Riga, 7-8 May 2009] (publication in preparation).

Dunn, L. C. (1991), *A short history of genetics: The development of some of the mainlines of Thought, 1864-1939.* Ames/Iowa: Iowa State University Press.

Essen-Möller, E. (1935), *Untersuchungen über die Fruchtbarkeit gewisser Gruppen von Geisteskranken (Schizophrenen, Manisch-Depressiven und Epileptikern).* Copenhagen: Munksgaard.

Essen-Möller, E. (1936), "Über die Fortpflanzung von Geisteskranken" in: Harmsen, H. and Lohse, F. (eds.), *Bevölkerungsfragen. Bericht des Internationalen Kongresses für Bevölkerungswissenschaft, Berlin, 26. August - 1.September 1935.* München: Lehmanns, 719-724.

Essen-Möller, E. (1939), "Die Persönlichkeitsdiagnose nach Sjöbring und ihre Bedeutung für die psychiatrische Erbdiagnose" *Allgemeine Zeitschrift für Psychiatrie*, vol. 112, 25-31.

Essen-Möller, E. (1941), *Psychiatrische Untersuchungen an einer Serie von Zwillingen*. Copenhagen: Munksgaard, 7-10.

Essen-Möller, E. (1955), "The calculation of morbid risk in parents of index cases, as applied to a family sample of schizophrenics" *Acta Genetica et Statistica Medica*, vol. 5, 334-342.

Essen-Möller, E. (1957), "Registration and subdivision of an entire population according to familial relationships" *Acta Genetica et Statistica Medica*, vol.7, 464-468.

Essen-Möller, E. (1958), "Das erbpsychiatrische Lebenswerk von Bruno Schulz" *Acta Pychiatrica et Neurologica Scandinavica*, vol. 34, 51-59.

Essen-Möller, E. (1970), "Twenty-one psychiatric cases and MZ co-twins: a thirty years follow-up" *Acta Geneticae Medicae et Gemellologiae*, vol. 19, 315-317.

Gershon, E. (1981), "The Historical Context of Franz Kallmann and Psychiatric Genetics" *Archiv für Psychiatrie und Nervenkrankheiten*, vol. 229, no.4 , 273-276.

Gottesman, I. (1993), "In Memoriam: Erik Essen-Möller (1901-1992)" *American Journal of Medical Genetics (Neuropsychiatric Genetics)*, vol. 48, no. 1, 4-5.

Gottesman, I. and McGuffin, P. (1996), "Eliot Slater and the birth of Psychiatric Genetics in Great Britain" in: Freeman, H. and Berrios, G. E. (eds.), *150 Years of British Psychiatry, Vol. II: The Aftermath*. London: Athlone, 537-548.

Hagnell, O. (1986), "The 25-year follow-up of the Lundby study: incidence and risk of alcoholism, depression, and disorders of the senium" in: Barrett, J. and Rose, R. M. (eds.), *Mental Disorders in the Community: Findings from Psychiatric Epidemiology*. New York: Guilford, 89-110.

Herndon, C. N. (1962), "Medical Genetics in the United States" in: Kallmann, F. (ed.), *Expanding Goals of Genetics in Psychiatry*. New York/ London: Grune & Stratton, 244-249.

Hurst, L. A. (1992), "Pursued by Genetics. An Autobiographical Disclosure" *Adler Museum Bulletin*, vol. 18, no. 2 (no pagination).

Hummel, K. (1981), "Laudatio" in: Hummel, K. and Gerchow, J. (eds.), *Biomathematical Evidence of Paternity. Festschrift for Erik Essen-Möller.* Berlin: Springer, 7-14.

Kallmann, F. (1936a), "Die Fruchtbarkeit der Schizophrenen" in: Harmsen, H. and Lohse, F. (eds.), *Bevölkerungsfragen. Bericht des Internationalen Kongresses für Bevölkerungswissenschaft, Berlin, 26. August - 1.September 1935*. München: Lehmanns, 725-729.

Kallmann, F. (1936b), "Erbprognose und Fruchtbarkeit bei den verschiedenen klinischen Formen der Schizophrenie. Vortrag auf der 1. Jahresversammlung der Gesellschaft Deutscher Neurologen und Psychiater, Dresden (1.-4. November 1935), vorgetragen von Bruno Schulz" *Allgemeine Zeitschrift für Psychiatrie*, vol. 104, 119-124.

Kallmann, F. (1938a), *The Genetics of Schizophrenia. A Study of Heredity and Reproduction in the Families of 1087 Schizophrenics*. New York: Augustin.

Kallmann, F. (1938b), "Heredity, Reproduction, and Eugenic Procedure in the Field of Schizophrenia" *Eugenical News*, vol. 23, 105-113.

Kallmann, F. (1938c), "Eugenic Birth Control in Schizophrenic Families" *Journal of Contraception*, vol. 3, no. 11, 195-199.

Kallmann, F. (1939), "Informal discussion" in: Moulton, F. R. (ed.), *Mental Health. Publication of the American Association for the Advancement of Science*. New York: Science Press, 143-145.

Kallmann, F. (1946), "The genetic theory of schizophrenia: An analysis of 691 schizophrenic twin index families" *American Journal of Psychiatry*, vol. 103, 309-322.

Kallmann, F. (1948), "Applicability of Modern Genetic Concepts in the Management of Schizophrenia" *Journal of Heredity*, vol. 39, 339-344.

Kallmann, F. (1953), *Heredity in Health and mental Disorder*. New York: W. Norton.

Kallmann, F. (1962), "Genetic Research and Counselling in the Mental Health Field, Present and Future" in: Kallmann, F. (ed.), *Expanding Goals of Genetics in Psychiatry*. New York/ London: Grune & Stratton, 250-255.

Kevles, D. (1985), *In the Name of Eugenics. Genetics and the Uses of Human Heredity*. Cambridge/ Mass.: Harvard University Press.

Koch, L. (2004), "The meaning of eugenics: Reflections on the government of genetic knowledge in the past and the present" *Science in Context*, vol. 17, no. 3, 315-331.

Kraepelin, E. (1908), "Zur Entartungfrage" *Zentralblatt für Nervenheilkunde und Psychiatrie* vol. 31, N. F. 19, 745-751.

Kühl, S. (1994), *The Nazi Connection. Eugenics, American Racism, and German National Socialism*. New York: Oxford University Press.

Lewis, A. (1931), "Genetic problems in psychiatry, and their solution by the study of twins" *Eugenics Review*, vol. 232, 119-125.

Lewis, A. (1934), "Inheritance of Mental Disorders" in: Blacker, C. P. (ed.), *The Chances of Morbid Inheritance*. London: Lewis, 86-133.

Luxenburger, H. (1928), "Vorläufiger Bericht über psychiatrische Serienuntersuchungen an Zwillingen" *Zeitschrift für die gesamte Neurologie und Psychiatrie*, vol. 116, 297-326.

Massin, B. (1999), "Anthropologie und Humangenetik im Nationalsozialismus oder: Wie schreiben deutsche Wissenschaftler ihre eigene Wissenschaftsgeschichte?" in: Kaupen-Haas, H. and Saller, C. (eds.), *Wissenschaftlicher Rassismus: Analysen einer Kontinuität in den Human- und Naturwissenschaften*. Frankfurt a. M.: Campus, 12-64.

Mazumdar, P. (1992), *Eugenics, Human Genetics, and Human Failings. The eugenics society, its sources and its critics in Britain*. London: Routledge.

Mildenberger, F. (2002), "Auf der Spur des 'scientific pursuit'. Franz Kallmann (1897-1965) und die rassenhygienische Forschung" *Medizinhistorisches Journal*, vol. 37, no. 2, 183-200.

Müller-Hill, B. (1988), *Murderous Science. Elimination by Scientific Selection of Jews, Gypsies, and Others. Germany 1933-1945*. Oxford: Oxford University Press.

Munk-Jörgensen, P. (2007), "Psychiatric epidemiology: a Scandinavian contribution to international psychiatry" *Acta Neuropsychiatrica*, vol. 19, no. 1, 6-10.

Nettelbladt, P.; Bogren, M. and Mattisson, C. et al. (2005), "Does it make sense to do repeated surveys? The Lundby Study, 1947-1997" *Acta Psychiatrica Scandinavica*, vol. 111, no. 6, 444-452.

Paul, D. (1995), *Controlling Human Heredity, 1865 to the Present*. Atlantic Highlands, NJ: Humanities Press.

Propping, P. (1989), *Psychiatrische Genetik. Befunde und Konzepte*. Berlin: J. Springer.

Propping, P.; Heuer, B. (1991), "Vergleich des 'Archiv für Rassen- und Gesellschaftsbiologie' (1904-1933) und des ‚Journal of Heredity' (1910-1939): Eine Untersuchung zu Hans Nachtsheims These von der Schwäche der Genetik in Deutschland" *Medizinhistorisches Journal*, vol. 26, 78-93.

Rado, S. (1962), "[Speech at Anniversary Banquet]" in: Kallmann, F. (ed.), *Expanding Goals of Genetics in Psychiatry*. New York: Grune & Stratton, 259-260.

Rainer, J. D. (1973), "Perspectives on the Genetics of Schizophrenia: A Re-evaluation of Kallmann's contribution, its influence and current relevance" in: Kolb, L. (ed.), *Progress in Psychiatric research and Education* (= *Psychiatric Quarterly of New York*), New York, 40-46.

Rainer, J. D. (1989), "Franz Kallmann's views on eugenics" *American Journal of Psychiatry*, vol. 146, 1361-1362.

Ritter, H. J. and Roelcke, V. (2005), "Psychiatric Genetics in Munich and Basel between 1925 and 1945: Programs – Practices – Cooperative Arrangements" *Osiris*, vol. 20, 263-288.

Roelcke, V. (1999), *Krankheit und Kulturkritik. Psychiatrische Gesellschaftsdiagnosen im bürgerlichen Zeitalter, 1790-1914*. Frankfurt/ Main: Campus.

Roelcke, V. (2000), "Psychiatrische Wissenschaft im Kontext nationalsozialistischer Politik und ‚Euthanasie': Zur Rolle von Ernst Rüdin und der Deutschen Forschungsanstalt/ Kaiser-Wilhelm-Institut für

Slater, E. (1939), “Über Begriff und Anwendbarkeit der Manifestationswahrscheinlichkeit” *Allgemeine Zeitschrift für Psychiatrie*, vol. 112, 148-152.

Slater, E. (1945), “Modern tendencies in eugenics” *Health Education Journal*, vol. 3, 182-185.

Slater, E. (1948), “Genetics, Medicine, and Practical Eugenics” *Eugenics Review*, vol. 40, 62-69.

Slater, E. (1953/1971), “Schizophrenia. Problems of Heterogeneity” reprinted in: Shields, J. and Gottesman, I. (eds.), *Man, Mind, and Heredity. Selected Papers of Eliot Slater on Psychiatry and Genetics*. Baltimore: Johns Hopkins University Press, 84-87.

Slater, E. (1960), “Galton’s heritage” *Eugenics Review*, vol. 52, 91-103.

Slater, E. (1965), “Obituary Notices: F. J. Kallmann, M.D” *British Medical Journal*, vol. 1, no. 5547 (29 May), 1440.

Slater, E. and Cowie, V. (1971), *The Genetics of Mental Disorders*. London: Oxford University Press.

Sommer, R. (1910), “Eine psychiatrische Abteilung des Reichsgesundheitsamtes” *Psychiatrisch-Neurologische Wochenschrift*, vol. 12, 295-298.

Stern, C. (1960), *Principles of Human Genetics, 2nd edition*. San Francisco/London: W. H. Freeman.

Strömgren, E. (1994), “Recent History of European Psychiatry – Ideas, Developments, and Personalities: The Annual Eliot Slater Lecture” *American Journal of Medical Genetics (Neuropyschiatric Genetics)*, vol. 54, 405-410.

Strous, R. D. and Edelman, M. C. (2007), “Eponyms and the Nazi Era: Time to Remember and Time for Change” *The Israel Medical Association Journal*, vol. 9, 207-214.

Turda, M. (2010), *Modernism and Eugenics*. Houndmills, Basingstoke: Palgrave.

Weber, M. M. (1991), "'Ein Forschungsinstitut für Psychiatrie ...': Die Entwicklung der Deutschen Forschungsanstalt für Psychiatrie in München zwischen 1917 und 1945" *Sudhoffs Archiv*, vol. 75, 74-89.

Weinberg, W. (1908/09), "Über Vererbungsgesetze beim Menschen" *Zeitschrift für induktive Abstammungs- und Vererbungsforschung*, vol. 1, 377-392 and 440-460.

Weinberg, W. (1910), "Statistik und Vererbung in der Psychiatrie" *Klinik für psychische und nervöse Krankheiten*, vol. 5, 34-43

Weinberg, W. (1912a), "Methoden der Vererbungsforschung beim Menschen" *Berliner Klinische Wochenschrift*, vol. 49, 646-649 and 697-701.

Weinberg, W. (1912b), "Weitere Beiträge zur Theorie der Vererbung" *Archiv für Rassen- und Gesellschaftsbiologie*, vol. 9, 165-174.

Weindling, P. J. (1989), *Health, Race, and German Politics between National Unification and Nazism, 1870-1945*. Cambridge: Cambridge University Press.

Weindling, P. (1999), "International Eugenics: Swedish Sterilization in Context" *Scandinavian Journal of History*, vol. 24, no. 2, 179-197.

Weindling, P. (2006), "Central Europe confronts German racial hygiene: Friedrich Hertz, Hugo Iltis and Ignaz Zollschan as critics of German racial hygiene" in: Turda, M. and Weindling, P. (eds.), *"Blood and Homeland": Eugenics and Racial Nationalism in Central and Southeast Europe, 1900-1940*. Budapest: Central European University Press, 263-282.

List of Contributors

Vsevolod Bashkuev is a research fellow at the Institute of Mongolian, Buddhist and Tibetan studies, Ulan-Ude, Russia.

Björn M. Felder is lecturer at the Seminar for Medieval and Modern History at the Georg-August-University Göttingen.

Arūnas Germanavičius is an associated professor at the Clinic of Psychiatry, Faculty of Medicine, Vilnius University.

Maciej Górny is an assistant professor at the Tadeusz Manteuffel Institute of History of the Polish Academy of Sciences in Warsaw.

Leiu Heapost is a senior researcher at the Institute of History, Tallinn University.

Ken Kalling is lecturer at the Institute of Public Health at the University of Tartu.

Vladimirs Kuznecovs is an associated professor at the Medical Faculty at the University of Latvia in Riga.

Ineta Lipša is a researcher at Institute of Latvian History at the University of Latvia in Riga.

Volker Roelcke is professor and director, Institute for the History of Medicine at the Giessen University

Maija Runcis is an associate professor in history at Stockholm University.

Paul J. Weindling is Wellcome Trust Professor in History of Medicine at the department for History, Philosophy and Religion at Oxford Brookes University.

www.ingramcontent.com/pod-product-compliance
Lightning Source LLC
LaVergne TN
LVHW020651300926
849161LV00004B/894
* 9 7 8 9 0 4 2 0 3 7 2 2 9 *